HOLT ESSENTIAL

MATHEMATICS

WELLS REMALEY ROBERTS THOMPSON

Holt, Rinehart and Winston, Publishers
New York • Toronto • Mexico City
London • Sydney • Tokyo

ABOUT THE AUTHORS

David W. Wells is a Mathematics teacher in St. Francis Xavier Junior High School, St. Joseph, Missouri, and a Lecturer in the Mathematics Department, Missouri Western State College, St. Joseph, Missouri.

Charlotte Evans Remaley is the Mathematics Curriculum Specialist, Hampton Public Schools, Hampton, Virginia.

Ralph E. Roberts is Mathematics Chairman, Pelham Memorial High School, Pelham, New York.

Frances McBroom Thompson is a Mathematics Consultant, K–12, Education Service Center, Region 10, Richardson, Texas.

Cover photo: D. Driscoll/Photo Researchers.

Photos at the Beginning of Each Chapter:

1. Lipsticks **2.** General Store, Cape Cod, MA, Newspaper Delivery **3.** Trombone Valve **4.** Computer Storage **5.** Los Angeles Memorial Coliseum **6.** Drill Bits **7.** Aerial View of Cars **8.** Pet Leashes **9.** Electrocardiogram Graph **10.** Computer Chip **11.** Spools of Dyed Synthetic Fibers **12.** Eggs in Container **13.** Fruit in Containers **14.** Railroad Tracks **15.** Lumber **16.** Computer Interior **17.** Aerial View of San Joaquin Valley **18.** Solar panels in Barstow, CA **19.** Stacks of Unbound Magazines **20.** Hotel in Bermuda

Photo Credits are on page 422.

Portions of this work were previously published under the title HOLT MATH 1000

Printed in the United States of America
ISBN: 0-03-006473-2
78901234 032 98765432

Contents

1 Addition of Whole Numbers

2 Subtraction of Whole Numbers

3 Problem Solving

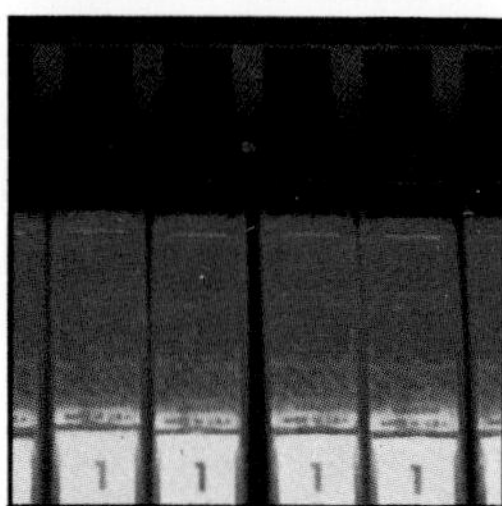

4 Multiplication of Whole Numbers

5 Division of Whole Numbers

6 *Addition and Subtraction of Decimals*

7 *Multiplication of Decimals*

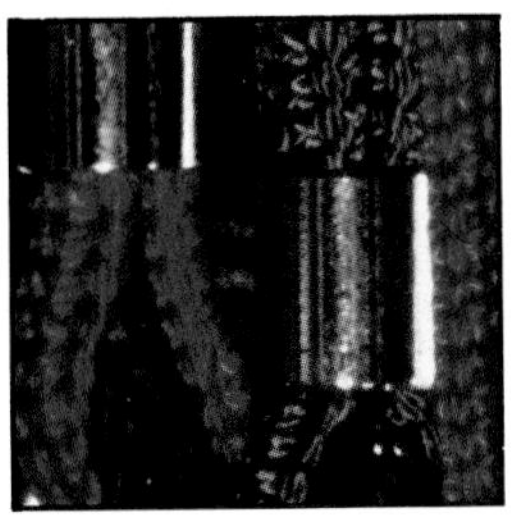

8 *Division of Decimals*

---

9 Measurement

10 The Meaning of Fractions

11 Addition and Subtraction of Fractions

12 Multiplication and Division of Fractions

13 Problem Solving with Formulas

14 Ratio and Proportion

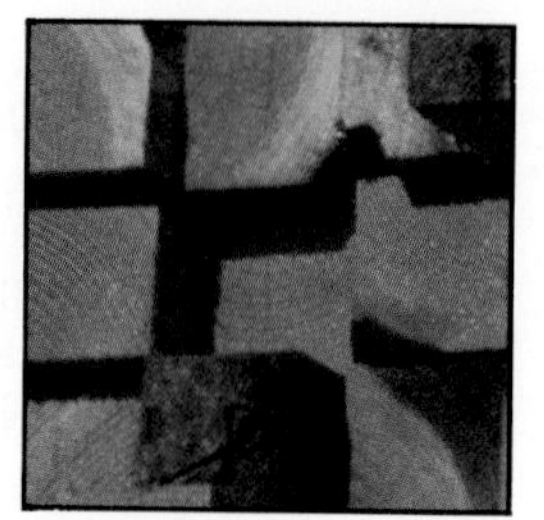

15 Geometry

16 Statistics

17 Percent

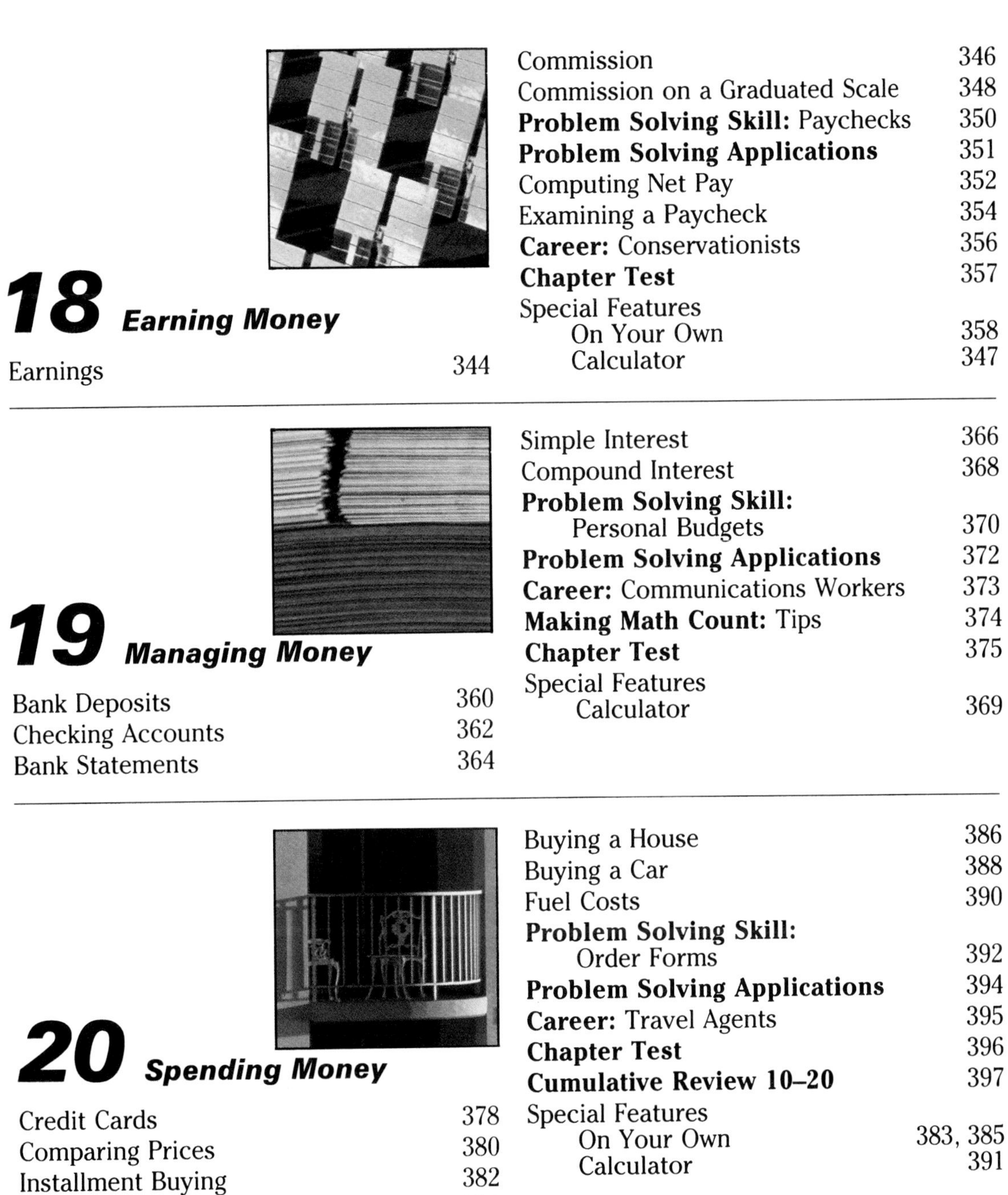

18 Earning Money

19 Managing Money

20 Spending Money

Addition of Whole Numbers

MAKING MATH COUNT

October

S	M	T	W	T	F	S
	1	2	3	4	5	6
7	8	9	10	11	12	13
14	15	16	17	18	19	20
21	22	23	24	25	26	27
28	29	30	31			

September

S	M	T	W	T	F	S
						1
2	3	4	5	6	7	8
9	10	11	12	13	14	15
16	17	18	19	20	21	22
23/30	24	25	26	27	28	29

November

S	M	T	W	T	F	S
				1	2	3
4	5	6	7	8	9	10
11	12	13	14	15	16	17
18	19	20	21	22	23	24
25	26	27	28	29	30	

EXAMPLE

Jill's school project began on September 23 and ended on October 8. How long did the project last?

A 1 wk 1 d **B** 1 wk 2 d
C 2 wk 1 d **D** 2 wk 4 d

Begin counting with the following day and end counting with the last date.
September 23 to October 7 is 2 wk.
October 7 to October 8 is 1 d.
So, **C** is the correct answer.

Choose the correct answer.

1. How long is it from October 14 to November 4?
A 2 wk 5 d **B** 2 wk 6 d
C 3 wk **D** 3 wk 1 d

2. How long is it from October 31 to November 30?
A 4 wk 2 d **B** 4 wk 1 d
C 4 wk **D** 3 wk 2 d

3. Ted started to build a model ship on September 9. He finished it on November 3. How long did it take to build the ship?
A 8 wk **B** 7 wk 6 d
C 7 wk 5 d **D** 7 wk 4 d

***4.** Marie ordered furniture on October 18. It arrived on December 7. How long did she wait?
A 7 wk **B** 7 wk 1 d
C 7 wk 2 d **D** 8 wk

ADDITION FACTS TEST A (3 MINUTES)

1.	7	5	4	2	9	7	9	3	1	6
	+3	+2	+2	+1	+2	+2	+0	+2	+2	+3
2.	9	2	8	6	6	4	3	7	1	2
	+3	+3	+2	+2	+4	+1	+4	+4	+3	+4
3.	5	8	4	3	5	4	5	9	4	8
	+3	+4	+3	+1	+1	+4	+4	+4	+5	+6
4.	6	1	7	9	2	3	7	8	9	1
	+5	+5	+6	+6	+5	+6	+5	+5	+5	+9
5.	2	4	5	5	9	1	2	6	7	7
	+6	+6	+6	+7	+7	+6	+7	+8	+1	+8
6.	6	4	6	7	8	6	3	9	9	3
	+1	+7	+9	+0	+7	+7	+7	+8	+1	+8
7.	4	1	5	3	2	0	9	3	8	8
	+9	+8	+9	+5	+8	+5	+9	+9	+9	+1

67 or more right? Go to page 7.

Less than 67 right? Go to page 4.

ADDITION FACTS TEST B (2 MINUTES)

1.	7	1	6	8	2	4	9	8	7	4
	+6	+5	+5	+6	+5	+5	+6	+5	+5	+6
2.	9	3	2	3	5	5	8	1	5	6
	+5	+5	+6	+6	+1	+6	+8	+6	+7	+8
3.	1	2	7	3	9	9	4	8	6	4
	+7	+7	+8	+7	+8	+7	+7	+7	+7	+8
4.	8	3	7	8	5	0	7	4	9	1
	+1	+9	+3	+9	+8	+8	+9	+9	+1	+8
5.	1	3	6	5	2	4	6	2	6	5
	+9	+8	+0	+9	+8	+8	+9	+9	+6	+4

48 or more right? Go to page 7.

Less than 48 right? See your teacher.

ADDITION FACTS

To learn the addition facts

- Try to use doubles.

EXAMPLE

Add. $\begin{array}{r} 7 \\ +8 \\ \hline \end{array}$ Suppose you know $\begin{array}{r} 7 \\ +7 \\ \hline 14. \end{array}$

Think: $\begin{array}{r} \mathbf{7} \\ \mathbf{+8} \\ \hline \end{array}$ → $\mathbf{7} + \mathbf{7} \rightarrow \mathbf{14}$, $\mathbf{1}$; $\begin{array}{r} \mathbf{14} \\ \mathbf{+\ 1} \\ \hline \mathbf{15} \end{array}$

So, $\begin{array}{r} 7 \\ +8 \\ \hline \mathbf{15.} \end{array}$

These addition facts are called **doubles:**

$\begin{array}{r} 1 \\ +1 \\ \hline 2 \end{array}$ $\begin{array}{r} 2 \\ +2 \\ \hline 4 \end{array}$ $\begin{array}{r} 3 \\ +3 \\ \hline 6 \end{array}$ $\begin{array}{r} 4 \\ +4 \\ \hline 8 \end{array}$ $\begin{array}{r} 5 \\ +5 \\ \hline 10 \end{array}$

$\begin{array}{r} 6 \\ +6 \\ \hline 12 \end{array}$ $\begin{array}{r} 7 \\ +7 \\ \hline 14 \end{array}$ $\begin{array}{r} 8 \\ +8 \\ \hline 16 \end{array}$ $\begin{array}{r} 9 \\ +9 \\ \hline 18 \end{array}$

GETTING READY

Add. Try to use doubles.

1. $\begin{array}{r} 7 \\ +6 \\ \hline \end{array}$ **2.** $\begin{array}{r} 5 \\ +8 \\ \hline \end{array}$ **3.** $\begin{array}{r} 4 \\ +7 \\ \hline \end{array}$ **4.** $\begin{array}{r} 2 \\ +5 \\ \hline \end{array}$ **5.** $\begin{array}{r} 5 \\ +7 \\ \hline \end{array}$ **6.** $\begin{array}{r} 3 \\ +8 \\ \hline \end{array}$

7. $\begin{array}{r} 4 \\ +2 \\ \hline \end{array}$ **8.** $\begin{array}{r} 4 \\ +5 \\ \hline \end{array}$ **9.** $\begin{array}{r} 3 \\ +1 \\ \hline \end{array}$ **10.** $\begin{array}{r} 9 \\ +7 \\ \hline \end{array}$ **11.** $\begin{array}{r} 5 \\ +6 \\ \hline \end{array}$ **12.** $\begin{array}{r} 2 \\ +6 \\ \hline \end{array}$

EXERCISES

Add.

1. $\begin{array}{r} 3 \\ +4 \\ \hline \end{array}$ **2.** $\begin{array}{r} 6 \\ +8 \\ \hline \end{array}$ **3.** $\begin{array}{r} 7 \\ +5 \\ \hline \end{array}$ **4.** $\begin{array}{r} 2 \\ +7 \\ \hline \end{array}$ **5.** $\begin{array}{r} 9 \\ +8 \\ \hline \end{array}$ **6.** $\begin{array}{r} 8 \\ +7 \\ \hline \end{array}$

7. $\begin{array}{r} 5 \\ +3 \\ \hline \end{array}$ **8.** $\begin{array}{r} 4 \\ +6 \\ \hline \end{array}$ **9.** $\begin{array}{r} 3 \\ +2 \\ \hline \end{array}$ **10.** $\begin{array}{r} 5 \\ +9 \\ \hline \end{array}$ **11.** $\begin{array}{r} 9 \\ +6 \\ \hline \end{array}$ **12.** $\begin{array}{r} 7 \\ +3 \\ \hline \end{array}$

Solve.

13. Rosa scored 6 points in the first half of the game and 4 points in the second half. How many points did she score in all?

14. Mark wrote 8 pages of a report in the morning and 6 pages in the afternoon. How many pages did he write in all?

MORE ADDITION FACTS

To learn the addition facts

- Try to make a 10.

EXAMPLE

Add. $\begin{array}{r} 9 \\ +5 \\ \hline \end{array}$

Remember the facts that make 10.

5 + 5 = 10	7 + 3 = 10	2 + 8 = 10
6 + 4 = 10	3 + 7 = 10	9 + 1 = 10
4 + 6 = 10	8 + 2 = 10	1 + 9 = 10

Each addition fact can be shown in several ways.

Example: 3 + 7, $\begin{array}{r} 3 \\ +7 \\ \hline \end{array}$, $\begin{array}{r} 7 \\ +3 \\ \hline \end{array}$, and 7 + 3 all equal 10.

Try to make a 10.

● ● ● ● ● 0 0 0 ● ● ● ● ● 0 0
● ● ● ● 0 0 ● ● ● ● 0 0 0

9 + 5 = 10 + 4 = 14

So, 9 + 5 = **14.**

GETTING READY

Add. Try to make a 10.

1. $\begin{array}{r} 9 \\ +8 \\ \hline \end{array}$ **2.** $\begin{array}{r} 7 \\ +4 \\ \hline \end{array}$ **3.** $\begin{array}{r} 6 \\ +9 \\ \hline \end{array}$ **4.** $\begin{array}{r} 9 \\ +4 \\ \hline \end{array}$ **5.** $\begin{array}{r} 5 \\ +7 \\ \hline \end{array}$ **6.** $\begin{array}{r} 8 \\ +3 \\ \hline \end{array}$ **7.** $\begin{array}{r} 6 \\ +5 \\ \hline \end{array}$

EXERCISES

Add.

1. $\begin{array}{r} 3 \\ +9 \\ \hline \end{array}$ **2.** $\begin{array}{r} 8 \\ +9 \\ \hline \end{array}$ **3.** $\begin{array}{r} 9 \\ +9 \\ \hline \end{array}$ **4.** $\begin{array}{r} 9 \\ +6 \\ \hline \end{array}$ **5.** $\begin{array}{r} 6 \\ +7 \\ \hline \end{array}$ **6.** $\begin{array}{r} 7 \\ +8 \\ \hline \end{array}$ **7.** $\begin{array}{r} 2 \\ +9 \\ \hline \end{array}$

8. $\begin{array}{r} 4 \\ +8 \\ \hline \end{array}$ **9.** $\begin{array}{r} 6 \\ +6 \\ \hline \end{array}$ **10.** $\begin{array}{r} 6 \\ +8 \\ \hline \end{array}$ **11.** $\begin{array}{r} 8 \\ +5 \\ \hline \end{array}$ **12.** $\begin{array}{r} 8 \\ +7 \\ \hline \end{array}$ **13.** $\begin{array}{r} 9 \\ +7 \\ \hline \end{array}$ **14.** $\begin{array}{r} 5 \\ +9 \\ \hline \end{array}$

15. List all of the addition facts that you missed on Facts Test A and practice them until you know them.

ADDING MULTIPLES

To add multiples of 10; 100; or 1,000

- Add in columns, starting from the right.
- Use addition facts.

EXAMPLE 1

Add.

$$\begin{array}{r} 50 \\ +30 \\ \hline \end{array} \qquad \begin{array}{r} 500 \\ +300 \\ \hline \end{array} \qquad \begin{array}{r} 5{,}000 \\ +3{,}000 \\ \hline \end{array}$$

Think: Each column is an addition fact.

$$\begin{array}{r} 50 \\ +30 \\ \hline \mathbf{80} \end{array} \qquad \begin{array}{r} 500 \\ +300 \\ \hline \mathbf{800} \end{array} \qquad \begin{array}{r} 5{,}000 \\ +3{,}000 \\ \hline \mathbf{8{,}000} \end{array}$$

> Multiples of a number are found by multiplying the number by 1, 2, 3, and so on.
> **Example:** Multiples of 10 are 10, 20, 30, and so on.

EXAMPLE 2

Add.

$$\begin{array}{r} 80 \\ +80 \\ \hline \end{array} \qquad \begin{array}{r} 800 \\ +800 \\ \hline \end{array} \qquad \begin{array}{r} 8{,}000 \\ +8{,}000 \\ \hline \end{array}$$

Use the facts.

$$\begin{array}{r} 80 \\ +80 \\ \hline \mathbf{160} \end{array} \qquad \begin{array}{r} 800 \\ +800 \\ \hline \mathbf{1{,}600} \end{array} \qquad \begin{array}{r} 8{,}000 \\ +8{,}000 \\ \hline \mathbf{16{,}000} \end{array}$$

GETTING READY

Add.

1. $\begin{array}{r} 40 \\ +30 \\ \hline \end{array}$ **2.** $\begin{array}{r} 700 \\ +100 \\ \hline \end{array}$ **3.** $\begin{array}{r} 2{,}000 \\ +5{,}000 \\ \hline \end{array}$ **4.** $\begin{array}{r} 80 \\ +90 \\ \hline \end{array}$ **5.** $\begin{array}{r} 400 \\ +500 \\ \hline \end{array}$

EXERCISES

Add.

1. $\begin{array}{r} 20 \\ +50 \\ \hline \end{array}$ **2.** $\begin{array}{r} 30 \\ +40 \\ \hline \end{array}$ **3.** $\begin{array}{r} 60 \\ +10 \\ \hline \end{array}$ **4.** $\begin{array}{r} 30 \\ +70 \\ \hline \end{array}$ **5.** $\begin{array}{r} 70 \\ +40 \\ \hline \end{array}$

6. $\begin{array}{r} 400 \\ +200 \\ \hline \end{array}$ **7.** $\begin{array}{r} 500 \\ +800 \\ \hline \end{array}$ **8.** $\begin{array}{r} 7{,}000 \\ +8{,}000 \\ \hline \end{array}$ **9.** $\begin{array}{r} 5{,}000 \\ +7{,}000 \\ \hline \end{array}$ **10.** $\begin{array}{r} 8{,}000 \\ +3{,}000 \\ \hline \end{array}$

Return to Facts Test B on page 3.

DIAGNOSTIC TEST: FORM A (18 MINUTES)

Add.

1.	2.	3.	4.	5.	6.
97	15	67	59	105	378
38	87	55	51	324	890
+11	+54	16	18	940	350
		+26	+43	997	94
				443	232
				+ 18	+376

7.	8.	9.	10.	11.	12.
392	599	249	246	7	23
781	350	1,611	1,448	43	37
60	4,622	51	61	61	42
99	585	3,805	107	31	50
18	1,087	251	277	57	95
+570	+ 87	+ 797	+6,408	9	10
				97	63
				+93	+60

11 or more right? Go to page 16. *Less than 11 right? Go to page 8.*

DIAGNOSTIC TEST: FORM B (18 MINUTES)

Add.

1.	2.	3.	4.	5.	6.
31	96	62	76	446	736
87	13	57	81	998	908
+28	+79	62	15	423	503
		+26	+63	206	873
				23	96
				+940	+323

7.	8.	9.	10.	11.	12.
582	789	347	3,836	6	24
91	530	2,610	637	34	11
199	1,262	71	971	72	36
37	855	508	1,599	43	38
19	2,087	152	102	58	76
+480	+ 93	+4,896	+ 937	8	24
				79	70
				+98	+98

11 or more right? Go to page 16. *Less than 11 right? See your teacher.*

SUMS OF ONE-DIGIT AND TWO-DIGIT NUMBERS

To find the sum of a one-digit and a two-digit number

- Write in columns.
- Use addition facts.

EXAMPLE 1

Add. 33 + 6

Each of the number symbols 0, 1, 2, 3, 4, 5, 6, 7, 8, and 9 is called a **digit.**

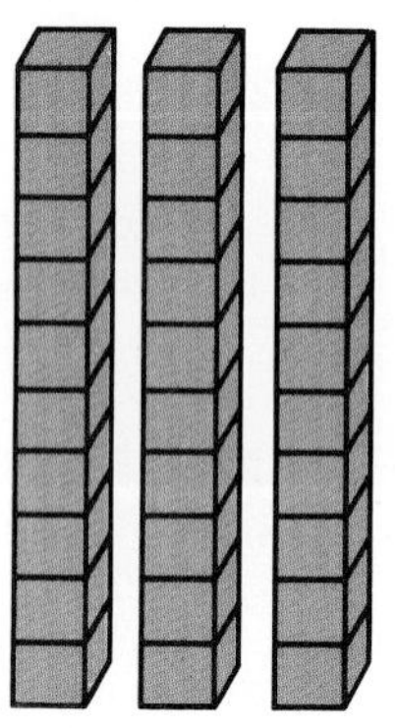

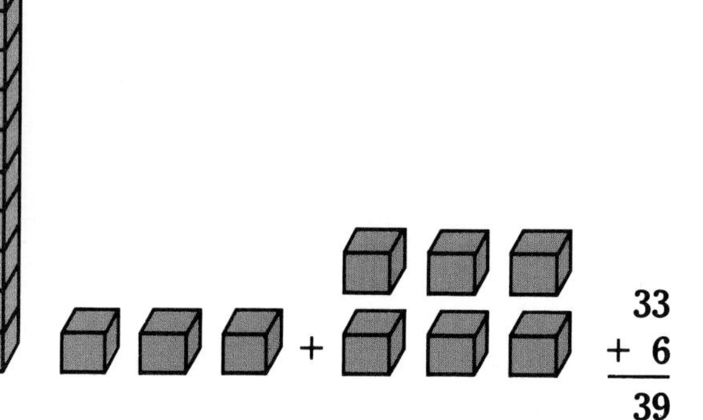

$$\begin{array}{r} 33 \\ +\ 6 \\ \hline 39 \end{array} \leftarrow \textbf{sum}$$

So, the sum is **39.**

EXAMPLE 2

Add. 58 + 4

Think:

$$\begin{array}{r} 58 \\ +\ 4 \\ \hline \end{array} \longrightarrow \begin{array}{r} \left\{\begin{array}{r} 50 \\ 8 \end{array}\right. \\ +\ 4 \\ \hline \end{array}$$

Use addition facts.

$$\begin{array}{r} 50 \\ \left.\begin{array}{r} 8 \\ +\ 4 \end{array}\right\} \\ \hline \end{array} \longrightarrow \begin{array}{r} 50 \\ +12 \\ \hline 62 \end{array}$$

So, the sum is **62.**

EXAMPLE 3

Add. 94 + 6

Think:

$$\begin{array}{r} 94 \\ +\ 6 \\ \hline \end{array} \longrightarrow \begin{array}{r} \left\{\begin{array}{r} 90 \\ 4 \end{array}\right. \\ +\ 6 \\ \hline \end{array} \quad \begin{array}{r} 90 \\ \left.\begin{array}{r} 4 \\ +\ 6 \end{array}\right\} \\ \hline \end{array} \longrightarrow \begin{array}{r} 90 \\ +10 \\ \hline 100 \end{array}$$

So, the sum is **100.**

GETTING READY

Add.

1. 56 + 3	**2.** 74 + 8	**3.** 51 + 8	**4.** 35 + 7	**5.** 95 + 9

EXERCISES

Add.

1. 23 + 5	**2.** 41 + 8	**3.** 35 + 3	**4.** 83 + 3	**5.** 14 + 3	**6.** 47 + 4
7. 61 + 5	**8.** 34 + 5	**9.** 46 + 1	**10.** 30 + 4	**11.** 52 + 7	**12.** 38 + 3
13. 47 + 2	**14.** 31 + 6	**15.** 94 + 4	**16.** 65 + 3	**17.** 63 + 1	**18.** 25 + 4
19. 47 + 5	**20.** 64 + 9	**21.** 27 + 7	**22.** 57 + 6	**23.** 18 + 9	**24.** 64 + 8
25. 78 + 5	**26.** 88 + 3	**27.** 81 + 9	**28.** 69 + 4	**29.** 37 + 3	**30.** 72 + 9
31. 96 + 9	**32.** 99 + 5	**33.** 96 + 6	**34.** 95 + 8	**35.** 95 + 6	**36.** 98 + 9

***37.** Mr. Adams paid $93 for car repairs. At the same time, he spent $8 for gasoline. How much did he spend altogether?

ON YOUR OWN

Complete the number patterns.

5, 5 + 7, 12 + 7, 19 + 7, ______, ______, ______, ______, ______

↓ ↓ ↓ ↓

5, 12, 19, ______, ______, ______, ______, ______, ______

4, 11, 18, 25, ______, ______, ______, ______, ______

1, 10, 19, 28, ______, ______, ______, ______, ______

ADDING A COLUMN OF NUMBERS

To find the sum of a column of one-digit numbers

- Look for facts that make 10.

EXAMPLE 1

Add. 3 + 5 + 7 + 5

Form a column.

Look for 10's.

3	● ● ●		
5	0 0 0 0 0		10
7	● ● ● ● ● ● ●		
+5	0 0 0 0 0		10
			20

So, the sum is **20.**

EXAMPLE 2

Add. 8 + 4 + 2 + 5 + 1 + 3

Form a column.

Make 10's and add.

8		
4	**10**	
2		**20**
5		
1	**10**	
+3		**+ 3**
		23

So, the sum is **23.**

EXAMPLE 3

Add. 8 + 7 + 3 + 5 + 9 + 4

The only facts that make 10 are 7 and 3. Add the remaining numbers beginning from the top.

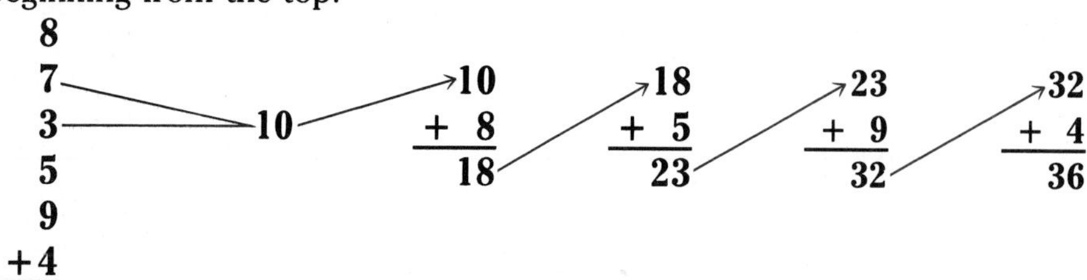

So, the sum is **36.**

GETTING READY

Add.

1.	2.	3.	4.	5.	6.	7.	8.
3	6	7	5	9	7	8	5
5	3	4	8	5	6	3	3
2	5	6	4	6	5	8	8
6	1	2	+6	4	4	5	4
+4	4	+3		+6	+9	9	7
	+1					+4	+9

EXERCISES

Add.

1.	2.	3.	4.	5.	6.	7.	8.	9.
3	5	7	9	3	5	2	7	6
7	4	2	5	4	8	9	6	9
2	1	6	1	8	2	4	3	5
6	7	1	4	5	4	3	0	8
+2	+3	3	2	2	5	3	9	2
		+1	+1	+3	+6	+5	+5	+6

10.	11.	12.	13.	14.	15.	*16.	*17.	*18.
4	3	4	8	5	3	9	1	8
6	2	6	1	7	8	5	7	3
8	5	3	5	2	1	3	8	9
3	7	1	7	0	2	8	4	4
6	2	7	4	4	6	6	5	8
+5	1	9	2	3	7	9	8	9
	8	2	3	9	4	6	7	5
	+2	+5	+6	+1	+5	+8	+4	+3

***19.** \$1 + \$3 + \$2 + \$4 + \$6 + \$2

***20.** \$3 + \$4 + \$7 + \$9 + \$8 + \$5

Solve.

21. Dorothy rode her bicycle 8 mi on Monday, 7 mi on Tuesday, and 5 mi on Wednesday. How far did she ride her bicycle in all?

22. Annie has a part-time job after school and on Saturday. This week she worked 2 h, 2 h, 3 h, 2 h, 3 h, and 5 h. How many hours did she work in all?

ADDING LARGE NUMBERS

To find the sum of large numbers

- Write each number in expanded form.
- Make columns and add.

EXAMPLE 1

Add.

$$\begin{array}{r} 37 \\ 64 \\ +52 \\ \hline \end{array}$$

Write in expanded form.

$$\begin{array}{rcrcrl} \mathbf{37} & = & \mathbf{30} & + & \mathbf{7} & \longleftarrow 3 \text{ tens} + 7 \text{ ones} \\ \mathbf{64} & = & \mathbf{60} & + & \mathbf{4} & \longleftarrow 6 \text{ tens} + 4 \text{ ones} \\ \mathbf{+52} & = & \mathbf{50} & + & \mathbf{2} & \longleftarrow 5 \text{ tens} + 2 \text{ ones} \\ \hline & & \mathbf{140} & + & \mathbf{13} & \end{array}$$

Numbers like 342 written as 300 + 40 + 2 are in **expanded form.**

Make columns and add.

$$\begin{array}{r} \mathbf{140} \\ \mathbf{+\ 13} \\ \hline \mathbf{153} \end{array}$$

So, the sum is **153**.

EXAMPLE 2

Add.

$$\begin{array}{r} 437 \\ 281 \\ +606 \\ \hline \end{array}$$

Write in expanded form.

$$\begin{array}{rcrcrcrl} \mathbf{437} & = & \mathbf{400} & + & \mathbf{30} & + & \mathbf{7} & \longleftarrow 4 \text{ hundreds} + 3 \text{ tens} + 7 \text{ ones} \\ \mathbf{281} & = & \mathbf{200} & + & \mathbf{80} & + & \mathbf{1} & \longleftarrow 2 \text{ hundreds} + 8 \text{ tens} + 1 \text{ one} \\ \mathbf{+606} & = & \mathbf{600} & + & \mathbf{0} & + & \mathbf{6} & \longleftarrow 6 \text{ hundreds} + 0 \text{ tens} + 6 \text{ ones} \\ \hline & & \mathbf{1{,}200} & + & \mathbf{110} & + & \mathbf{14} & \end{array}$$

Make columns and add.

$$\begin{array}{r} \mathbf{1{,}200} \\ \mathbf{110} \\ \mathbf{+\ \ 14} \\ \hline \mathbf{1{,}324} \end{array}$$

So, the sum is **1,324.**

GETTING READY

Add. Use expanded form.

1. $\begin{array}{r} 26 \\ 51 \\ +\ 80 \\ \hline \end{array}$ **2.** $\begin{array}{r} 52 \\ 45 \\ +\ 17 \\ \hline \end{array}$ **3.** $\begin{array}{r} 125 \\ 302 \\ +412 \\ \hline \end{array}$ **4.** $\begin{array}{r} 324 \\ 170 \\ +\ 681 \\ \hline \end{array}$ **5.** $\begin{array}{r} 463 \\ 531 \\ +\ 187 \\ \hline \end{array}$

EXERCISES

Add. Use expanded form.

1.	**2.**	**3.**	**4.**	**5.**	**6.**
34	43	78	61	82	79
25	27	41	19	35	60
+17	+30	+27	+53	+74	+18

7.	**8.**	**9.**	**10.**	**11.**	**12.**
26	19	33	47	55	98
65	73	19	50	34	53
+71	+20	+56	+93	+21	+45

13.	**14.**	**15.**	**16.**	**17.**	**18.**
48	53	28	35	57	64
32	47	52	15	22	12
+ 4	+ 6	+ 7	+48	+43	+36

19.	**20.**	**21.**	**22.**	**23.**	**24.**
56	52	27	35	85	77
14	37	41	72	8	15
+20	+ 1	+54	+89	+63	+23

25.	**26.**	***27.**	**28.**	***29.**	**30.**
84	59	351	109	873	106
23	64	76	318	542	894
+55	+96	+180	+452	+ 74	+109

31.	**32.**	***33.**	**34.**	***35.**	***36.**
952	445	555	147	783	529
200	381	4	375	803	58
+380	+611	+ 79	+224	+ 50	+831

Copy and complete. Write the sum of each row in the box to the right. Write the sum of each column in the box at the bottom.

37.

10	5	
60	3	

38.

45	7	
36	5	

39.

20	16	
30	6	

40.

50	14	
18	9	

ON YOUR OWN

The picture shows 5 nails with a total length of 60 mm. They form 3 rectangles. Make a similar sketch of 6 nails with a total length of 81 mm forming 5 triangles.

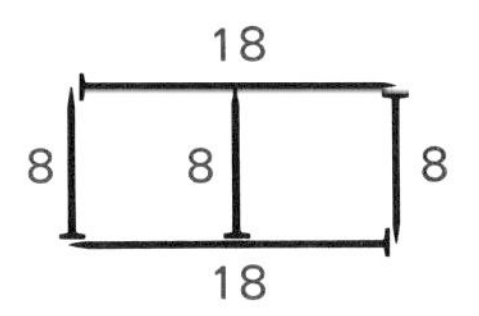

ADDING LARGE NUMBERS BY RENAMING

To find the sum of large numbers

- Add each column of numbers from right to left.
- When the sum of a column is more than 9, use renaming.

EXAMPLE 1

Add.

$$\begin{array}{r} 32 \\ 15 \\ +47 \\ \hline \end{array}$$

Write each number in expanded form.

$$\begin{array}{rcll} 32 &=& 30 + 2 & \leftarrow 3 \text{ tens} + 2 \text{ ones} \\ 15 &=& 10 + 5 & \leftarrow 1 \text{ ten} + 5 \text{ ones} \\ +47 &=& 40 + 7 & \leftarrow 4 \text{ tens} + 7 \text{ ones} \\ &=& 80 + 14 & \\ &=& 80 + 10 + 4 & \\ &=& 90 + 4 & \\ &=& \mathbf{94} & \end{array}$$

Another way to add is to use renaming.

Add the **ones** column first.

$$\begin{array}{r} 2 \\ 5 \\ +7 \\ \hline 14 \end{array}$$

Move this ten → to the top of the **tens** column. →

Now add the **tens** column.

$$\begin{array}{r} {}^{1} \\ 32 \\ 15 \\ +47 \\ \hline 94 \end{array}$$

So, the sum is **94.**

> When the sum of a column is more than 9, add the tens digit to the next column. This is called **renaming.**

EXAMPLE 2

Add.

$$\begin{array}{r} 165 \\ 382 \\ 517 \\ +264 \\ \hline \end{array}$$

Add the ones column first. Use renaming.

$$\begin{array}{r} {}^{1} \\ 165 \\ 382 \\ 517 \\ +264 \\ \hline 8 \end{array} \qquad \begin{array}{r} {}^{2\,1} \\ 165 \\ 382 \\ 517 \\ +264 \\ \hline 28 \end{array} \qquad \begin{array}{r} {}^{2\,1} \\ 165 \\ 382 \\ 517 \\ +264 \\ \hline 1{,}328 \end{array}$$

So, the sum is **1,328.**

GETTING READY

Add. Use renaming.

1.	2.	3.	4.
56	78	21	315
97	41	53	178
+ 8	+59	74	462
		+89	+281

EXERCISES

Add. Use renaming.

1.	2.	3.	4.	5.	6.
16	22	201	769	701	1,131
59	11	736	62	465	649
16	78	783	932	2,560	588
56	48	783	666	849	3,551
12	63	303	811	1,228	977
43	85	746	659	+ 358	+ 584
30	3	797	965		
+41	+14	+ 24	+827		

7.	8.	9.	10.	11.	12.
95	67	194	54	330	246
34	8	322	245	269	333
43	13	744	15	5,212	1,223
44	16	886	176	560	731
13	5	841	22	463	673
50	68	118	69	3,486	399
+37	+40	+902	+469	+ 233	+4,292

Check. Correct if needed.

***13.** 1,326 + 83 + 574 + 657 = 2,537

***14.** 405 + 8,026 + 2,003 + 51 = 10,475

Solve.

***15.** Mr. Edwards budgets a total of $669 to pay five monthly bills. He allows $47 for telephone, $78 for electricity, $365 for mortgage, and $27 for water. What does he budget for heat?

Return to Form B on page 7.

ROUNDING NUMBERS

To round a number to the nearest 10 or 100

- Find the 10 or 100 closest to that number.

EXAMPLE 1

Round 36 to the nearest 10.
36 is between 30 and 40.
Since 36 is closer to 40,
36 rounds up to 40.
So, 36 rounded to the nearest 10 is **40.**

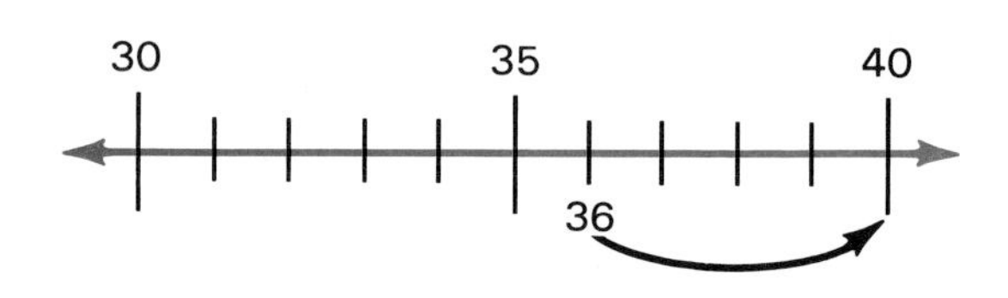

EXAMPLE 2

Round 842 to the nearest 100.
842 is between 800 and 900.
Since 842 is closer to 800,
842 rounds down to 800.
So, 842 rounded to the nearest 100 is **800.**

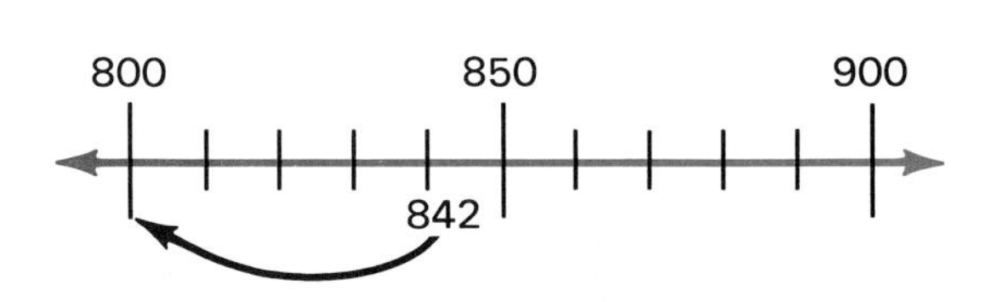

EXAMPLE 3

Round 384 to the nearest 10.
384 is between 380 and 390. **3 8 (4)**
The digit in the **ones** place is 4. ↓
Round down. **3 8 0**
So, 384 rounded to the nearest 10 is **380.**

> To round a number to the nearest 10, look at the digit in the **ones** place. If the digit is 5 or more, round up. If the digit is less than 5, round down.

EXAMPLE 4

Round 2,369 to the nearest 100.
2,369 is between 2,300 and 2,400. **2, 3 (6) 9**
The digit in the **tens** place is 6. ↓
Round up. **2, 4 0 0**
So, 2,369 rounded to the nearest 100 is **2,400.**

> To round a number to the nearest 100, look at the digit in the **tens** place. If the digit is 5 or more, round up. If the digit is less than 5, round down.

GETTING READY

Round to the nearest 10.

1. 84 **2.** 567 **3.** 385 **4.** 8,943

Round to the nearest 100.

5. 281 **6.** 654 **7.** 945 **8.** 8,199

EXERCISES

Round to the nearest 10.

1. 77 **2.** 9 **3.** 84 **4.** 361 **5.** 473

6. 179 **7.** 643 **8.** 355 **9.** 91 **10.** 1,563

11. 865 **12.** 2,472 ***13.** 98 ***14.** 197 ***15.** 1,396

Round to the nearest 100.

16. 229 **17.** 864 **18.** 149 **19.** 582 **20.** 722

21. 257 **22.** 356 **23.** 144 **24.** 3,379 **25.** 5,809

26. 187 **27.** 937 ***28.** 87 ***29.** 968 ***30.** 2,961

CALCULATOR

One famous set of numbers is called Pascal's Triangle. Use a calculator to find the sum of each row of numbers. Try to make two **more** rows for the triangle. Find their sums too!

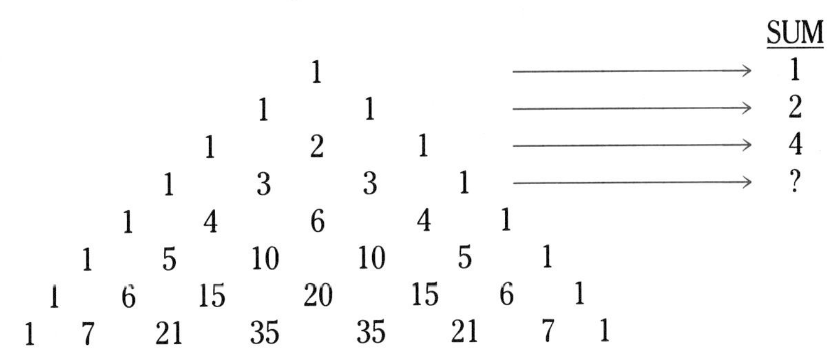

ESTIMATING SUMS

To estimate sums

- Round each number to the same place.
- Add the rounded numbers.

EXAMPLE 1

Estimate the sum of 84 + 67 + 35. Round each number to the nearest 10.

Round each number.	**84** ⟶	**80**	
	67 ⟶	**70**	
Add.	**+35** ⟶	**40**	
		190	So, the estimated sum is **190**.

EXAMPLE 2

About how much is the total bill for items costing \$3.38, \$2.49, and \$0.83? Round each cost to the nearest dollar.

Round to the nearest dollar.	**\$3.38** ⟶	**\$3**	
	2.49 ⟶	**2**	
Add.	**+0.83** ⟶	**1**	
		\$6	So, the total is about **\$6.**

EXAMPLE 3

Estimate the sum of 572 + 347 + 55.

Round to the nearest 100.	**572** ⟶	**600**
	347 ⟶	**300**
Add.	**+ 55** ⟶	**100**
		1,000

Notice that 55 is the smallest number. The largest place in 55 is tens.

To get a closer estimate, you can round to the nearest 10.

Round to the nearest 10.	**572** ⟶	**570**
	347 ⟶	**350**
Add.	**+ 55** ⟶	**60**
		980

So, **1,000** is the estimate to the nearest 100, and **980** is the estimate to the nearest 10.

> To get a closer estimate of a sum, round each number to the largest place in the smallest number.

GETTING READY

Estimate.

1. 23 + 46 **2.** 179 + 260 + 413 **3.** 109 + 25 + 17 **4.** $1.39 + $3.54

EXERCISES

Estimate. Round each number to the nearest 10.

1. 18 + 25 + 51 **2.** 34 + 75 + 62 ***3.** 18 + 27 + 56 + 3

Estimate. Round each number to the nearest 100.

4. 320 + 185 + 407 **5.** 276 + 542 + 109 **6.** 89 + 235 + 674 ***7.** 231 + 34 + 91

Estimate. Round each amount to the nearest dollar.

8. $2.29 + $1.89 **9.** $0.39 + $1.25 + $4.99 **10.** $2.17 + $0.73 + $4.10

11. $\begin{array}{r} \$3.50 \\ 1.98 \\ 0.29 \\ +4.33 \\ \hline \end{array}$ **12.** $\begin{array}{r} \$0.79 \\ 3.85 \\ 4.22 \\ +0.57 \\ \hline \end{array}$ **13.** $\begin{array}{r} \$5.01 \\ 1.46 \\ 0.75 \\ +2.18 \\ \hline \end{array}$

Estimate.

14. $\begin{array}{r} 155 \\ 347 \\ 611 \\ +294 \\ \hline \end{array}$ **15.** $\begin{array}{r} 549 \\ 267 \\ 54 \\ +333 \\ \hline \end{array}$ **16.** $\begin{array}{r} 213 \\ 586 \\ 674 \\ +\ 37 \\ \hline \end{array}$

17. $\begin{array}{r} 1{,}236 \\ 725 \\ 2{,}861 \\ +\ \ 84 \\ \hline \end{array}$ ***18.** $\begin{array}{r} 85 \\ 5 \\ 43 \\ +\ 8 \\ \hline \end{array}$ ***19.** $\begin{array}{r} \$101.49 \\ 21.13 \\ 14.21 \\ +\ 15.42 \\ \hline \end{array}$

Solve by estimating.

20. The Davidson family used 597 kilowatt hours (kW·h) of electricity in June, 624 kW·h in July and 721 kW·h in August. Estimate the total kilowatt hours used for the three-month period.

***21.** About how far is it from Marshall to Johnson? Round to the nearest 10 mi.

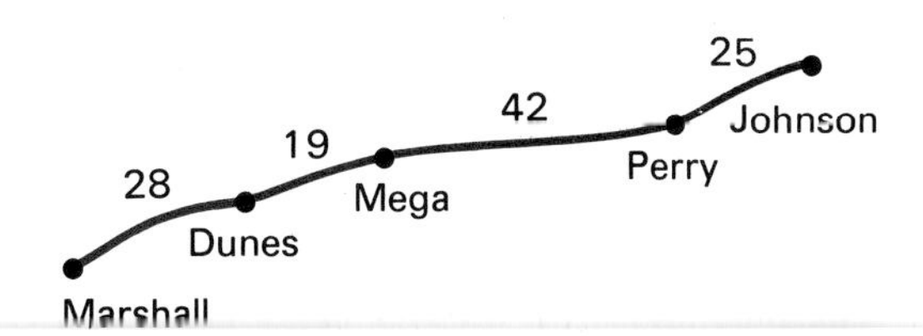

ADDING MENTALLY

To find sums mentally

- Use numbers that are multiples of 10.
- With dollars and cents, try to make whole dollar amounts.

EXAMPLE 1

Find 98 + 99 mentally.

Use 100.
98 is 2 less than 100.
99 is 1 less than 100.

$$\begin{array}{r} \mathbf{98} = \mathbf{100 - 2} \\ \underline{\mathbf{+99}} = \underline{\mathbf{100 - 1}} \\ \mathbf{200 - 3 = 197} \end{array}$$

So, the sum is **197.**
Check by adding in columns.

> To add mentally, choose the multiple of 10 closest to all of the numbers.

EXAMPLE 2

Find 37 + 42 + 43 mentally.

Use 40.
37 is 3 less than 40.
42 is 2 more than 40.
43 is 3 more than 40.

$$\begin{array}{r} \mathbf{37} = \mathbf{40 - 3} \\ \mathbf{42} = \mathbf{40 + 2} \\ \underline{\mathbf{+43}} = \underline{\mathbf{40 + 3}} \\ \mathbf{120 + 2 = 122} \end{array}$$

So, the sum is **122.**
Check by adding in columns.

EXAMPLE 3

Find \$1.25 + \$0.79 + \$0.15 mentally.

Think:

\$1.25	**\$1.25** → **0.75**	**\$2.00**
0.79 → 0.75 + 0.04		
+0.15	**0.04** + **0.15**	**0.19**
		\$2.19

So, the sum is **\$2.19.**
Check by adding in columns.

> To add dollars and cents mentally, try to make whole dollar amounts.

GETTING READY

Find the sum mentally. Check by adding in columns.

1. 96 + 98 **2.** 27 + 31 + 30 **3.** \$2.25 + \$1.75

4. \$0.89 + \$0.41 **5.** \$4.50 + \$0.99 + \$0.31 **6.** \$1.85 + \$0.17 + \$3.20

EXERCISES

Find the sum mentally. Check by adding in columns.

1. 98 + 101 **2.** 99 + 97 + 100 **3.** 45 + 35 + 42 **4.** 151 + 149

5. 103 + 94 **6.** 16 + 21 + 22 **7.** 47 + 53 + 46 **8.** 155 + 147

9. 8 + 12 + 10 + 15 + 9

10. 89 + 88 + 93

11. \$55 + 45 + 50 + 48

12. 39 + 36 + 42 + 45 + 40

13. \$0.75 + \$0.50 **14.** \$1.20 + \$0.89 **15.** \$4.89 + \$0.31

16. \$2.28 + 0.12 + 0.40

17. \$0.98 + 0.42 + 1.50

18. \$7.50 + 0.99 + 0.08

***19.** Patti budgets \$5.00 a week for her school lunches. One week she paid the following: \$1.02, \$0.94, \$1.08, \$0.95, and \$1.04. Was she within her budget?

CALCULATOR

There is a special sequence of numbers in which each new number is found by adding the two numbers just before it. It is called a Fibonacci sequence.

Use the calculator to find some missing numbers of the sequence.

1, 1, 2, 3, 5, 8, ___, ___, ___, ___, ___, ___, ___, ___, . . .

How many numbers in the sequence are less than 1,000 in value?

Use the calculator to find the sum of these particular numbers.

Problem Solving Skill

To read and interpret a bar graph

- Find the bar you need.
- Think of a line from the top of the bar to the scale.
- Now read the scale.

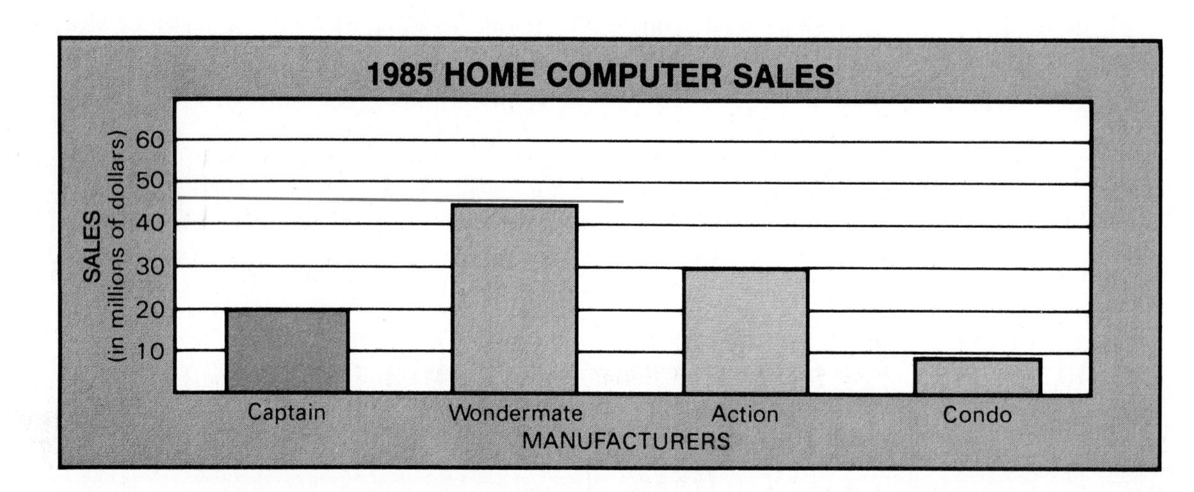

EXAMPLE

What were the total sales of Wondermate computers in 1985?

Find the bar representing Wondermate.
Think of a line from the top of the bar to the scale.
Notice that the line is halfway between 40 and 50.

So, Wondermate computer sales were **$45 million.**

Answer each question about the bar graph.

1. What does the bar graph describe?

2. What does the scale show?

3. What kind of units does the scale represent?

4. What were the sales of Action computers in 1985?

5. By about how much more were the sales of Wondermate computers than Captain computers?

6. Which company had the greatest amount of sales in 1981?
7. Whose sales were lowest in 1981?
*8. About how many times greater were the sales of Action computers than Condo computers?

The graph on page 22 is an example of a **vertical** bar graph. Below you see a **horizontal** bar graph. It can be read in the same way, but notice that the scale is along the bottom of the graph.

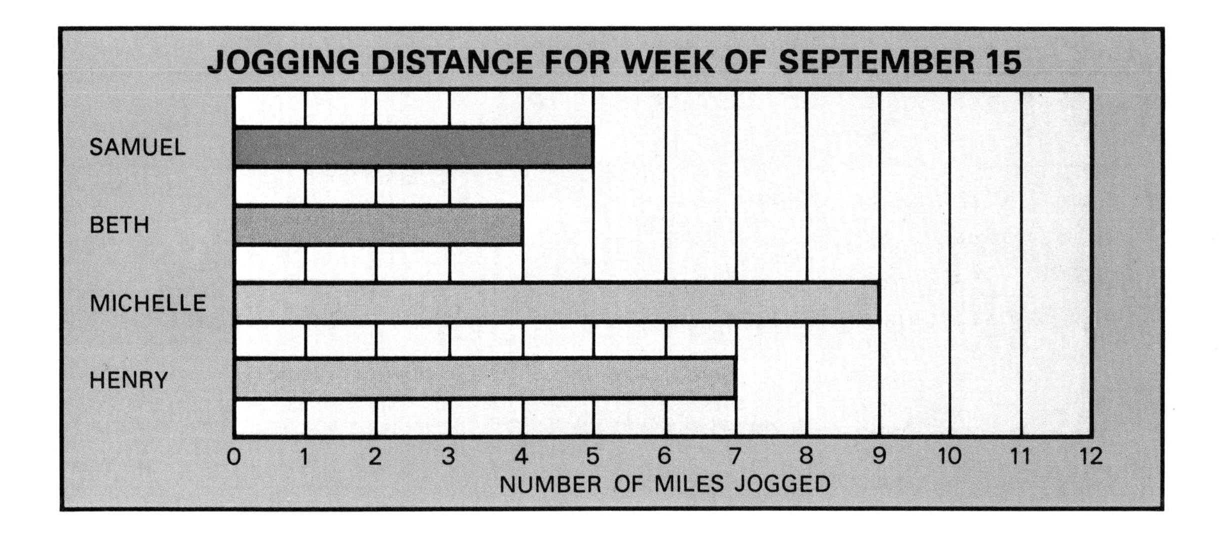

Answer each question about the bar graph.

9. What is the title of the bar graph?
10. Who jogged more miles than Samuel?
11. Who jogged the most miles during the week?
12. Find the number of miles jogged for the week by the entire group.
13. How many more miles did Henry jog during the week than Beth?
*14. During the week of September 8, Samuel jogged 8 mi. How many more miles would he have had to jog the week of the 15th to equal 8 mi?

Problem Solving Applications

READ • PLAN • SOLVE • CHECK

The Morris Career Development Center offers a 2-year program in cosmetology to high school juniors and seniors. The students attend classes, see demonstrations of basic techniques, and practice these techniques in a special laboratory.

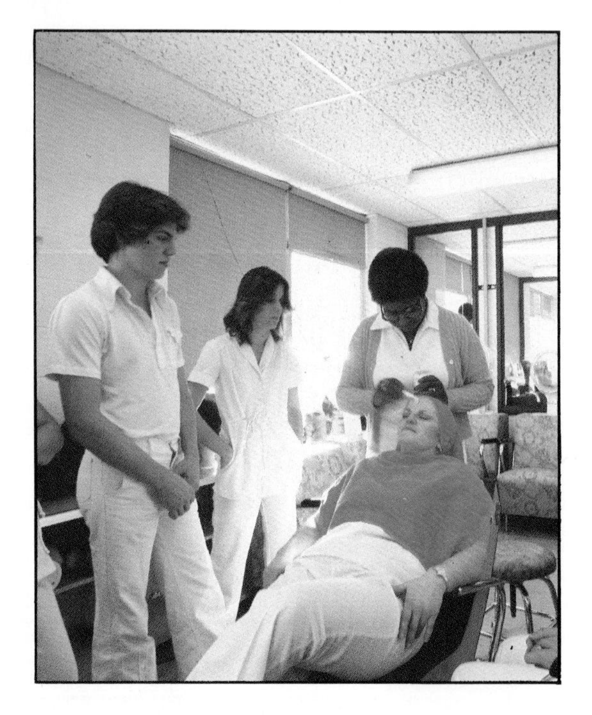

AS YOU READ

- activator—a chemical used to make other chemicals mix
- developer—a chemical used in mixing a bleach
- peroxide—a bleaching chemical

Answer these questions about the cosmetology program at Morris.

1. Morris offers 3 sessions: morning, afternoon, and evening. This year's 3 sessions enrolled 20, 24, and 25 students. How many students are enrolled in the program?

2. To mix a bleach, 2 oz of oil *developer,* 2 oz of cream *peroxide*, and 2 oz of clear peroxide are added to 2 packets of *activator* powder. How much peroxide and developer are added in all?

3. To enter the program, students must complete a health form and pay a $10 registration fee. They must also pay $78 for a kit, $19 for a uniform, and $15 for shoes. Find the total expenses.

4. Bill is preparing shampoo. He adds 7 parts water to 1 part shampoo concentrate. If he uses 1 pt of shampoo concentrate, how many pints of shampoo is he preparing?

5. The laboratory contains 25 hairdressing stations and 18 manicure tables. How many work areas are there in all?

6. To frost a person's hair, 1 oz of color is mixed with 2 oz of peroxide. How many ounces of the color-peroxide mixture are needed for 8 people?

Career: Cosmetologists

READ • PLAN • SOLVE • CHECK

Cosmetologists work in beauty salons, barber and styling shops, department stores, hospitals, and hotels. Owners of large salons employ hairdressers, shampooists, manicurists, and receptionists. Training in cosmetology can last from 6 months to a year.

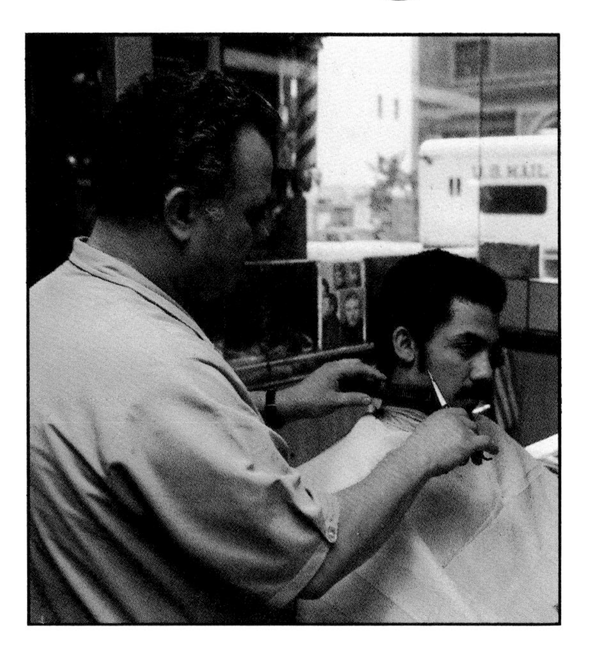

AS YOU READ

- manicurist—a person who cares for the fingernails
- setting lotion—a lotion used in setting hair on rollers

Solve.

1. Mario, owner of Mario's Beautique, is preparing *setting lotion.* He mixes 16 oz of concentrate with 112 oz of water. How many ounces of lotion are there?

2. Anita Furman operates the beauty salon in a hospital. She keeps the salon open 4 h a day, 5 d a week. How long is the salon open each week?

3. Mario charges the following: permanent, \$35; color tint, \$18; bleaching, \$35; and haircut, \$9. How much would a customer be charged for a haircut and color tint?

4. Marlene Bennett works part-time as a *manicurist.* She earns \$4.50 for every \$6 manicure she gives. If she gave 4 manicures on Thursday, how much did she earn that day?

5. Jerry and Lisa Winston are co-owners of the Eastland Barber Shop. Jerry gave 35 haircuts on Tuesday, 30 on Wednesday, 44 on Thursday, 41 on Friday, and 50 on Saturday. If Lisa gave 200 haircuts that week, did she give more haircuts than Jerry?

6. Rudy Porter owns his own salon. This week he ordered the following supplies: 2 pairs of scissors for \$20 each, 2 cases of shampoo for \$24 each, and 1 case of hair spray for \$54. How much did these supplies cost?

CHAPTER TEST 1

Add.

1.	2.	3.	4.	5.	6.
257	651	377	819	1,760	689
390	45	69	542	432	3,044
235	220	190	108	2,105	162
880	404	444	385	49	4,306
588	56	746	27	+ 381	95
+341	+731	+ 43	455		+ 417
			16		
			+577		

Round to the nearest 10.

7. 53 **8.** 785

Round to the nearest 100.

9. 248 **10.** 83

Solve.

11. Estimate the total votes cast if the results were as follows: Ann Davey, 3,271; Jim Matsel, 1,564; Paula Ramirez, 899; and Melvin Comet, 320. (Round each result to the nearest 100 votes.)

12. Marion paid $35 for a permanent, $18 for a color tint, and $9 for a haircut. How much did she pay in all?

Use the bar graph for **13–14.**

13. During which year were the most cars sold?

14. Estimate how many millions of new cars were sold in 1979.

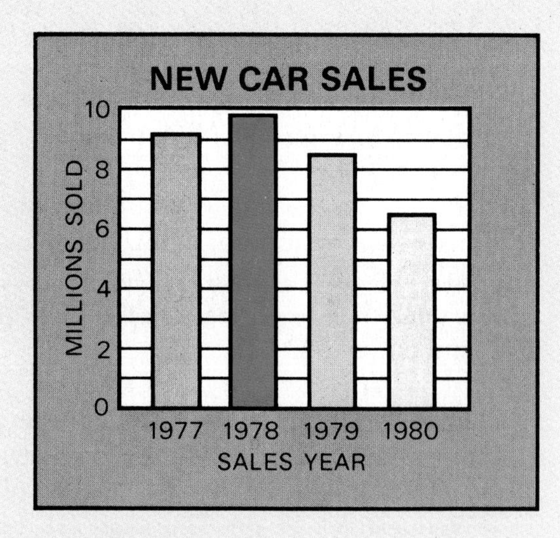

15. Estimate the sum by rounding each amount to the nearest dollar.
$3.75 + $0.99 + $2.10 + $1.85 + $3.26

Subtraction of Whole Numbers

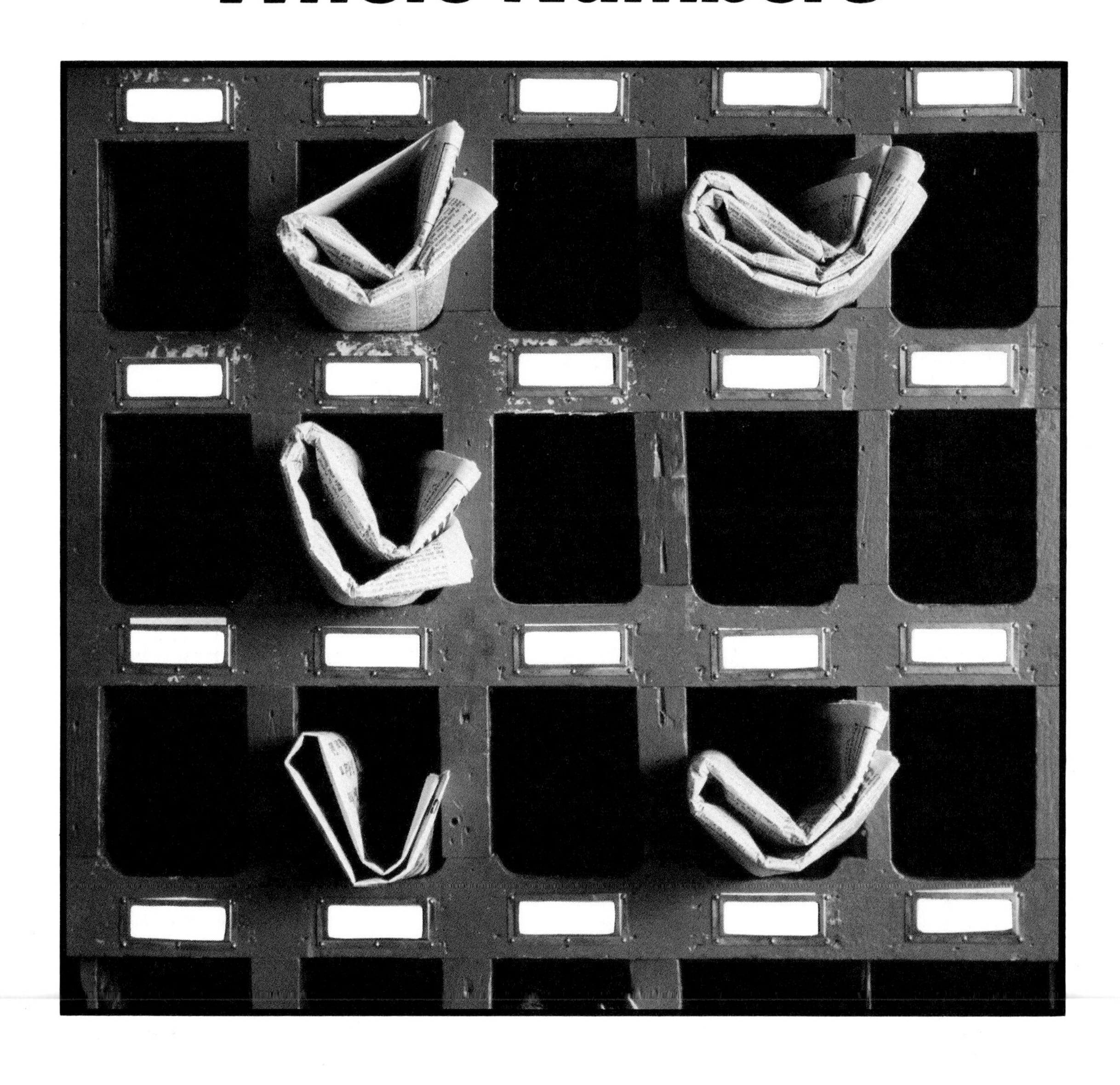

SUBTRACTION FACTS TEST A (3 MINUTES)

1.	$6 - 5$	$4 - 3$	$3 - 1$	$9 - 8$	$10 - 1$	$8 - 3$	$11 - 9$	$2 - 1$	$8 - 2$	$9 - 6$
2.	$10 - 9$	$5 - 2$	$5 - 1$	$3 - 2$	$4 - 1$	$7 - 6$	$6 - 1$	$8 - 1$	$6 - 2$	$17 - 9$
3.	$7 - 1$	$4 - 2$	$11 - 6$	$10 - 5$	$5 - 4$	$10 - 3$	$17 - 8$	$14 - 9$	$6 - 3$	$10 - 4$
4.	$15 - 9$	$7 - 2$	$16 - 7$	$12 - 9$	$9 - 9$	$14 - 7$	$10 - 6$	$10 - 8$	$5 - 0$	$8 - 4$
5.	$11 - 2$	$12 - 6$	$10 - 7$	$18 - 9$	$9 - 2$	$12 - 4$	$15 - 8$	$9 - 4$	$9 - 7$	$7 - 5$
6.	$13 - 7$	$12 - 8$	$13 - 8$	$9 - 5$	$9 - 3$	$14 - 6$	$8 - 0$	$12 - 5$	$13 - 9$	$7 - 7$
7.	$14 - 8$	$13 - 5$	$7 - 4$	$11 - 4$	$5 - 3$	$12 - 3$	$15 - 6$	$15 - 7$	$13 - 4$	$11 - 5$

67 or more right? Go to page 33. *Less than 67 right? Go to page 29.*

SUBTRACTION FACTS TEST B (2 MINUTES)

1.	$10 - 5$	$12 - 9$	$6 - 3$	$11 - 9$	$10 - 3$	$10 - 2$	$14 - 9$	$10 - 4$	$11 - 2$	$16 - 8$
2.	$18 - 9$	$10 - 8$	$12 - 6$	$2 - 2$	$14 - 7$	$10 - 7$	$16 - 7$	$13 - 7$	$12 - 8$	$15 - 8$
3.	$13 - 8$	$12 - 4$	$5 - 0$	$14 - 8$	$12 - 5$	$11 - 6$	$14 - 6$	$3 - 3$	$11 - 4$	$8 - 5$
4.	$15 - 7$	$5 - 3$	$17 - 8$	$15 - 6$	$11 - 5$	$16 - 9$	$8 - 3$	$11 - 8$	$9 - 6$	$13 - 4$
5.	$17 - 9$	$5 - 0$	$12 - 3$	$11 - 7$	$13 - 5$	$13 - 6$	$12 - 7$	$11 - 3$	$14 - 5$	$16 - 7$

48 or more right? Go to page 33. *Less than 48 right? See your teacher.*

SUBTRACTION FACTS

To learn the subtraction facts

- Think of removing objects from a group.
- Use related addition facts.

EXAMPLE 1

Subtract. $\begin{array}{r} 12 \\ -\ 6 \\ \hline \end{array}$

Think of 12 circles. Cross out 6.
How many are left?
So, $12 - 6 =$ **6.**

○○⊗⊗⊗⊗
○○○○⊗⊗

> Each subtraction fact can be shown in two ways.
> **Example:** $\begin{array}{r} 12 \\ -\ 6 \\ \hline \end{array}$ or $12 - 6$.

EXAMPLE 2

Subtract. $\begin{array}{r} 17 \\ -\ 9 \\ \hline \end{array}$

Think: What do you add to 9 to get 17?
The related addition fact is $9 + 8 = 17$.
So, $17 - 9 =$ **8.**

GETTING READY

Subtract. Try to use related addition facts.

1. $\begin{array}{r} 16 \\ -\ 7 \\ \hline \end{array}$ **2.** $\begin{array}{r} 13 \\ -\ 8 \\ \hline \end{array}$ **3.** $\begin{array}{r} 15 \\ -\ 6 \\ \hline \end{array}$ **4.** $\begin{array}{r} 12 \\ -\ 7 \\ \hline \end{array}$

5. List all of the subtraction facts that you missed on Facts Test A and practice them until you know them.

EXERCISES

Subtract.

1. $\begin{array}{r} 17 \\ -\ 8 \\ \hline \end{array}$ **2.** $\begin{array}{r} 16 \\ -\ 9 \\ \hline \end{array}$ **3.** $\begin{array}{r} 15 \\ -\ 7 \\ \hline \end{array}$ **4.** $\begin{array}{r} 14 \\ -\ 6 \\ \hline \end{array}$

5. $14 - 9$ **6.** $13 - 7$ **7.** $18 - 9$ **8.** $15 - 8$

***9.** Ted read 6 pages of a driver's education manual containing 13 pages. How many more are left to read?

FINDING DIFFERENCES WITH THE FACTS

To find the difference of a one-digit and a two-digit number

- Subtract in columns, starting from the right.
- Use subtraction facts.
- Write multiples of 10 as a sum.

EXAMPLE 1

Subtract. $\begin{array}{r} 26 \\ -\ 3 \\ \hline \end{array}$

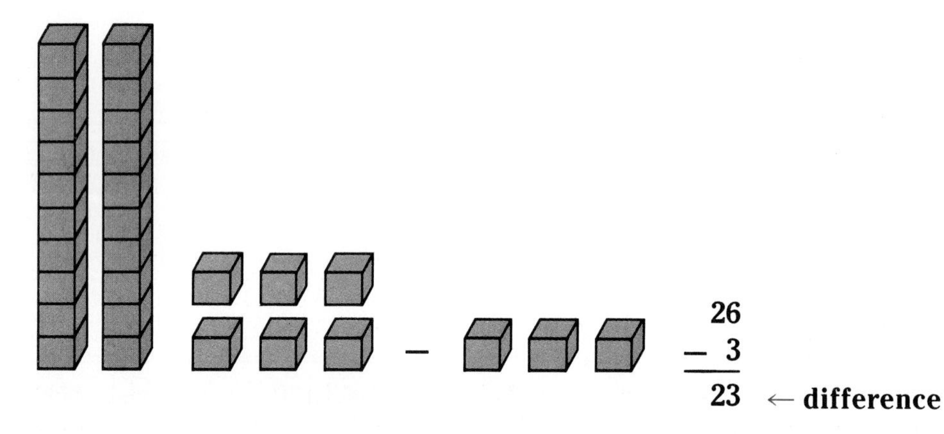

$\begin{array}{r} 26 \\ -\ 3 \\ \hline 23 \end{array}$ ← **difference**

So, the difference is **23.**

EXAMPLE 2

Subtract. $\begin{array}{r} 48 \\ -\ 8 \\ \hline \end{array}$

Use subtraction facts. $\begin{array}{r} \mathbf{48} \\ \mathbf{-\ 8} \\ \hline \mathbf{40} \end{array}$

So, the difference is **40.**

EXAMPLE 3

Subtract. $\begin{array}{r} 50 \\ -\ 9 \\ \hline \end{array}$

You cannot subtract 9 from 0.
Write 50 as a sum.

$\begin{array}{r} \mathbf{50} \\ \mathbf{-\ 9} \\ \hline \end{array} \longrightarrow \begin{array}{r} \mathbf{40 + 10} \\ \mathbf{-\qquad 9} \\ \hline \mathbf{40 + \ \ 1} \end{array} \mathbf{= 41}$

So, the difference is **41.**

GETTING READY

Subtract.

1. 47 − 3	**2.** 76 − 4	**3.** 29 − 5	**4.** 37 − 5	**5.** 28 − 8	**6.** 30 − 2
7. 46 − 4	**8.** 63 − 3	**9.** 40 − 7	**10.** 60 − 5	**11.** 50 − 8	**12.** 57 − 3

EXERCISES

Subtract.

1. 39 − 7	**2.** 84 − 2	**3.** 97 − 5	**4.** 39 − 6	**5.** 75 − 4	**6.** 88 − 4
7. 99 − 5	**8.** 67 − 3	**9.** 78 − 5	**10.** 99 − 2	**11.** 83 − 3	**12.** 95 − 5
13. 89 − 9	**14.** 74 − 4	**15.** 87 − 7	**16.** 60 − 4	**17.** 70 − 6	**18.** 50 − 9
19. 80 − 7	**20.** 70 − 8	**21.** 87 − 5	**22.** 66 − 4	**23.** 88 − 8	**24.** 90 − 4
25. 27 − 7	**26.** 84 − 3	**27.** 78 − 5	**28.** 60 − 9	**29.** 80 − 8	**30.** 75 − 5

Solve.

***31.** Ed spent $50 for paint and supplies. He returned an unused brush and received a $6 refund. How much did his project cost?

CALCULATOR

Make a guess for the answer. Then use the calculator to get the actual answer. Was your guess larger or smaller than the calculator's answer?

1. 278,015 − 83,699 **2.** 546,397 − 392,748

3. 8,009,007 − 5,998,099 **4.** 45,329,855 − 8,900,376

SUBTRACTING MULTIPLES

To subtract multiples of 10; 100; or 1,000

- Subtract in columns, starting from the right.
- Use subtraction facts.

EXAMPLE

Subtract.

$$\begin{array}{r} 170 \\ -\ 90 \\ \hline \end{array} \qquad \begin{array}{r} 1,700 \\ -\ 900 \\ \hline \end{array} \qquad \begin{array}{r} 17,000 \\ -\ 9,000 \\ \hline \end{array}$$

Look for subtraction facts.

$$\begin{array}{r} 170 \\ -\ 90 \\ \hline \mathbf{80} \end{array} \qquad \begin{array}{r} 1,700 \\ -\ 900 \\ \hline \mathbf{800} \end{array} \qquad \begin{array}{r} 17,000 \\ -\ 9,000 \\ \hline \mathbf{8,000} \end{array}$$

GETTING READY

Subtract.

1. $\begin{array}{r} 90 \\ -50 \\ \hline \end{array}$ **2.** $\begin{array}{r} 500 \\ -200 \\ \hline \end{array}$ **3.** $\begin{array}{r} 130 \\ -\ 90 \\ \hline \end{array}$ **4.** $\begin{array}{r} 1,600 \\ -\ 800 \\ \hline \end{array}$ **5.** $\begin{array}{r} 900 \\ -200 \\ \hline \end{array}$

EXERCISES

Subtract.

1. $\begin{array}{r} 80 \\ -20 \\ \hline \end{array}$ **2.** $\begin{array}{r} 110 \\ -\ 70 \\ \hline \end{array}$ **3.** $\begin{array}{r} 100 \\ -\ 50 \\ \hline \end{array}$ **4.** $\begin{array}{r} 160 \\ -\ 90 \\ \hline \end{array}$ **5.** $\begin{array}{r} 170 \\ -\ 80 \\ \hline \end{array}$

6. $\begin{array}{r} 60 \\ -40 \\ \hline \end{array}$ **7.** $\begin{array}{r} 40 \\ -30 \\ \hline \end{array}$ **8.** $\begin{array}{r} 130 \\ -\ 70 \\ \hline \end{array}$ **9.** $\begin{array}{r} 140 \\ -\ 80 \\ \hline \end{array}$ **10.** $\begin{array}{r} 110 \\ -\ 60 \\ \hline \end{array}$

11. $\begin{array}{r} 700 \\ -400 \\ \hline \end{array}$ **12.** $\begin{array}{r} 200 \\ -200 \\ \hline \end{array}$ **13.** $\begin{array}{r} 1,300 \\ -\ 400 \\ \hline \end{array}$ **14.** $\begin{array}{r} 1,500 \\ -\ 900 \\ \hline \end{array}$ **15.** $\begin{array}{r} 1,000 \\ -\ 400 \\ \hline \end{array}$

16. $\begin{array}{r} 9,000 \\ -3,000 \\ \hline \end{array}$ **17.** $\begin{array}{r} 11,000 \\ -\ 5,000 \\ \hline \end{array}$ **18.** $\begin{array}{r} 15,000 \\ -\ 6,000 \\ \hline \end{array}$ **19.** $\begin{array}{r} 14,000 \\ -\ 9,000 \\ \hline \end{array}$

***20.** Miss McQuire withdrew $600 from a savings account containing $13,000. How much remained in her account?

Return to Facts Test B on page 28.

DIAGNOSTIC TEST: FORM A (20 MINUTES)

Subtract.

1. 45 − 6	**2.** 63 − 5	**3.** 82 − 29	**4.** 72 − 23
5. 831 − 84	**6.** 824 − 45	**7.** 947 − 899	**8.** 9,596 − 6,825
9. 9,845 − 8,249	**10.** 71,117 − 5,555	**11.** 54,793 − 32,895	**12.** 78,905 − 18,737
13. 60,801 − 3,625	**14.** 302,402 − 91,559	**15.** 303,504 − 113,486	**16.** 40,050 − 8,512
17. 40,000 − 329	**18.** 60,000 − 1,086	**19.** 20,000 − 78	**20.** 27,342 − 7,407

19 or more right? Go to page 38. *Less than 19 right? Go to page 34.*

DIAGNOSTIC TEST: FORM B (20 MINUTES)

Subtract.

1. 83 − 6	**2.** 33 − 5	**3.** 81 − 28	**4.** 73 − 35
5. 741 − 76	**6.** 851 − 46	**7.** 947 − 889	**8.** 9,684 − 7,741
9. 9,633 − 8,439	**10.** 82,237 − 6,666	**11.** 45,683 − 32,785	**12.** 68,804 − 28,646
13. 30,702 − 4,435	**14.** 802,406 − 80,667	**15.** 702,405 − 222,376	**16.** 30,070 − 7,463
17. 30,000 − 436	**18.** 60,000 − 2,078	**19.** 40,000 − 97	**20.** 26,632 − 6,805

19 or more right? Go to page 38. *Less than 19 right? See your teacher.*

RENAMING WITH TWO-DIGIT NUMBERS

To subtract a one-digit or a two-digit number from a two-digit number

- Subtract in columns, starting from the right.
- Rename tens as ones.

EXAMPLE 1

Subtract. $\begin{array}{r} 50 \\ -\ 8 \\ \hline \end{array}$

Look at the ones column. $\begin{array}{r} 5\ 0 \\ -\ \ 8 \\ \hline \end{array}$ ⟵ You cannot subtract 8 from 0.

Rename 50 as 4 tens and 10 ones.

Subtract. $\begin{array}{r} \scriptstyle 4\ 10 \\ \not{5}\ \not{0} \\ -\ \ 8 \\ \hline 4\ 2 \end{array}$

So, the difference is **42.**

EXAMPLE 2

Subtract. $\begin{array}{r} 34 \\ -\ 7 \\ \hline \end{array}$

Look at the ones column. $\begin{array}{r} 3\ 4 \\ -\ \ 7 \\ \hline \end{array}$ ⟵ You cannot subtract 7 from 4.

Rename 34 as 2 tens and 14 ones.

Subtract. $\begin{array}{r} \scriptstyle 2\ 14 \\ \not{3}\ \not{4} \\ -\ \ 7 \\ \hline 2\ 7 \end{array}$

So, the difference is **27.**

EXAMPLE 3

Subtract and check. $\begin{array}{r} 73 \\ -55 \\ \hline \end{array}$

Rename and subtract. $\begin{array}{r} \scriptstyle 6\ 13 \\ \not{7}\ \not{3} \\ -\ 5\ 5 \\ \hline 1\ 8 \end{array}$

Check by adding. $18 + 55 = 73.$

So, the difference is **18.**

> To check subtraction use addition.
> **Example:**
> $\begin{array}{r} 54 \\ -\ 6 \\ \hline 48 \end{array}$ ⟶ $\begin{array}{r} 48 \\ +\ 6 \\ \hline 54 \end{array}$ ✓

GETTING READY

Subtract and check.

1. $80 - 6$	**2.** $45 - 8$	**3.** $94 - 17$	**4.** $62 - 58$	**5.** $81 - 18$

EXERCISES

Subtract and check.

1. $50 - 9$	**2.** $60 - 5$	**3.** $40 - 3$	**4.** $70 - 8$	**5.** $90 - 3$
6. $47 - 9$	**7.** $53 - 8$	**8.** $62 - 7$	**9.** $46 - 9$	**10.** $82 - 8$
11. $36 - 8$	**12.** $43 - 6$	**13.** $52 - 4$	**14.** $86 - 7$	**15.** $58 - 9$
16. $30 - 18$	**17.** $50 - 27$	**18.** $40 - 17$	**19.** $60 - 12$	**20.** $80 - 29$
21. $70 - 28$	**22.** $80 - 34$	**23.** $90 - 53$	**24.** $70 - 54$	**25.** $60 - 27$
26. $84 - 16$	**27.** $92 - 35$	**28.** $77 - 19$	**29.** $64 - 58$	**30.** $42 - 37$
31. $48 - 29$	**32.** $74 - 28$	**33.** $45 - 18$	**34.** $53 - 47$	**35.** $28 - 19$
36. $91 - 77$	**37.** $83 - 69$	**38.** $47 - 19$	**39.** $58 - 39$	**40.** $27 - 18$
41. $63 - 48$	**42.** $72 - 47$	**43.** $67 - 38$	**44.** $84 - 65$	**45.** $82 - 58$
46. $84 - 57$	**47.** $77 - 28$	**48.** $53 - 37$	**49.** $64 - 29$	**50.** $56 - 27$

Solve.

***51.** Mr. Fisher's checkbook balance was \$54. After writing a check for \$29 to pay for a plumbing bill, what was his checkbook balance?

RENAMING WITH LARGE NUMBERS

To find the difference of large numbers

- Subtract in columns, starting from the right.
- Use renaming, if necessary.

EXAMPLE 1

Subtract.

$$\begin{array}{r} 254 \\ -\ 27 \\ \hline \end{array}$$

Rename 54.
Subtract.

$$\begin{array}{r} {}^{4\ 14} \\ 2\not{5}\not{4} \\ -\ 27 \\ \hline 227 \end{array}$$

So, the difference is **227.**

EXAMPLE 2

Subtract.

$$\begin{array}{r} 832 \\ -\ 54 \\ \hline \end{array}$$

Rename 32.

$$\begin{array}{r} {}^{2\ 12} \\ 8\not{3}\not{2} \\ -\ 54 \\ \hline 8 \end{array}$$

Rename in the hundreds and tens column.
Then subtract.

$$\begin{array}{r} {}^{7\ 12} \\ {}^{\not{2}\ 12} \\ \not{8}\not{3}\not{2} \\ -\ 54 \\ \hline 778 \end{array}$$

So, the difference is **778.**

EXAMPLE 3

Subtract and check.

$$\begin{array}{r} 5{,}000 \\ -\ 436 \\ \hline \end{array}$$

Rename as necessary.

$$\begin{array}{r} {}^{4\ 9\ 9\ 10} \\ \not{5}\not{0}\not{0}\not{0} \\ -\ 436 \\ \hline 4{,}564 \end{array}$$

Check by adding. $4{,}564 + 436 = 5{,}000$

So, the difference is **4,564.**

GETTING READY

Subtract and check.

1. 326 − 17	**2.** 753 − 85	**3.** 4,000 − 654	**4.** 3,800 − 941

EXERCISES

Subtract and check.

1. 352 − 27	**2.** 485 − 39	**3.** 642 − 36	**4.** 843 − 28	**5.** 456 − 48
6. 927 − 54	**7.** 835 − 61	**8.** 563 − 72	**9.** 845 − 62	**10.** 795 − 97
11. 356 − 29	**12.** 478 − 89	**13.** 727 − 58	**14.** 471 − 29	**15.** 563 − 72
16. 235 − 87	**17.** 718 − 578	**18.** 111 − 88	**19.** 846 − 97	**20.** 703 − 46
21. 808 − 329	**22.** 2,811 − 734	**23.** 5,736 − 1,999	**24.** 4,906 − 788	**25.** 8,777 − 5,987
26. 9,472 − 9,466	**27.** 32,351 − 5,470	**28.** 53,854 − 9,208	**29.** 15,460 − 7,923	**30.** 10,030 − 8,879
31. 80,302 − 46,783	**32.** 65,347 − 28,939	**33.** 83,006 − 61,747	**34.** 50,105 − 23,137	**35.** 50,000 − 8,000
36. 30,000 − 60	**37.** 70,000 − 83	**38.** 50,000 − 805	**39.** 50,000 − 8,005	**40.** 90,000 − 6,000
41. 70,000 − 988	**42.** 40,000 − 681	**43.** 60,000 − 2,568	**44.** 40,000 − 9,106	**45.** 75,000 − 67,472

***46.** *Subtract.* 63,922 − 48,256

***47.** The Census Bureau reported the population of the United States to be 179,323,175 in 1960. In 1970, it reported a population of 203,235,298. What was the increase?

Return to Form B on page 33.

MAKING CHANGE

To make change correctly by counting

- Count from the amount owed to the amount given.
- Subtract any coins given from the cost and amount given.

EXAMPLE 1

An item costs $1.89 and the buyer pays with a $5 bill. How much change is to be returned to the buyer?

Count:	$1.89 to $1.90	is	1 cent	⟶	$0.01
	$1.90 to $2.00	is	1 dime	⟶	0.10
	$2.00 to $5.00	is	3 dollars	⟶	3.00
			Total change to be returned:		**$3.11**

EXAMPLE 2

Buyers often pay part of their bills with coins, particularly pennies.

The cost of an item is $3.59. The buyer uses a $5 bill and 4 pennies as payment. What is the amount of change?

Mentally subtract $0.04 from the cost and from the amount given.

$3.59 − $0.04 = $3.55 $5.04 − $0.04 = $5.00

Count:	$3.55 to $3.75	is	2 dimes	⟶	$0.20
	$3.75 to $4.00	is	1 quarter	⟶	0.25
	$4.00 to $5.00	is	1 dollar	⟶	1.00
			Total change to be returned:		**$1.45**

GETTING READY

How much change should the buyer receive?

1. Change for $1.75 out of $5

2. Change for $4.98 out of $10

3. Change for $8.43 out of $20

4. Change for $6.63 out of $10.63

5. Change for $3.47 out of $5.02

6. Change for $12.86 out of $20.06

EXERCISES

How much change should the buyer receive from a $5 bill?

1. ball-point pen, $0.89

2. 1 pair of socks, $2.43

3. toothpaste, $1.34

4. compact, $3.19

How much change should the buyer receive from a $10 bill?

5. hand calculator, $8.76

6. video cassette, $7.29

7. tote bag, $9.98

8. tennis balls, $5.98

How much change should the buyer receive from a $20 bill?

9. wallet, $12.89

10. scissors, $18.45

11. slacks, $14.68

12. charm bracelet, $15.94

How much change should a cashier give?

13. Change for $5.43 out of $10.43

14. Change for $3.97 out of $5.02

15. Change for $12.65 out of $20.15

16. Change for $4.31 out of $10.06

17. Change for $1.65 out of $20.15

18. Change for $3.97 out of $20.47

ON YOUR OWN

Mrs. Gray, Mrs. Brown, and Mrs. Green are working together. One woman is wearing a gray dress, one a brown dress, and the other a green dress, but no one is wearing a dress whose color matches her name. How is each woman dressed?

SUBTRACTING MENTALLY

To find differences mentally

- Count from the smaller number to the larger number.
- Count to the next ten and so on.

EXAMPLE 1

Find 3,400 − 1,976 mentally.

Count from 1,976	1,976		
to the next ten,	1,980	⟶	**4**
then to the next hundred.	2,000	⟶	**20**
Then complete the count.	3,400	⟶	**1,400**
			1,424

So, 3,400 − 1,976 = **1,424.** Check: 1,976 + 1,424 = 3,400

EXAMPLE 2

Find \$15.04 − \$9.89 mentally.

Count from \$9.89	\$9.89		
to the next dollar.	10.00	⟶	**\$0.11**
Then count in whole dollars.	15.00	⟶	**5.00**
Add the \$0.04 last.	15.04	⟶	**0.04**
			\$5.15

So, \$15.04 − \$9.89 = \$5.15. Check: \$5.15 + \$9.89 = \$15.04

GETTING READY

Find the difference mentally. Check by using addition.

1. 57 − 35 **2.** 84 − 38 **3.** 216 − 150

4. \$2.16 − \$0.79 **5.** \$8.10 − \$5.87 **6.** \$27.26 − \$18.16

EXERCISES

Find the difference mentally. Check by using addition.

1. 80 − 30 **2.** 35 − 15 **3.** 74 − 34 **4.** \$96 − \$75

5. \$560 − \$400 **6.** 1,200 − 900 **7.** 660 − 160 **8.** \$870 − \$650

9. 359 − 129 **10.** 128 − 80 **11.** \$389 − \$345 **12.** \$125 − \$89

13. \$7.00 − \$2.50 **14.** \$46.20 − \$26.00 **15.** \$5.09 − \$3.49 **16.** \$55.80 − \$47.80

Problem Solving Skill

To read a double bar graph

- Notice the two types of bars used.
- Read like a single bar graph.

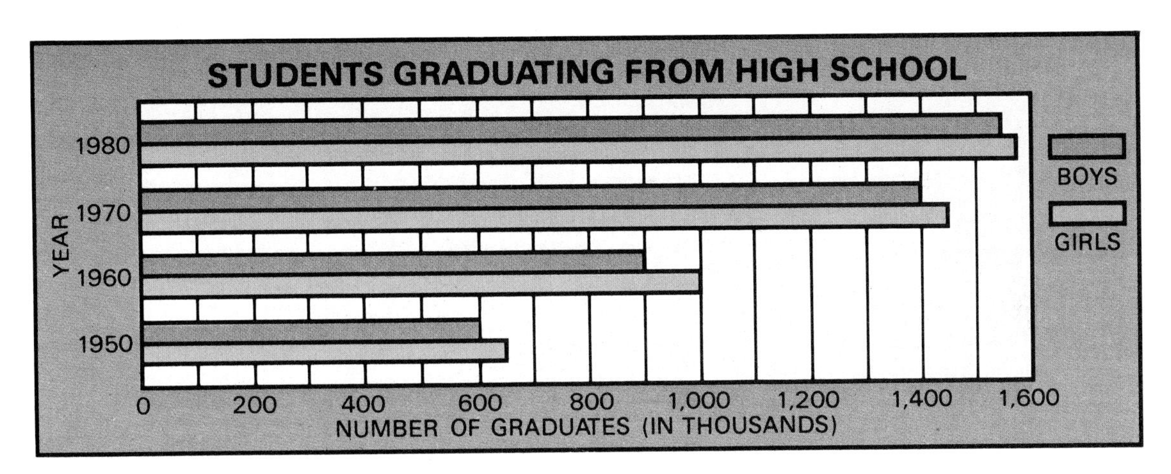

EXAMPLE

How many more girls graduated from high school in 1960 than boys? Find the bars representing 1960. Imagine a line from the end of each bar to the scale at the bottom of the graph.

1,000 × 1,000 = 1,000,000 ⟶	The number of girls is 1,000 thousands.
900 × 1,000 = 900,000 ⟶	The number of boys is 900 thousands.
1,000,000 − 900,000 = 100,000	So, **100,000 more girls** graduated in 1960.

Answer each question about the double bar graph.

1. What does the scale represent?

2. What is the range of years shown on the graph?

3. How many students graduated in 1970?

4. How many girls graduated in 1950?

5. In which year(s) did exactly 600,000 boys graduate?

6. In which year(s) did more boys graduate than girls?

7. How many more boys graduated in 1980 than in 1960?

8. How many more girls graduated in 1970 than boys?

9. In which year(s) did more than twice as many girls graduate as boys?

ESTIMATING DIFFERENCES

To estimate differences

- Round each number to the same place.
- Subtract the rounded numbers.

EXAMPLE 1

Estimate. 674 − 498

Round to the nearest 100.

$$\begin{array}{r} 674 \\ -498 \\ \hline \end{array} \longrightarrow \begin{array}{r} \mathbf{700} \\ \mathbf{-500} \\ \hline \mathbf{200} \end{array}$$

So, the estimated difference is **200.**

EXAMPLE 2

Estimate. 48,761 − 6,250

Round to the nearest 1,000.

$$\begin{array}{r} 48{,}761 \\ -\ 6{,}250 \\ \hline \end{array} \longrightarrow \begin{array}{r} \mathbf{49{,}000} \\ \mathbf{-\ 6{,}000} \\ \hline \mathbf{43{,}000} \end{array}$$

So, the estimated difference is **43,000.**

When estimating differences, round each number to the largest place value in the smaller number.

EXAMPLE 3

Mrs. Barnes has $8,140 in stocks and $7,920 in bonds. About how much more money does she have in stocks than in bonds?

Both numbers are in the thousands. They could each be rounded to $8,000. Then the difference would be 0.

To estimate, round to hundreds.

$$\begin{array}{r} \$8{,}140 \\ -7{,}920 \\ \hline \end{array} \longrightarrow \begin{array}{r} \mathbf{\$8{,}100} \\ \mathbf{-7{,}900} \\ \hline \mathbf{\$200} \end{array}$$

So, she has about **$200** more in stocks.

EXAMPLE 4

Estimate. $46.37 − $29.80

Round to the nearest $10.

$46.37	→	**$50**
− 29.80	→	**− 30**
		$20

So, the estimated difference is **$20.**

GETTING READY

Estimate.

1. 587 − 143
2. 1,245 − 769
3. $2,314 − 1,810
4. 5,286 − 942
5. 2,420 − 968
6. $3,095 − $1,500
7. 17,805 − 5,447
8. $13.25 − $9.89
9. $24.87 − $10.15
10. $324.40 − $103.99

EXERCISES

Estimate.

1. 847 − 398
2. 286 − 74
3. 501 − 92
4. $9,555 − $542
5. 8,299 − 317
6. $4,650 − 1,523
7. 115 − 36
8. 1,915 − 899
9. $20.00 − $14.98
10. $10.00 − $4.98
11. $16.25 − $12.37
12. $14.79 − $9.99
13. $50.00 − 39.50
14. $100.00 − 75.23
15. $130.00 − 79.65
16. $75.06 − 43.85

17. Carter High School has 1,765 students. On Tuesday, 837 of the students ate hot lunches. About how many did not eat hot lunches?

*18. Rodney bought items costing the following: $0.79, $2.09, $1.48, $0.29, and $0.43. If he has $5, estimate to determine if he has enough money.

Problem Solving Applications

READ • PLAN • SOLVE • CHECK

Some of the students at Bridgeport High School have a delivery service. They are hired by businesses to deliver advertising circulars, newspapers, and bills door to door. Sometimes the students volunteer to deliver items for community groups.

AS YOU READ

- brochure—a small booklet
- flier—an advertising circular
- packet—a small bundle

Answer these questions about the student delivery service.

1. During "Help Beautify Bridgeport" week, Sara helped deliver *fliers* for the mayor. She worked 4 h on Monday, 6 h on Wednesday, 3 h on Thursday, and 5 h on Friday. Find how many hours she worked in all.

2. Each week Ruth collects for her newspaper route. This week she collected only $390.25. If all her accounts total $437.50 a week, how much does she still need to collect?

3. Recently Beth helped a friend hand out political campaign *brochures*. On Saturday Beth handed out 1,500 and her friend handed out 1,010. How many more brochures did Beth hand out?

4. Chris delivers newspapers for the Gazette. On his old route he delivered 198 papers a day. On his new route he delivers 265 papers a day. How many more papers does he deliver on his new route?

5. The community Welcome Wagon uses the students to deliver gift *packets* to new families in town. Kirk delivered 27 packets and Jeff delivered 35. How many packets did they deliver altogether?

*6. Claude delivers the monthly bills for a local business. It takes him 1 h 20 min to walk the route. This month he rode his bicycle and it took 55 min. How much time did he save by biking?

Career: Postal Workers

READ • PLAN • SOLVE • CHECK

The United States Postal Service employs mail carriers, window clerks, mail handlers, and postmasters. In order to work for the Postal Service, a person must pass a Civil Service Examination.

AS YOU READ

- case—to separate by address numbers
- parcel—a package

Solve.

1. The post office charges $0.75 for a $25 money order and $1.10 for a $25.01 money order. Find the difference in cost.

2. A mail carrier punched in on the time clock at 6:45 am and punched out at 3:15 pm. How long did the carrier work?

3. Each morning Richard Holt, a mail carrier, *cases* mail by address numbers. One day he cased 1,580 letters and 647 magazines into address slots. How many more letters than magazines did he case for delivery that day?

4. Arthur Louis, a mail handler, unloaded 10 mail bags and 13 *parcels* on Monday, 14 bags and 25 parcels on Tuesday, and 25 bags and 19 parcels on Wednesday. How many more parcels than bags did he unload during the 3-day period?

5. Today on her mail route, Marian Harbison collected the following amounts of postage due on forwarded mail: $0.50, $0.75, $1.50, $0.20, $0.95, and $1.20. How much was collected in all in postage due?

6. To send a letter by express mail costs $9.35. If the letter is mailed by 5 pm on Monday, it is guaranteed to be delivered by 3 pm the next day. Within how many hours will the delivery be made?

CHAPTER TEST 2

Subtract and check.

1. $\begin{array}{r} 1{,}122 \\ -\ \ 705 \\ \hline \end{array}$

2. $\begin{array}{r} 60{,}000 \\ -\ 9{,}633 \\ \hline \end{array}$

3. $\begin{array}{r} 78{,}456 \\ -24{,}739 \\ \hline \end{array}$

4. $\begin{array}{r} 351{,}004 \\ -\ \ 8{,}177 \\ \hline \end{array}$

5. $25 − $13

6. 20,000 − 91

7. $75,965 − $865

Solve.

8. Mt. McKinley is about 6,095 m high. Pikes Peak is about 4,234 m high. How much higher is Mt. McKinley?

9. Jupiter's diameter is about 143,179 km. Mercury's diameter is about 4,858 km. What is the difference in the diameters?

How much change should be given?

10. Change for $2.87 out of $5

11. Change for $8.48 out of $20

12. Change for $3.71 out of $5.01

Estimate.

13. Carol earned $137.93. Her deductions amounted to $15.36. About how much is her take-home pay?

14. Weisman's had 1,128 shirts in stock on Monday. By Wednesday, 272 shirts had been sold. About how many shirts are left in stock?

Use the double bar graph for **15–16.**

15. On which test did Jo Ann's score and Len's score differ the least?

16. What score did each person make on Test I?

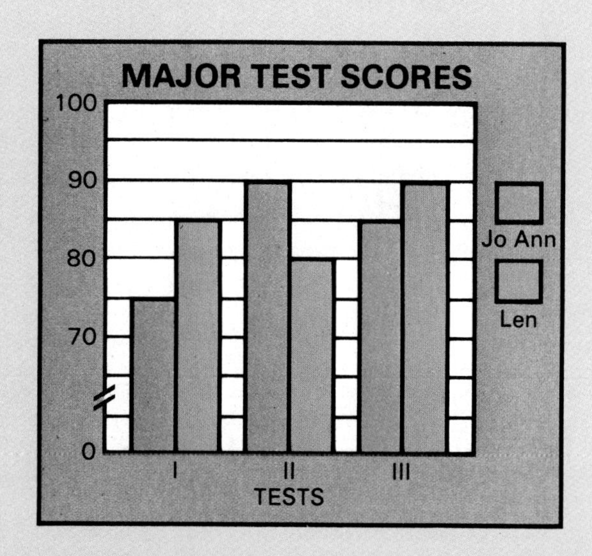

Problem Solving

MAKING MATH COUNT

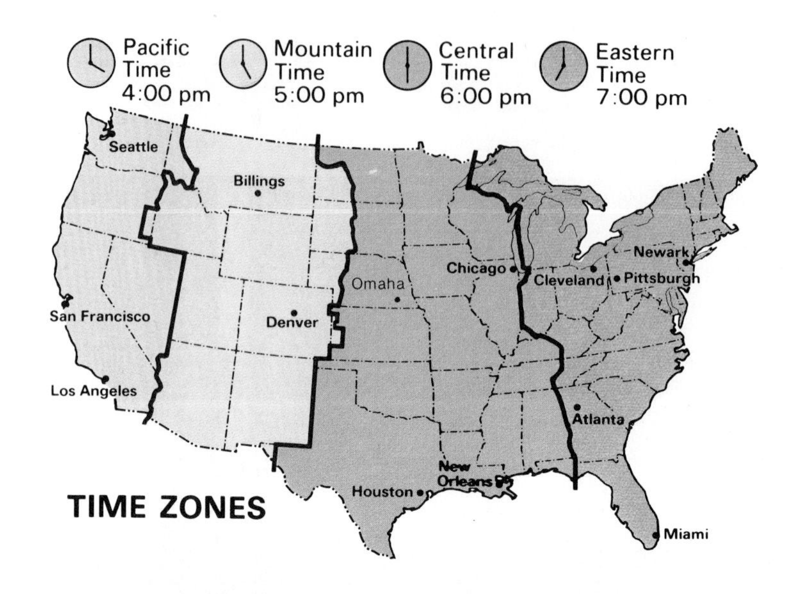

EXAMPLE

If it is 9:30 am in Pittsburgh, what time is it in San Francisco?

A 5:30 am **B** 6:30 am
C 7:30 am **D** 8:30 am

Look at the map.
Pittsburgh is in the Eastern time zone.
San Francisco is in the Pacific time zone.
Now look at the clocks shown.
It is 3 h earlier in San Francisco.

So, **B** is the correct answer.

Choose the correct answer.

1. If it is 10:00 am in Seattle, what time is it in Omaha?
A 11:00 am **B** 12:00 pm
C 1:00 pm **D** 2:00 pm

2. If it is 8:50 pm in Newark, what time is it in Billings?
A 6:50 pm **B** 7:50 pm
C 9:50 pm **D** 10:50 pm

3. At 11:10 am in New Orleans, Bill called his mother in Miami. What time was it in Miami?
A 10:10 am **B** 11:10 am
C 12:10 am **D** 12:10 pm

***4.** Carlotta's flight from Los Angeles landed in Atlanta at 1:15 am. What time was it in Los Angeles?
A 4:15 am **B** 10:15 am
C 4:15 pm **D** 10:15 pm

DIAGNOSTIC TEST: FORM A (35 MINUTES)

Read each problem and answer the questions.

The students at Carver High School are selling raffle tickets. The sophomores sold 3,278; the juniors 4,117; and the seniors 6,059. How many tickets were sold in all?

1. What are the students selling?

2. How many tickets did the seniors sell?

3. Who sold 3,278 tickets?

4. What are you to find?

There are 35 mopeds and 18 go-carts in Joe's Motor Shop. How many more mopeds are there than go-carts?

5. What kind of shop does Joe have?

6. How many mopeds does he have?

7. How many go-carts does he have?

8. What are you to find?

Write what is given and what you are to find in each problem.

9. Joan planted apple trees. If each row contains 20 trees and there are 25 rows, how many trees are there?

10. Frank was born in 1967. In what year will he be 18?

What other information do you need to know to solve each problem?

11. Ted has read 208 pages in his new book. How many more pages does he need to read to finish it?

12. It is 896 km to New York. How many liters of gas must you buy for the trip?

What is the extra information in each problem?

13. On the Ellises' 200-acre farm there are 95 cows, 20 sheep, and 46 pigs. How many more cows than pigs are there?

14. Fay's Music Shop sells guitars for \$169.99 and trumpets for \$175.99. Donna buys a guitar with a down payment of \$50. How much does she owe?

Decide what operation to use to solve each problem.

15. Cary is earning money for the band trip. The cost is \$125. He has \$59. How much more money does he need?

16. A factory packed 25,905 cans of tuna in 5 days. About how many cans per day were packed?

17. There are 30 desks in each classroom in Memorial High School. There are 45 classrooms. How many desks are there in all?

16 or more right? Go to page 58.

Less than 16 right? Go to page 51.

DIAGNOSTIC TEST: FORM B (35 MINUTES)

Read each problem and answer the questions.

There are 583 juniors and 491 seniors at Mt. Clair High School. How many more juniors are there than seniors?

1. What is the name of the school?
2. How many seniors are in the school?
3. How many juniors?
4. What are you to find?

Barry and Cam collect matchbook covers. Barry has 916 and Cam has 731. How many do they have altogether?

5. What do Barry and Cam collect?
6. Who has 916 covers?
7. Who has 731 covers?
8. What are you to find?

Write what is given and what you are to find in each problem.

9. Toni delivers the paper. She collects Friday afternoons and Saturdays. She collected $28 on Friday and $42 on Saturday. How much did she collect?
10. Ly Thi sleeps 8 h each day and spends 6 h in school. How many more hours does he sleep than he spends in school?

What other information do you need to know to solve each problem?

11. A video game system usually sells for $149.97. How much will you save if you buy at the sale price?
12. Z Mart is selling a 10-speed bike for $125. Bargain Town is selling the same bike. How much will you save if you buy the bike at Z Mart?

What is the extra information in each problem?

13. Tony gained 710 yd in 10 games and Earl gained 691 yd in 12 games this season. Together how many yards have they gained?
14. A store owner bought 320 kerosene heaters for a special sale. She sold 107 the first week, 73 the second week, and 88 the third week. How many were sold altogether?

Decide what operation to use to solve each problem.

15. Calculators that are regularly priced at $54.99 are selling for $25 off. What is the sale price?
16. Ann bought a notebook for $1.19, notebook paper for $1.98, a ball-point pen for $0.79, and a pencil for $0.15. How much did she spend?
17. During a trip, the Washingtons used 22 gal of gasoline to go 410 mi. How many miles per gallon did they get?

16 or more right? Go to page 58.

Less than 16 right? See your teacher.

READING FOR INFORMATION

To recognize the information that is needed to solve a word problem

- Identify what is given and what you are to find.

EXAMPLE

Write what is given and what you are to find.

READ A 3-lb can of coffee sells for $8.97. How much does the coffee cost per pound?

Write what is given.
amount of coffee ⟶ **3 lb**
cost ⟶ **$8.97**

Write what you are to find.
cost per pound ⟶ **cost of 1 lb**

GETTING READY

Write what is given and what you are to find.

Mr. Harold put a roast in the oven at 2:00 pm. The roast should cook for 3 h, 20 min. At what time should he remove the roast from the oven?

EXERCISES

Write what is given and what you are to find in each problem.

1. Mary's scout troop had a pennant sale. They sold 261 mini pennants and 197 standard pennants. How many did they sell altogether?

2. The main floor in a theater seats 1,282 people and the balcony seats 1,089. How many people does the entire theater seat?

3. When Maria left home, the odometer on her car read 18,931 mi. When she arrived in Chicago, it read 19,328 mi. How far did she drive?

4. Paul had $3,175 in his savings account. He bought a used car for $2,500. How much does he have left in his account?

5. Dennis has been dieting and now weighs 128 lb. He has lost 29 lb. How much did he weigh when he began his diet?

6. Cheryl bought 3 cassette tapes for $17.88. How much did each cassette tape cost?

***7.** Solve problems 1–6.

READING TO UNDERSTAND

To understand a word problem

- Read the problem carefully.
- Be sure you can answer any questions about it.

EXAMPLE

Read the problem and answer the questions.

READ Monica plans to drive a distance of 1,946 km in 4 days. If she drives 1,515 km in 3 days, how far must she drive on the fourth day?

a. Who is driving? **Monica**
b. What is the total distance? **1,946 km**
c. What measure is used for the distance? **kilometers**
d. What is the total amount of time? **4 days**
e. What is the measure used for the time? **days**
f. What is the distance driven in 3 days? **1,515 km**
g. What are you to find? **distance on fourth day**

GETTING READY

Read the problem and answer the questions.

The total attendance at the homecoming football game was 2,146 people. If 1,462 of them were students, how many were not students?

a. How many people attended?
b. What did they attend?
c. How many students attended?
d. What are you to find?

EXERCISES

Read each problem and answer the questions.

1. George bought 5 doz cookies to serve at a party. If 4 doz were eaten, how many dozen were left?

a. How many cookies were bought?
b. Who bought the cookies?
c. How many cookies were eaten?
d. What are you to find?

2. In last night's Star Path video game, Mildred scored 48 points and Allison scored 35. How many more points did Mildred score than Allison?

   a. How many points did Mildred score?

   b. How many points did Allison score?

   c. What are you to find?

3. Each day Bruce delivers 45 newspapers before school and 28 after school. How many newspapers does he deliver each day?

   a. When does he deliver the newspapers?

   b. How many newspapers are delivered before school?

   c. How many newspapers are delivered after school?

   d. What are you to find?

4. Mrs. Rodriguez used 597 kilowatt hours (kW·h) of electricity in June and 624 kW·h in July. How many kilowatt hours did she use in both months?

   a. What does kW·h stand for?

   b. How many kilowatt hours were used in June?

   c. How many kilowatt hours were used in July?

   d. What are you to find?

*5. Solve problems 1–4.

CALCULATOR

Use your calculator.

1. Pick a two-digit number. Find the sum of all the whole numbers beginning with 1 up to and including the number you picked.

2. Now try this. Take the number you picked, multiply it by one more than itself, and then divide the product by 2.

3. What happens?

4. Try this with another two-digit number.

CHOOSING THE CORRECT OPERATION

To choose the correct operation to solve a word problem

EXAMPLE

Decide what operation to use to solve this problem.

A pizza costs $5.49. How much change would you receive from a $10.00 bill?

PLAN Think:

amount given − cost = change

↓ ↓ ↓

$10.00 − $5.49 = ?

So, you would **subtract**.

GETTING READY

Decide what operation to use to solve each problem.

1. Mr. Chan purchased toothpaste for $1.69, aspirin for $1.18, a razor for $4.99, and shampoo for $3.52. How much did he spend altogether?
2. Ground beef costs $2.25 a pound. Mrs. Wilbur needs $2\frac{1}{2}$ lb of ground beef to make hamburgers for a cookout. What is the total cost?
3. Ms. King drove for 8 h in one day. She averaged 60 km each hour. How many kilometers did she travel that day?
4. Five buses will carry the same number of senior citizens on a field trip. Two hundred people are going. How many people will ride on each bus?
5. Flashlight batteries cost 3 for $4.59. How much does one battery cost?

EXERCISES

Decide what operation to use to solve each problem.

1. Sandi's bowling scores were 141, 128, and 136. What was her total score for the three games?
2. Hannah bought a tape deck for $147, speakers for $125, and an amplifier for $165. How much did she spend altogether?
3. Records are on sale for $1 off. If the original price was $6.98, what is the sale price?
4. Bananas cost 75¢ a kilogram. How much do 3 kg cost?

5. Mary is $18\frac{1}{2}$ y old. Her brother Todd is 3 y younger. How old is Todd?

6. Kevin spent $8.13 for groceries. He gave the clerk a $20 bill. How much change did he get?

7. A recipe calls for $\frac{3}{4}$ c of wheat flour. If you double the recipe, how much wheat flour is needed?

8. Audra studies $\frac{1}{2}$ h before dinner and 2 h after dinner. How long does she study?

9. The students in Enrico's homeroom voted 18 to 13 to go to Hope Park for their picnic. How many students voted?

10. There are 6 coins in each of 7 display cases. How many coins are there in all?

11. A plane flew 2,000 km in $2\frac{1}{2}$ h. How many kilometers per hour did it fly?

12. Kenneth bought 12 tomato plants for $15.48. How much did each plant cost?

13. Ruth, Frank, Lisa, Sean, Joan, and Bill shared the expenses of a party. The party cost $30.72. How much was each one's share?

14. Apples were selling at 3 kg for $0.99. Mr. Aurie bought 1 kg of apples. How much did he pay?

***15.** Solve problems 1–14.

ON YOUR OWN

In professional football, the scoring is as follows:

safety	2 points
field goal	3 points
touchdown	6 points
touchdown with conversion	7 points

Every whole number larger than 1 is a possible total score. For example, 12 could be scored as 6 + 6, 7 + 3 + 2, or 3 + 3 + 3 + 3.

Find at least three ways of reaching each of these total scores.

1. 9
2. 20
3. 6
4. 14
5. 48
6. 31

MISSING AND EXTRA INFORMATION

To identify missing information that is needed to solve a word problem
To identify extra information in a word problem

EXAMPLE 1

Some word problems may not have all the information that is needed to solve them.

What other information do you need to know to solve this problem?

A coat is on sale for $15 off. What is the sale price of the coat?

PLAN Think: original price − amount off = sale price

$$\begin{array}{ccccc} \text{original price} & - & \text{amount off} & = & \text{sale price} \\ \downarrow & & \downarrow & & \downarrow \\ \textbf{?} & - & \textbf{\$15} & = & \textbf{?} \end{array}$$

So, you need to know the **original price** of the coat.

EXAMPLE 2

Some word problems may contain extra information.

What is the extra information in this problem?

John made a deposit of $150. Then he wrote checks for $21, $85, and $16. What was the total amount of the checks?

PLAN Think: You are to find the total amount of the checks.

$21 + $85 + $16 = total amount

You do *not* need to know that John made a deposit of $150.

So, the **deposit of $150** is extra information.

GETTING READY

What other information do you need to know to solve this problem?

1. Randi's plane ticket cost $144. How much change did she get?

What is the extra information in this problem?

2. A bookstore sold 756 paperback books, 289 hardback tour books and 69 hardback cookbooks. How many hardback books were sold?

EXERCISES

What other information do you need to know to solve each problem?

1. Tomato juice is sold in 6-can packages. How much would 12 cans cost?

2. A metal pipe is 2 m long. How many pieces can be cut from it?

What is the extra information in each problem?

3. A space mission was delayed a week. It was to last ten days, but had to be shortened by four days. How long was the mission?

4. Mark and Eloise were each given 25 tickets to sell. Mark sold 18 of his tickets and Eloise sold 20 of hers. Find the total number of tickets sold.

5. Carol's nine-year-old cat weighs 12 lb and her one-year-old puppy weighs 28 lb. Find the difference in weight between the cat and the puppy.

6. A restaurant served 3,700 dinners during one week. A dinner cost $7.95 for adults and $3.95 for children. If 2,450 children's dinners were served, how many adults were served?

7. Mr. Moto's rent is $350 per month. His food averages $120 per month and his commuting expenses are $60 a month. How much are his monthly rent and food expenses combined?

8. Otis left his home at 7:00 am. He drove at an average of 56 km/h. If he arrived at the ski lodge at 1:00 pm, how long did the trip take?

***9.** Solve problems 3–8.

Return to Form B on page 50.

ON YOUR OWN

Choose the most reasonable answer.

1. Alfredo bought the following at Main's Smart Shop: sport coat, $59.95; slacks, $33.98; shirt, $15.00; tie, $11.00. How much did he spend altogether?
   A $100 to $150 **B** $150 to $200 **C** More than $200

2. Estelle purchased a new camera on credit. She paid $20 down. If the cost of the camera was $127, what is a reasonable amount for her to pay per month for 6 months?
   A $10.00 **B** $20.00 **C** $30.00 **D** $100.00

SOLVING PROBLEMS

To solve a word problem

- Identify what is given and what you are to find.
- Choose the correct operation to solve the problem.
- Do the arithmetic.

EXAMPLE

A bookstore sold 82 books on Friday and 159 books on Saturday. How many more books were sold on Saturday than on Friday?

You know the number of books sold on Friday and the number of books sold on Saturday.
You want to find *how many more* were sold on Saturday, so you will subtract.

books sold on Saturday − books sold on Friday = how many more?

↓ ↓ ↓

159 − 82 = ?

SOLVE Solve by subtracting.
159 − 82 = 77

So, **77** more books were sold on Saturday.

GETTING READY

Solve.

1. A record store sold 842 rock albums, 375 jazz albums, and 58 country and western albums. How many albums were sold?

2. Abbott's jewelry store had 371 watches in stock. Of these, 125 were quartz. After a sale there were 285 watches in stock. How many were sold?

3. The largest city in the state has 979,654 people. The next largest city has 589,320. What is the difference in population between the two cities?

4. Jan's letter weighs 3 oz. If a first-class letter costs 20¢ for the first ounce, and 17¢ for each additional ounce, how much will it cost to mail Jan's letter first-class?

5. At a sale, a large tube of conditioner is selling for $2.39 and a small tube for $1.05. How much will four large tubes cost?

EXERCISES

Solve.

1. The swimming pool in Phil's yard is 18 m long. Phil swam the length 25 times. How many meters did he swim?

2. The previous attendance record for Memorial Stadium was 82,573. This year the attendance was 84,321. By how much was the record broken?

3. The trip from Birmingham, Alabama, to Washington, D.C., is 648 mi. It is 726 mi from Washington, D.C., to Chicago. How far is it from Birmingham to Chicago by way of Washington, D.C.?

4. The temperature reading this morning was 24° F. The predicted temperature for this afternoon is 52° F. What is the predicted rise in temperature?

5. Mr. Andston started on a trip with $285. When he returned he had $47. How much did he spend during the trip?

6. Louise collects aluminum cans for recycling. If she has collected 105 so far this week and her goal is 300, how many more must she collect?

7. Jack and Henry were each given 125 raffle tickets to sell. Jack sold all of his tickets and Henry sold 22 of his. How many tickets were sold in all?

8. Alex's three younger brothers weigh 56 lb, 72 lb, and 103 lb. What is the difference between the lowest and highest weights?

***9.** Mrs. Arnold cashed a paycheck for $320. She paid a car repair bill for $129.86 and a utility bill for $56.92. How much did she have left from her paycheck?

***10.** During each hour, station WAVM broadcasts 10 min of news, $13\frac{1}{2}$ min of commercials, and 7 min of community notes. If the rest is music, how many minutes of music are broadcast each hour?

CHECKING THE SOLUTION

To check the solution of a word problem

- Recheck the arithmetic.
- Obtain an estimate and compare the estimate with your solution.

EXAMPLE 1

Solve and check.

A rocket had 29,750 lb of fuel. Four minutes into the flight it had 16,949 lb of fuel. How much fuel had burned in four minutes?

Subtract to find the amount of fuel burned.

29,750 − 16,949 = 12,801; or 12,801 lb

CHECK You can check the subtraction by adding:

16,949 + 12,801 = 29,750

So, **12,801 lb** of fuel had burned.

EXAMPLE 2

Estimate the answer, then solve the problem. Check by comparing the estimate with your solution.

How many calories (cal) are contained in this meal: roast lamb, 206; fresh peas, 66; boiled potato, 122; slice of bread, 72; apple, 117; glass of milk, 176?

Round each number to the nearest ten to obtain an estimate of the total number of calories.

lamb	**206**	⟶	**210**
peas	**66**	⟶	**70**
potato	**122**	⟶	**120**
bread	**72**	⟶	**70**
apple	**117**	⟶	**120**
milk	**176**	⟶	**+180**
			770

So, the estimated total is **770 cal.**

Find the actual total.

206 + 66 + 122 + 72 + 117 + 176 = 759; or 759 cal

CHECK Is 759 cal a reasonable answer? To check, compare this solution with your estimate.
Since 759 and 770 are very close, 759 is a reasonable answer.

So, **759 cal** are contained in the meal.

GETTING READY

Solve and check.

1. In the student election, Robert received 1,332 votes and Mark received 978 votes. How many more votes did Robert receive than Mark?

2. The Drake family used 6,207 cubic units of gas in February and 4,450 cubic units in March. How much gas did they use in the two months?

Estimate the answer, then solve each problem. Check by comparing the estimate with your solution.

3. Bethel High School presented three performances of *West Side Story* this year. There were 1,058 people at the opening performance, and 993 and 1,196 people at the next two performances. What was the total attendance? (To estimate, round to thousands.)

EXERCISES

Solve and check.

1. A baker made 60 doz bran muffins. He sold 52 doz. How many were left?

2. Frank paid $4.74 for lunch. How much change will he get from a $10 bill?

3. Harry scored 121 points, Susan 175 points, and Amy 111 points. How many points did they score in all?

4. The odometer on Helen's car reads 5,269 mi. What will it read after a trip of 884 mi?

Estimate the answer, then solve each problem. Check by comparing the estimate with your solution.

5. The Record Shop had 2,217 records on Tuesday. By Friday, it had 982 records left. How many records were sold? (To estimate, round to hundreds.)

6. What is the cost of Barbara's lunch if she bought a cheeseburger for $1.69, a large order of fries for $0.72, and a shake for $0.96? (To estimate, round to nearest dollar.)

7. It is 103 km from Leigh to Surrey and 179 km from Surrey to Blackstone. How far is it from Leigh to Blackstone by way of Surrey? (To estimate, round to hundreds.)

USING THE 4-STEP METHOD

To solve word problems

- Use the 4 steps: Read, Plan, Solve, Check.

EXAMPLE 1

Solve by the 4-step method.

Jonathan sells shoes. On Friday, he sold 35 pairs of shoes and on Saturday he sold 41 pairs. How many pairs did he sell altogether?

READ What information is given?
35 pairs sold on Friday
41 pairs sold on Saturday

What are you to find?
total pairs sold

PLAN Which operation should you use?
You are to find the total, so you *add*.

SOLVE Add. **35 + 41 = 76, or 76 pairs**

CHECK Check your arithmetic.

$$\begin{array}{r} \mathbf{35} \\ \mathbf{+41} \\ \hline \mathbf{76} \end{array}$$

So, Jonathan sold **76 pairs** of shoes.

EXAMPLE 2

Solve by the 4-step method.

Lisa is taking a 3,200-mi trip across the country in her car, which gets 32 mi to the gallon. She covered 965 mi in the first two days and 512 mi the next day. How far does Lisa still have to go?

READ What information is given?

total mileage of trip	3,200
miles covered in 2 days	965
miles covered next day	512
car gets 32 mpg	

What are you to find?
miles that Lisa still has to go

PLAN Which operation(s) should you use?
Add to find total miles covered in 3 days.
Subtract from 3,200 to get the number of miles remaining.

SOLVE Before solving, estimate the answer.
Round to hundreds.

$$\begin{array}{rcr} \mathbf{965} & \longrightarrow & \mathbf{1{,}000} \\ \mathbf{512} & \longrightarrow & \mathbf{+\ \ 500} \\ \hline & & \mathbf{1{,}500} \end{array} \quad \mathbf{3{,}200 - 1{,}500 = 1{,}700}$$

estimate ↓ (1,700)

Find the actual number of miles still to go.

Add. **965 + 512 = 1,477** Subtract. **3,200 − 1,477 = 1,723**

(Note that 32 mpg is extra information that is *not* needed to solve the problem.)

CHECK Compare this solution with your estimate. Since 1,723 and 1,700 are very close, 1,723 is a reasonable answer.

So, Lisa still has **1,723 mi** to go.

GETTING READY

Solve using the 4-step method.

1. A basketball team won 18 of the first 24 games played. The team lost 15 of the next 21 games played. How many games did they play?

2. Chuck started jogging at 6:55 pm and jogged until 8:20 pm. How long did he jog?

EXERCISES

Solve using the 4-step method.

1. Shawn, Chris, and Shanda washed cars on Saturday. They washed 8 more cars in the morning than in the afternoon. If they washed 17 cars in the morning, how many did they wash in the afternoon?

2. A rock group is planning a concert at a local stadium. If 13,000 tickets are available, and 8,732 tickets have been sold so far, how many tickets are still left?

3. Craig had $15.25 in savings. He earned $24.50 last week and then spent $35.50 on clothing. How much money did he have left?

4. Ching bought a tablet for $1.19, typing paper for $1.50, a typewriter ribbon for $1.98, and an eraser for $.25. How much did he spend?

5. One out of every three seniors at Washington High School takes mathematics, and one out of five takes chemistry. There are 510 seniors. How many seniors take mathematics?

6. Bargain Town is having a sale on audio equipment. They are selling speakers at two for $89, tape decks for $115 each, and cassette tapes for $6.95 each. What is the cost of a tape deck and two speakers?

7. Jane would like to buy running shoes that cost $32. She has saved $19. How much more money does she need to buy the shoes?

***8.** The 210-member marching band is going by bus to the state band contest. If each bus holds 58 people, how many buses are needed?

Problem Solving Applications

READ • PLAN • SOLVE • CHECK

The students at East High School are preparing for their spring musical production. Art and woodshop students are working on the scenery and props, and homemaking students are making costumes. The band and choral students will also perform.

AS YOU READ

- complimentary—given free
- performance—a public presentation
- rights—use permitted for a fee

Answer these questions about the musical production.

1. The *rights* for the musical cost $500, the material for costumes costs $475, and the supplies needed for the scenery cost $250. What are the total expenses?

2. The students are selling grapefruit to help with the expenses. If the profit on each box is $1.20, how much will they make if they sell 250 boxes?

3. There are 14 dancers and 65 singers. It will take 3 yd of material for each of the dancers' skirts. How many yards of material are needed?

4. The length of the performance is 2 h 30 min, and there will be two 15-min intermissions. If the performance ends at 11:00 pm, when should it begin?

5. The band will play for 20 min before the beginning of the *performance*. If each selection takes about 5 min, how many selections will they play?

6. The auditorium seats 1,500. Tickets cost $1.25 each. What will be the receipts if there is a full house including 100 *complimentary* tickets?

Career: Musicians

READ • PLAN • SOLVE • CHECK

Musicians perform for the enjoyment of others. Some play in dance bands, rock groups, or jazz combos. Others may play in community symphony orchestras or in armed forces bands.

AS YOU READ

- audition—a trial performance to judge a performer's talents
- soprano—a person having the highest singing voice

Solve.

1. The musical group Change consists of Jon, Nadine, Eric, and Andie. Last night the group was paid \$125. If Jon received \$50 and the others shared equally what was left, how much did Nadine receive?

2. Barbara Coleman is a *soprano* in the St. Luke's choral group. The group performed for a community fund raiser. The total receipts were \$4,000. If the group charged \$750, how much money was raised?

3. Change plans to record on a 30-min tape. If the average length of a song is 4 min, at most how many songs can they record on the tape?

4. Tim Barnes repairs musical instruments. It will take a week to get a part he needs to repair a trumpet and 5 days for the repairs. How many days will it take?

5. Joseph Mantini is flying to Los Angeles for an *audition*. One flight departs at 1:30 pm and arrives at 6:45 pm, and another departs at 8:00 am and arrives at 1:45 pm. Which flight is the faster?

6. Friday and Saturday nights Tomika Jones plays piano at the Lion Inn. This Friday she played from 8:00 pm until 1:00 am with two 15-min breaks. How long did she play?

CHAPTER TEST 3

Read the problem and answer the questions.

A compact stereo with cassette deck is on sale for $100 off the regular price of $369.95. What is the sale price of the stereo?

1. What is on sale?
2. What is the regular price?
3. How much will you save?
4. What are you to find?
5. Which operation would you use?

Write what is given and what you are to find in each problem.

6. Tennis balls sell for $2.91 a can. There are 3 tennis balls in a can. What is the price per tennis ball?
7. A cyclist travels 8 mi in an hour. In 6 h, how far will she travel?

Decide what operation to use to solve each problem.

8. A restaurant uses 130 eggs each day. How many days will a supply of 520 eggs last?
9. Bruce spends 45 min a day practicing the piano. How many minutes does he practice in a week?

What other information do you need to know to solve this problem?

10. A suit is on sale for $159.95. How much would you save by buying at the sale price?

What is the extra information in this problem?

11. Seven of the twelve girls on the swim team and five of the nine boys are going to state championships. How many swimmers are going?

Estimate the answer, then solve this problem. Check by comparing the estimate with your solution.

12. Ski mittens that regularly sell for $43.98 are on sale for $21.99. How much would you save on the mittens by buying them at the sale price? (To estimate, round to the nearest dollar.)

Solve using the 4-step method.

13. Sandi is laying a brick walk. She needs 1,200 bricks for the walk and 135 bricks for the edges. How many bricks does Sandi need altogether?

Multiplication of Whole Numbers

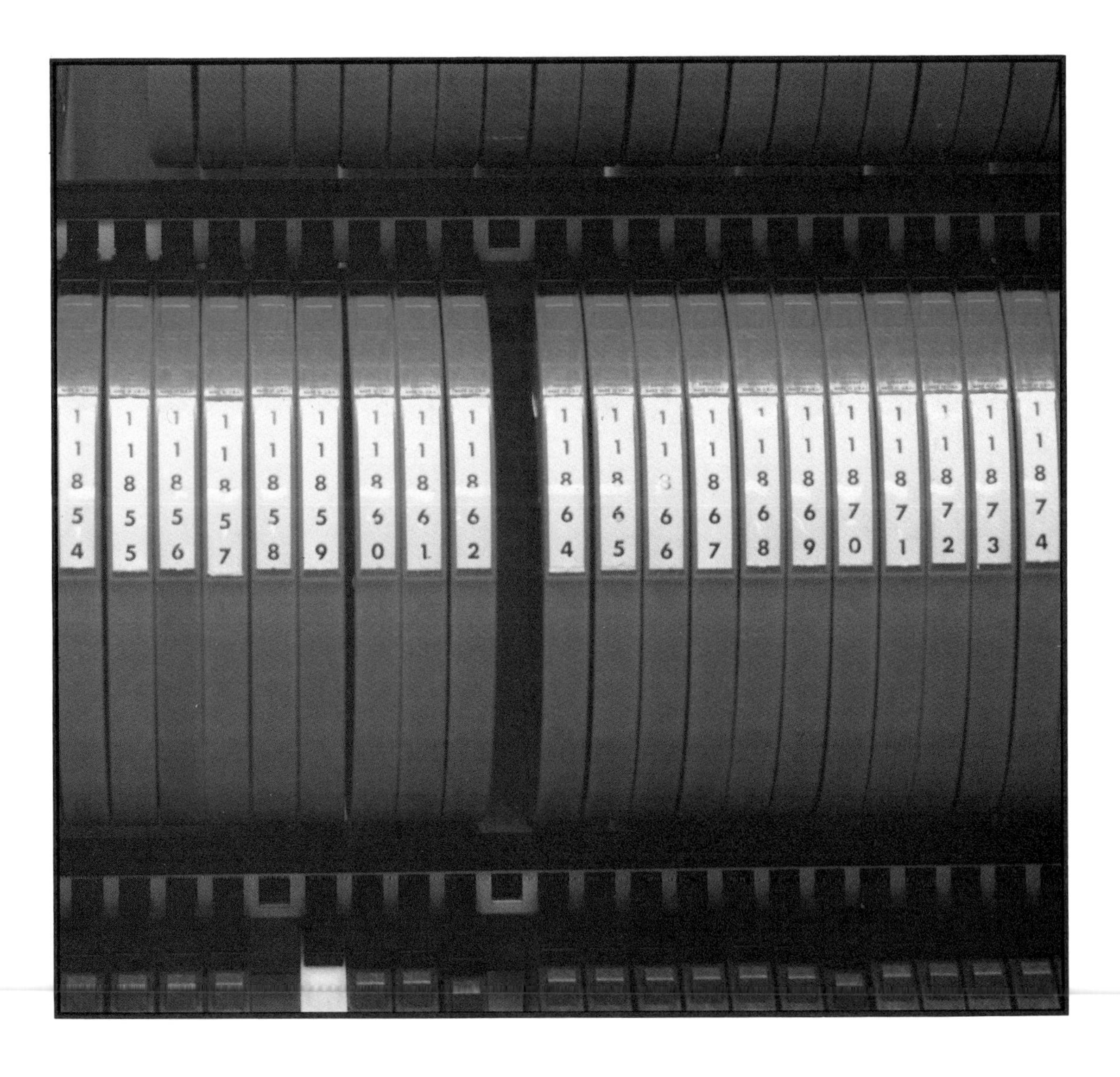

MAKING MATH COUNT

EXAMPLE

A bus leaves Midway at 10:45 am and arrives in Tipton at 1:25 pm. How long is the trip?

A 1 h 40 min **B** 2 h 15 min
C 2 h 25 min **D** 2 h 40 min

From 10:45 am to 12 pm → 1 h 15 min
From 12 pm to 1:25 pm → + 1 h 25 min
2 h 40 min

So, **D** is the correct answer.

Choose the correct answer.

1. Jean left on her bicycle at 11:30 am and returned at 2:00 pm. How long was she gone?
A 30 min **B** 1 h 30 min
C 2 h 30 min **D** 9 h

2. Mr. Spencer leaves for work at 7:30 am. He returns home at 5:30 pm. How long is he gone?
A 2 h **B** 4 h 30 min
C 9 h **D** 10 h

***3.** Randy starts work at 8:45 am and works 5 h 15 min. What time does he finish?
A 12:45 pm **B** 1:45 pm
C 1:00 pm **D** 2:00 pm

***4.** Mrs. Bono left her house at 6:30 am and arrived at work at 7:20 am. In the afternoon, she left work at 4:30 pm and arrived home at 5:35 pm. How long did it take her to commute to and from work?
A 55 min **B** 1 h 55 min
C 2 h 55 min **D** 3 h 5 min

MULTIPLICATION FACTS TEST A (3 MINUTES)

1.	8×3	4×2	2×1	3×2	2×5	1×2	8×2	6×3	2×4	6×9
2.	7×4	2×3	3×4	4×1	6×4	5×4	3×1	4×4	5×2	7×3
3.	7×2	1×1	9×2	6×2	9×3	5×3	8×4	9×9	4×3	7×9
4.	3×3	9×4	6×5	1×5	6×6	7×6	6×0	5×5	9×6	2×0
5.	5×6	8×5	2×7	9×5	3×6	4×5	8×6	7×5	4×6	3×5
6.	8×7	2×9	4×8	1×9	2×8	8×9	5×8	3×9	8×1	0×9
7.	5×9	7×8	4×9	3×8	9×1	9×8	6×7	7×7	4×7	8×8

67 or more right? Go to page 74. *Less than 67 right? Go to page 70.*

MULTIPLICATION FACTS TEST B (2 MINUTES)

1.	6×5	6×8	7×7	4×8	1×8	5×1	6×6	7×1	4×7	2×8
2.	9×1	3×6	5×5	5×7	1×6	8×9	6×9	4×5	8×6	8×7
3.	7×5	5×6	7×2	4×6	8×5	3×5	7×9	8×1	6×7	9×9
4.	2×7	2×9	5×9	4×9	2×6	1×5	7×8	8×8	1×9	3×8
5.	9×5	7×6	6×1	9×6	1×7	9×7	5×8	2×5	9×8	3×7

48 or more right? Go to page 74. *Less than 48 right? See your teacher.*

MULTIPLICATION FACTS

To learn the multiplication facts

- Think of rows of objects.
- Use repeated addition.
- Use facts that you know.

EXAMPLE 1

Multiply. $\begin{array}{r} 9 \\ \times 7 \\ \hline \end{array}$

$\begin{array}{r} 9 \\ \times 7 \\ \hline \end{array}$ can be shown as 7 rows of 9 objects.

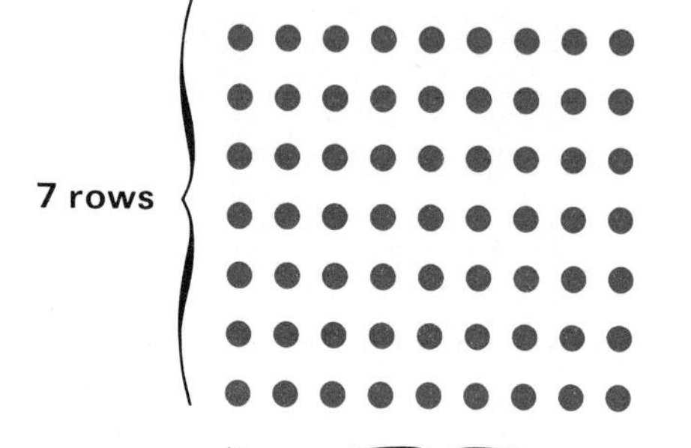

Count all of the objects.
There are 63.
So, $\begin{array}{r} 9 \\ \times 7 \\ \hline \mathbf{63.} \end{array}$ ← **product**

EXAMPLE 2

Multiply. $\begin{array}{r} 2 \\ \times 6 \\ \hline \end{array}$

Think of repeated addition.
6×2 means six 2's, or
$\mathbf{2 + 2 + 2 + 2 + 2 + 2 = 12}$
So, $6 \times 2 =$ **12.**

> Each multiplication fact can be written in different ways.
> **Example:** $\begin{array}{r} 3 \\ \times 5, \\ \hline \end{array}$ $\begin{array}{r} 5 \\ \times 3, \\ \hline \end{array}$
> 5×3, and 3×5 all equal 15.

EXAMPLE 3

Find the product. $\begin{array}{r} 7 \\ \times 8 \\ \hline \end{array}$

Use facts that you know.
Suppose you know that $\begin{array}{r} 7 \\ \times 5 \\ \hline 35. \end{array}$

Since 8×7 means eight 7's,
you need 3 more 7's, or $\begin{array}{r} \mathbf{7} \\ \mathbf{\times 3} \\ \hline \mathbf{21.} \end{array}$

So, the product is **56.**

Now add five 7's and three 7's.
$\mathbf{35 + 21 = 56}$

GETTING READY

Multiply. Try to use repeated addition, or facts that you know.

1. 9×5 **2.** 5×7 **3.** 8×3 **4.** 5×5 **5.** 4×2 **6.** 9×4 **7.** 8×5

8. 7×5 **9.** 6×8 **10.** 9×9 **11.** 8×4 **12.** 7×7 **13.** 7×8 **14.** 6×5

15. List all of the facts that you missed on Facts Test A and practice them until you know them.

EXERCISES

Multiply.

1. 9×3 **2.** 4×7 **3.** 6×6 **4.** 4×5 **5.** 3×7 **6.** 3×6 **7.** 4×4

8. 6×7 **9.** 4×6 **10.** 2×9 **11.** 6×9 **12.** 5×0 **13.** 7×9 **14.** 6×2

15. 3×5 **16.** 2×8 **17.** 8×8 **18.** 3×4 **19.** 9×1 **20.** 5×3 **21.** 9×0

Solve.

22. The first 7 rows in a parking lot have 2 spaces each for the handicapped. How many spaces for the handicapped are there?

23. There are 4 duplicating machines on each of the 9 floors of an office building. How many are there altogether?

CALCULATOR

In the number sentence $2^4 = 16$, the small numeral 4 is called an **exponent.** It tells you to use 2 four times in multiplying. That is, $2^4 = 2 \times 2 \times 2 \times 2 = 16$. The result, 16, is called a **power** of 2.

Use your calculator to find the following powers.

1. 5^3 **2.** 3^4 **3.** 7^4 **4.** 6^5 **5.** 4^3 **6.** 2^6 **7.** 5^4

USING THE FACTS

To multiply multiples of 10, 100, or 1,000 by a one-digit number

- Find the number of 10's, 100's, or 1,000's.
- Use the multiplication facts.

EXAMPLE

Multiply. $\begin{array}{r} 30 \\ \times 4 \\ \hline \end{array}$ $\begin{array}{r} 300 \\ \times 4 \\ \hline \end{array}$ $\begin{array}{r} 3{,}000 \\ \times 4 \\ \hline \end{array}$

Think: 30 means 3 tens, 300 means 3 hundreds, and 3,000 means 3 thousands.

$\begin{array}{r} \textbf{3 tens} \\ \times \textbf{4} \\ \hline \textbf{12 tens} \end{array}$ $\begin{array}{r} \textbf{3 hundreds} \\ \times \textbf{4} \\ \hline \textbf{12 hundreds} \end{array}$ $\begin{array}{r} \textbf{3 thousands} \\ \times \textbf{4} \\ \hline \textbf{12 thousands} \end{array}$

So, $\begin{array}{r} 30 \\ \times 4 \\ \hline \textbf{120} \end{array}$, $\begin{array}{r} 300 \\ \times 4 \\ \hline \textbf{1,200} \end{array}$, and $\begin{array}{r} 3{,}000 \\ \times 4 \\ \hline \textbf{12,000} \end{array}$.

GETTING READY

Multiply.

1. $\begin{array}{r} 20 \\ \times 8 \\ \hline \end{array}$ **2.** $\begin{array}{r} 60 \\ \times 4 \\ \hline \end{array}$ **3.** $\begin{array}{r} 400 \\ \times 2 \\ \hline \end{array}$ **4.** $\begin{array}{r} 5{,}000 \\ \times 4 \\ \hline \end{array}$ **5.** $\begin{array}{r} 700 \\ \times 6 \\ \hline \end{array}$

EXERCISES

Multiply.

1. 90×4 **2.** 200×7 **3.** 80×8 **4.** 800×6 **5.** 4×50

6. $\begin{array}{r} 2{,}000 \\ \times 6 \\ \hline \end{array}$ **7.** 40×7 **8.** $\begin{array}{r} 4{,}000 \\ \times 7 \\ \hline \end{array}$ **9.** $\begin{array}{r} 50 \\ \times 8 \\ \hline \end{array}$ **10.** $\begin{array}{r} 200 \\ \times 9 \\ \hline \end{array}$

11. $\begin{array}{r} 800 \\ \times 7 \\ \hline \end{array}$ **12.** $\begin{array}{r} 7{,}000 \\ \times 5 \\ \hline \end{array}$ **13.** $\begin{array}{r} 300 \\ \times 8 \\ \hline \end{array}$ **14.** $\begin{array}{r} 800 \\ \times 5 \\ \hline \end{array}$ **15.** $\begin{array}{r} 7{,}000 \\ \times 9 \\ \hline \end{array}$

Solve.

***16.** Ms. Feeney agreed to repay a loan in 6 monthly payments of $70 each. How much was she to repay?

MULTIPLYING BY A ONE-DIGIT NUMBER

To multiply by a one-digit number with no renaming

- Use the multiplication facts.

EXAMPLE 1

Multiply. 52×3

$$\begin{array}{r} 52 \\ \times 3 \\ \hline 6 \\ 150 \\ \hline \mathbf{156} \end{array} \qquad \begin{array}{l} \\ \\ \leftarrow 3 \times 2 \\ \leftarrow 3 \times 50 \\ \end{array}$$

> When multiplying in vertical form, the number with more digits is placed on top.

EXAMPLE 2

Here is a shorter way to multiply.

Multiply. $2{,}024 \times 2$

$$\begin{array}{r} \mathbf{2{,}024} \\ \mathbf{\times 2} \\ \hline \mathbf{8} \end{array} \qquad \begin{array}{r} \mathbf{2{,}024} \\ \mathbf{\times 2} \\ \hline \mathbf{48} \end{array} \qquad \begin{array}{r} \mathbf{2{,}024} \\ \mathbf{\times 2} \\ \hline \mathbf{048} \end{array} \qquad \begin{array}{r} \mathbf{2{,}024} \\ \mathbf{\times 2} \\ \hline \mathbf{4{,}048} \end{array}$$

So, $2{,}024 \times 2 =$ **4,048.**

GETTING READY

Multiply.

1. 43×3 **2.** 92×4 **3.** 823×2 **4.** $9{,}011 \times 5$

EXERCISES

Multiply.

1. $\begin{array}{r} 84 \\ \times 2 \\ \hline \end{array}$ **2.** $\begin{array}{r} 51 \\ \times 7 \\ \hline \end{array}$ **3.** $\begin{array}{r} 303 \\ \times 3 \\ \hline \end{array}$ **4.** $\begin{array}{r} 740 \\ \times 2 \\ \hline \end{array}$ **5.** $\begin{array}{r} 901 \\ \times 9 \\ \hline \end{array}$

6. $\begin{array}{r} 743 \\ \times 2 \\ \hline \end{array}$ **7.** $\begin{array}{r} 6{,}221 \\ \times 4 \\ \hline \end{array}$ **8.** $\begin{array}{r} 9{,}033 \\ \times 3 \\ \hline \end{array}$ **9.** $\begin{array}{r} 8{,}004 \\ \times 2 \\ \hline \end{array}$ **10.** $\begin{array}{r} 7{,}233 \\ \times 3 \\ \hline \end{array}$

Solve.

11. Tim bought 3 pairs of jeans for $31 each. How much did he pay altogether?

***12.** Lisa pays $202 every 3 months for insurance. How much does she pay a year?

Return to Facts Test B on page 69.

DIAGNOSTIC TEST: FORM A (30 MINUTES)

Multiply.

1. 429 × 26	**2.** 595 × 43	**3.** 417 × 35	**4.** 118 × 36
5. 495 × 76	**6.** 278 × 93	**7.** 8,617 × 37	**8.** 5,293 × 21
9. 9,215 × 75	**10.** 7,450 × 89	**11.** 802 × 205	**12.** 307 × 415
13. 439 × 727	**14.** 913 × 240	**15.** 409 × 770	**16.** 480 × 803
17. 612 × 151	**18.** 7,939 × 168	**19.** 5,673 × 916	**20.** 8,500 × 245
21. 691 × 10	**22.** 326 × 123	**23.** 9,732 × 69	**24.** 8,342 × 108

23 or more right? Go to page 80. *Less than 23 right? Go to page 75.*

DIAGNOSTIC TEST: FORM B (25 MINUTES)

Multiply.

1. 543 × 56	**2.** 274 × 65	**3.** 568 × 44	**4.** 955 × 14
5. 336 × 23	**6.** 493 × 76	**7.** 1,706 × 89	**8.** 6,423 × 38
9. 6,803 × 32	**10.** 7,946 × 65	**11.** 137 × 123	**12.** 843 × 486
13. 924 × 850	**14.** 209 × 104	**15.** 807 × 501	**16.** 306 × 145
17. 803 × 250	**18.** 5,675 × 912	**19.** 6,072 × 516	**20.** 4,109 × 307

19 or more right? Go to page 80. *Less than 19 right? See your teacher.*

RENAMING IN MULTIPLICATION

To multiply by a one-digit number with renaming

- Use the multiplication facts.
- Rename when necessary.

EXAMPLE

Multiply. 3×457

A short way to multiply is to use renaming.

$$\begin{array}{r} {}^{2} \\ 457 \\ \times 3 \\ \hline 1 \end{array}$$

$3 \times 7 = 21$
Put the 2 over the **tens** column.

$$\begin{array}{r} {}^{1\,2} \\ 457 \\ \times 3 \\ \hline 71 \end{array}$$

$3 \times 5 = 15$
$15 + 2 = 17.$
Put the 1 over the **hundreds** column.

$$\begin{array}{r} {}^{1\,2} \\ 457 \\ \times 3 \\ \hline 1{,}371 \end{array}$$

$3 \times 4 = 12$
$12 + 1 = 13.$

So, $3 \times 457 =$ **1,371.**

GETTING READY

Multiply.

1. 4×43 **2.** 6×139 **3.** $7 \times 2{,}423$ **4.** $3 \times 1{,}274$

EXERCISES

Multiply.

1. 3×48 **2.** 5×37 **3.** 7×69 **4.** 4×70 **5.** 2×84

6. $\begin{array}{r} 135 \\ \times 4 \\ \hline \end{array}$ **7.** $\begin{array}{r} 243 \\ \times 3 \\ \hline \end{array}$ **8.** $\begin{array}{r} 350 \\ \times 2 \\ \hline \end{array}$ **9.** $\begin{array}{r} 342 \\ \times 5 \\ \hline \end{array}$ **10.** $\begin{array}{r} 366 \\ \times 6 \\ \hline \end{array}$

11. 7×451 **12.** 5×506 **13.** 8×714 **14.** 3×328 **15.** 4×622

16. $\begin{array}{r} 4{,}321 \\ \times 4 \\ \hline \end{array}$ **17.** $\begin{array}{r} 4{,}290 \\ \times 5 \\ \hline \end{array}$ **18.** $\begin{array}{r} 5{,}321 \\ \times 6 \\ \hline \end{array}$ **19.** $\begin{array}{r} 6{,}543 \\ \times 6 \\ \hline \end{array}$ **20.** $\begin{array}{r} 2{,}453 \\ \times 7 \\ \hline \end{array}$

Solve.

21. Adam sells about $3,965 worth of sporting goods each week. About how much is this in 6 weeks?

***22.** Marjorie makes $295 a week. How much is this a year? (Use 52 wk = 1 y.)

MULTIPLYING BY A TWO-DIGIT NUMBER

To multiply by a two-digit number

- Multiply by each digit.
- Add the partial products.

EXAMPLE 1

Multiply. 56 × 23

Multiply by each digit.

Add the two partial products.

$$\begin{array}{r} \mathbf{56} \\ \underline{\mathbf{\times 23}} \\ \mathbf{168} \\ \underline{\mathbf{1\ 120}} \\ \mathbf{1{,}288} \end{array} \quad \begin{array}{l} \\ \\ \longleftarrow 3 \times 56 = 168 \\ \longleftarrow 2 \text{ tens} \times 56 = 112 \text{ tens.} \\ \end{array}$$

So, the product is **1,288.**

EXAMPLE 2

Multiply. 237 × 40

Multiply by each digit.

Add.

$$\begin{array}{r} \mathbf{237} \\ \underline{\mathbf{\times 40}} \\ \mathbf{0} \\ \underline{\mathbf{9\ 480}} \\ \mathbf{9{,}480} \end{array} \quad \begin{array}{l} \\ \\ \longleftarrow 0 \times 237 \\ \longleftarrow 4 \text{ tens} \times 237 \\ \end{array}$$

So, the product is **9,480.**

EXAMPLE 3

Multiply. 37 × 4,351

$$\begin{array}{r} \mathbf{4{,}351} \\ \underline{\mathbf{\times 37}} \\ \mathbf{30\ 457} \\ \underline{\mathbf{130\ 530}} \\ \mathbf{160{,}987} \end{array} \quad \begin{array}{l} \\ \\ \longleftarrow 7 \times 4{,}351 \\ \longleftarrow 3 \text{ tens} \times 4{,}351 \\ \end{array}$$

So, the product is **160,987.**

GETTING READY

Multiply.

1. 18 × 26 **2.** 30 × 54 **3.** 14 × 236 **4.** 53 × 5,107

EXERCISES

Multiply.

1. 23 × 49 **2.** 76 × 98 **3.** 89 × 21 **4.** 59 × 37 **5.** 64 × 28

6. 56 × 38 **7.** 80 × 42 **8.** 90 × 75 **9.** 72 × 40 **10.** 67 × 30

11. 60 × 20 **12.** 88 × 77 **13.** 64 × 36 **14.** 85 × 73 **15.** 86 × 79

16. 432 × 33 **17.** 516 × 42 **18.** 236 × 95 **19.** 561 × 26 **20.** 874 × 35

21. 25 × 903 **22.** 13 × 807 **23.** 88 × 704 **24.** 72 × 600 **25.** 59 × 931

26. 880 × 33 **27.** 777 × 54 **28.** 555 × 67 **29.** 841 × 50 **30.** 406 × 70

31. 21 × 3,256 **32.** 53 × 5,642 **33.** 37 × 5,947 **34.** 64 × 8,721

35. 7,400 × 83 **36.** 8,246 × 30 **37.** 1,907 × 16 **38.** 1,037 × 90 **39.** 4,095 × 77

Solve.

***40.** Patterson City has 35 level-two employees. Each gets $4,500 in benefits. How much does the city pay for all level-two benefits?

***41.** A truck contains 350 cartons. Each carton weighs 49 lb. How much does the load weigh?

ON YOUR OWN

If the pages of a book are numbered 1 through 100, how many times will each of the digits 0–9 be used in the numbering of the pages?

MULTIPLYING BY A THREE-DIGIT NUMBER

To multiply by a three-digit number

- Multiply by each digit.
- Add the partial products.

EXAMPLE 1

Multiply. 408 × 348

Multiply by each digit.

Add.

408		
×348		
3 264	⟵	8 × 408
16 320	⟵	4 tens × 408
122 400	⟵	3 hundreds × 408
141,984		So, the product is **141,984.**

EXAMPLE 2

Multiply. 234 × 5,376

5,376		
×234		
21 504	⟵	4 × 5,376
161 280	⟵	3 tens × 5,376
1 075 200	⟵	2 hundreds × 5,376
1,257,984		So, the product is **1,257,984.**

EXAMPLE 3

Multiply. 831 × 907

831		
×907		
5 817	⟵	7 × 831
0	⟵	0 tens × 831
747 900	⟵	9 hundreds × 831
753,717		So, the product is **753,717.**

GETTING READY

Multiply.

1. 241 × 352 **2.** 364 × 507 **3.** 186 × 1,343 **4.** 250 × 3,814

EXERCISES

Multiply.

1. 209 × 384 **2.** 327 × 234 **3.** 723 × 342 **4.** 695 × 313 **5.** 349 × 407

6. 516 × 728 **7.** 504 × 254 **8.** 641 × 400 **9.** 753 × 247 **10.** 903 × 345

11. 432 × 311 **12.** 819 × 536 **13.** 900 × 437 **14.** 715 × 200 **15.** 807 × 304

16. 653 × 245 **17.** 901 × 438 **18.** 875 × 193 **19.** 732 × 423 **20.** 641 × 452

21. 3,207 × 234 **22.** 4,409 × 325 **23.** 4,503 × 300 **24.** 5,610 × 432 **25.** 6,741 × 652

26. 5,832 × 364 **27.** 8,923 × 542 **28.** 9,064 × 543 **29.** 9,246 × 657 **30.** 4,791 × 267

31. 1,704 × 214 **32.** 5,351 × 800 **33.** 7,066 × 106 **34.** 3,298 × 615 **35.** 4,425 × 200

Solve.

***36.** Every day 435 passengers travel one route of the Urban Bus Lines. How many passengers are carried in one year? (Use 365 days = 1 y.)

***37.** The average weekly salary for the employees of a manufacturing company is $237. If there are 304 employees, what is the total amount paid in salary by the company each week?

Return to Form B on page 74

MULTIPLYING BY 10, 100, OR 1,000

To multiply by 10, 100, or 1,000

- Count the number of zeros.
- Attach zeros to the product of the nonzero digits.

EXAMPLE 1

Multiply 1,000 by 10.

Count the number of zeros in 1,000.
Count the number of zeros in 10.
Multiply 1 × 1.
Attach 4 zeros.

$$\begin{array}{r l r} 1{,}000 & \longleftarrow & 3 \text{ zeros} \\ \times 10 & \longleftarrow & +1 \text{ zero} \\ \hline 10{,}000 & \longleftarrow & 4 \text{ zeros} \end{array}$$

So, the product is **10,000.**

EXAMPLE 2

Multiply. 100 × 6,000

Count the number of zeros in 100 and in 6,000.
Then add to find the total number of zeros. 2 zeros + 3 zeros = 5 zeros
Multiply the nonzero digits. 1 × 6 = 6
Attach 5 zeros to the product 6. 600,000
So, 100 × 6,000 = **600,000.**

EXAMPLE 3

Multiply. 500 × 300

Count the number of zeros in each factor.
Then add. 2 zeros + 2 zeros = 4 zeros
Multiply the nonzero digits. 5 × 3 = 15
So, the product is **150,000.**

EXAMPLE 4

Multiply 672 by 10, by 100, and by 1,000.

$$\begin{array}{r l} 672 & \\ \times 10 & \longleftarrow 1 \text{ zero} \\ \hline \underbrace{6{,}72}_{672 \times 1}0 & \end{array} \qquad \begin{array}{r l} 672 & \\ \times 100 & \longleftarrow 2 \text{ zeros} \\ \hline \underbrace{67{,}2}_{672 \times 1}00 & \end{array} \qquad \begin{array}{r l} 672 & \\ \times 1{,}000 & \longleftarrow 3 \text{ zeros} \\ \hline \underbrace{672}_{672 \times 1}{,}000 & \end{array}$$

So, the products are **6,720; 67,200;** and **672,000.**

GETTING READY

Multiply.

1. 10×20 **2.** 70×100 **3.** 500×900 **4.** $800 \times 7,000$

Multiply by 10, by 100, and by 1,000.

5. 25 **6.** 326 **7.** 408 **8.** 1,297

EXERCISES

Multiply.

1. 10×10 **2.** 10×100 **3.** 100×100 **4.** $1,000 \times 100$ **5.** $10,000 \times 10$

6. 300×10 **7.** 900×100 **8.** 100×30 **9.** $1,000 \times 6$ **10.** $8,000 \times 100$

11. 60×40 **12.** 70×30 **13.** 50×20 **14.** $4,000 \times 300$ **15.** 700×700

16. 30×20 **17.** 80×50 **18.** 90×40 **19.** $5,000 \times 400$ **20.** 400×300

21. $5,000 \times 800$ **22.** $7,000 \times 2,000$ **23.** $9,000 \times 700$ **24.** $9,000 \times 400$ **25.** $8,000 \times 9,000$

Multiply by 10, by 100, and by 1,000

26. 47 **27.** 360 **28.** 871 **29.** 1,002 **30.** 4,205

Solve.

31. Arthur bought 10 raffle tickets for $1 each. How much did he spend?

32. Esther received 100 phone orders for a magazine. The magazine costs $27 a year. How much is this altogether?

***33.** How many zeros would you attach to the product of the nonzero digits in a million times a thousand?

ESTIMATING PRODUCTS

To estimate products

- Round each number to its largest place.
- Then multiply the rounded numbers.

EXAMPLE 1

Estimate the product of 249 × 461.

The largest place in both numbers is hundreds.
Round to the nearest hundred.

$$\begin{array}{r} 461 \\ \times 249 \\ \hline \end{array} \begin{array}{c} \longrightarrow \\ \longrightarrow \\ \end{array} \begin{array}{r} \mathbf{500} \\ \mathbf{\times 200} \\ \hline \mathbf{100{,}000} \end{array}$$

So, the estimated product is **100,000.**

EXAMPLE 2

Estimate the product of 85 × 113.

The largest place in 113 is hundreds.
The largest place in 85 is tens.

$$\begin{array}{r} 113 \\ \times 85 \\ \hline \end{array} \begin{array}{c} \longrightarrow \\ \longrightarrow \\ \end{array} \begin{array}{r} \mathbf{100} \\ \mathbf{\times 90} \\ \hline \mathbf{9{,}000} \end{array}$$

So, the estimated product is **9,000.**

EXAMPLE 3

Angelo ordered 18 spring jackets from the factory. Each jacket cost $12.93. About how much did he pay for them?

Round.
Multiply.

$$\begin{array}{r} \$12.93 \\ \times 18 \\ \hline \end{array} \begin{array}{c} \longrightarrow \\ \longrightarrow \\ \end{array} \begin{array}{r} \mathbf{\$10} \\ \mathbf{\times 20} \\ \hline \mathbf{\$200} \end{array}$$

The jackets cost about **$200.**

You can get a closer estimate by rounding $12.93 to the nearest dollar.
Round.

$$\begin{array}{r} \$12.93 \\ \times 18 \\ \hline \end{array} \begin{array}{c} \longrightarrow \\ \longrightarrow \\ \end{array} \begin{array}{r} \mathbf{\$13} \\ \mathbf{\times 20} \\ \hline \mathbf{\$260} \end{array}$$

So, rounding to the nearest $10, the estimated cost is **$200.** Rounding to the nearest $1, the estimated cost is **$260.**

GETTING READY

Estimate.

1. 63×205
2. 155×390
3. $13 \times \$11.69$
4. $11 \times \$17.26$

EXERCISES

Estimate.

1. 11×89
2. 18×315
3. 27×32
4. 28×41
5. 185×119
6. 76×69
7. 731×507
8. 54×468
9. 87×591
10. 85×899
11. $\$19.76 \times 16$
12. $12 \times \$114.95$
13. $1{,}209 \times 5$
14. $\$4.95 \times 3$

Solve.

15. Jordan bought 8 L of gas for 36¢ a liter. About how much did the gas cost?

***16.** Amy has only a $20 bill. She wants to buy 48 party favors for 39¢ each. Estimate the cost to determine if she has enough money.

CALCULATOR

Use a calculator and follow the instructions in order to complete the following record.

Instructions	Record
1. Find 1,365 + 5,292. Record the answer.	**1.** ________
2. Subtract 3,214 from line 1. Record.	**2.** ________
3. Find 69 × 47. Record.	**3.** ________
4. Compare line 2 and line 3. Record the larger number.	**4.** ________
5. Multiply line 4 by 763. Record.	**5.** ________
6. Subtract line 3 from line 2. Record.	**6.** ________
7. Divide line 6 by 200. Record.	**7.** ________
8. Subtract 1 from line 7. Record.	**8.** ________

Problem Solving Skill

To solve problems about shapes and patterns

- Draw a diagram.

EXAMPLE 1

You have a sheet of plywood 8 ft long and 4 ft wide. You need to cut from it a 2-ft by 4-ft piece and two 5-ft by 2-ft pieces. How many cuts will you have to make? What is the size of the leftover piece?

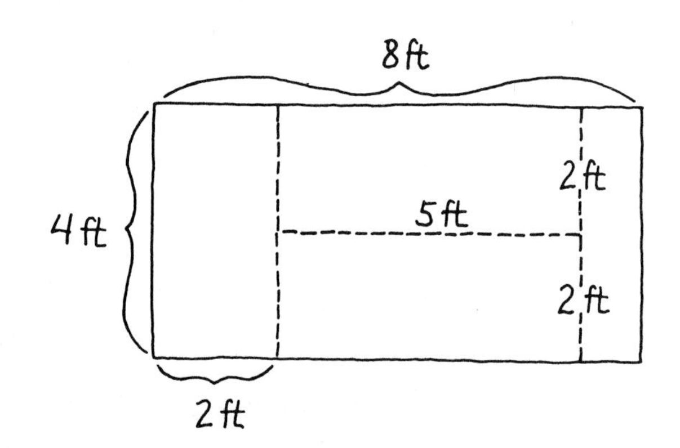

Sketch the pieces within a rectangle.
Label the lengths you know.
Look at your sketch.
You will need to make **3** cuts.
The length of 8 ft had 7 ft removed (2 ft + 5 ft).

So, the leftover piece is **1 ft by 4 ft.**

EXAMPLE 2

You are fencing a 9-ft by 5-ft garden. A post is needed at every corner and at least one post every 3 ft. How many posts will you need?

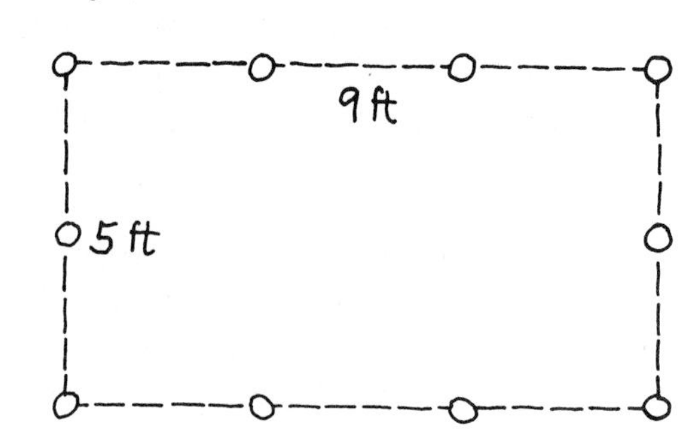

Sketch a rectangle.
Label the dimensions.
Mark the corner posts first.
You need at least one post along the width.
If you space the posts evenly, there will be 3 ft between them along the length.

So, you will need **10** posts.

EXAMPLE 3

Each square panel in a patchwork quilt contains 2 red squares and 2 blue squares. How many different panels can there be?

Draw the possible patterns.

R	R
B	B

R	B
R	B

R	B
B	R

B	B
R	R

B	R
B	R

B	R
R	B

So, there are **6** kinds of panels.

EXERCISES

Solve. Draw a diagram.

1. How many pieces of plywood, each 3 ft by 1 ft, can you cut from a sheet that is 4 ft square (a square that measures 4 ft on a side)?

2. A fence 25 ft long needs posts no more than 6 ft apart, and of course at the ends. How many posts are needed?

3. You are covering a floor that is 11 ft by 9 ft with carpet tiles that are 2 ft on a side. You can cut the tiles into smaller pieces if needed. What is the smallest number of tiles you could use to complete the job?

4. Maureen uses a 4-ring red and white target for a dart board. The center ring is white, the next one red, and so on. In how many ways can she throw 4 darts and have 3 land on red and 1 on white?

5. You walk 1 mi north, 1 mi east, 2 mi south, 1 mi east, and 1 mi north. How far and in what direction are you from your starting point?

*6. A robot is standing with its back against a fence. It is facing a wall 24 ft away. It walks half the distance to the wall. Then it turns toward the fence and walks half that distance. It does this one more time. How far is it from the fence? How far is it from the wall?

Problem Solving Applications

READ • PLAN • SOLVE • CHECK

The Counseling Center at Dalton High School provides many services to students. Course scheduling, permanent-record storage, peer tutoring, career guidance, and personal counseling are just a few of the services offered. Students work in the center each class period as peer tutors and aides.

AS YOU READ

- seminar—a meeting to give and discuss information

Answer these questions about the Counseling Center.

1. Jim does filing for the counselors. At the end of the first semester, he filed copies of report cards for 63 freshmen, 81 sophomores, 27 juniors, and 55 seniors. Find the total number of copies he filed.

2. Three student aides work in the Counseling Center each period, running errands. If there are 7 class periods, how many student aides in all work in the center each day?

3. In December the Counseling Center staff sponsored a canned-food drive. Twelve boxes holding 48 cans each were collected. How many cans of food were collected in all?

4. For five days Mr. Anderson, the career guidance counselor, met with English classes to discuss career choices. He talked with four classes each day. How many classes did Mr. Anderson talk with altogether?

***5.** The clerk in the center keeps a records file on all graduating seniors. If there are 378 seniors and 5 different record forms for each, how many record forms in all will the clerk have to file for the senior class?

***6.** Each year the counselors take a group of students to an engineering *seminar* at Austin State College. Forty-nine students went last year, and twice as many students went this year. How many students went this year?

Career: Employment Counselors

READ • PLAN • SOLVE • CHECK

Large employment offices employ file clerks, typists, interviewers or counselors, and office managers. In very small offices, the same person may have to combine several jobs, including typing and filing, along with interviewing. Training varies from a high school education to a college degree.

AS YOU READ

- applicant—a person who applies for something
- client—a person who gets advice or services from someone else
- screen—to divide into groups

Solve.

1. Last year a local employment office helped Tamcoat, Inc., hire 200 new employees. If there were 15 *applicants* for each job, how many people did the employment office *screen*?

2. One month Robert Davis, an employment interviewer, placed the following number of persons in permanent jobs: 15, 12, 18, and 23. How many jobs was he able to fill that month?

3. Each week a small state employment office processes about 125 claims for continued benefits. Find the number processed in a 4-wk period.

4. Marilyn Wallace had to file 23 employment aptitude-test forms, 57 claim reports, and 18 job-training applications. How many items did she file in all?

*5. Martha Levitas, an interviewer for the State Employment Commission, receives about 8 new claims each working day for unemployment benefits. About how many claims will she receive in a month? (Use 20 d = 1 mo.)

6. Annette Sims works for a private employment agency. A *client* is charged one month's salary for job placement. If the weekly salary is $180, how much would the agency collect for its services? (Use 4 wk = 1 mo.)

CHAPTER TEST 4

Multiply.

1. 4×39
2. 3×57
3. 7×24
4. 95×121
5. 338×41
6. 734×65
7. $8{,}561 \times 894$
8. $6{,}522 \times 370$
9. $8 \times 2{,}000$
10. $76 \times 3{,}000$
11. 80×300
12. $1{,}000 \times 100$
13. $3{,}087 \times 706$
14. $8{,}101 \times 500$
15. $7{,}040 \times 650$
16. $5{,}091 \times 303$

Estimate.

17. 89×34
18. 315×497
19. $2{,}046 \times 518$
20. About how much will 18 portable fans cost at $29.95 each?
21. If the paper service charges $2.27 per week, estimate the cost per year.

Solve by drawing a diagram.

22. You walk 3 blocks east, 1 block south, 2 blocks west, 2 blocks north, and 1 block west. How far and in what direction are you from your starting point?
23. How many ways can you make a 3-striped flag if you use only 3 colors and a different color for each stripe?

Solve.

24. One delivery box holds five bags of frozen french fries. One bag can fill two deep-fry cooking baskets. How many baskets of fries can the cook get from one delivery box?
25. Jay's Bar-B-Q uses a daily order of 8 packages of dinner rolls from the local bakery. There are 12 rolls per package. How many rolls are used each day at Jay's?
26. Multiply by 10, by 100, and by 1,000.

    1,285

*27. About how much will 11 oranges cost at 23¢ apiece?

Division of Whole Numbers

DIVISION FACTS TEST A (4 MINUTES)

1.	$2\overline{)16}$	$2\overline{)2}$	$2\overline{)14}$	$2\overline{)8}$	$1\overline{)0}$	$2\overline{)4}$	$3\overline{)0}$	$2\overline{)18}$	$2\overline{)10}$	$1\overline{)1}$
2.	$2\overline{)6}$	$2\overline{)12}$	$1\overline{)7}$	$3\overline{)3}$	$3\overline{)9}$	$4\overline{)16}$	$3\overline{)6}$	$2\overline{)0}$	$3\overline{)12}$	$4\overline{)32}$
3.	$6\overline{)0}$	$6\overline{)6}$	$3\overline{)27}$	$4\overline{)24}$	$4\overline{)36}$	$3\overline{)24}$	$4\overline{)12}$	$3\overline{)15}$	$4\overline{)4}$	$3\overline{)21}$
4.	$3\overline{)18}$	$6\overline{)36}$	$4\overline{)20}$	$5\overline{)45}$	$4\overline{)0}$	$4\overline{)28}$	$4\overline{)8}$	$5\overline{)35}$	$6\overline{)24}$	$5\overline{)5}$
5.	$6\overline{)12}$	$5\overline{)15}$	$5\overline{)10}$	$5\overline{)0}$	$1\overline{)3}$	$5\overline{)20}$	$1\overline{)4}$	$5\overline{)30}$	$8\overline{)0}$	$5\overline{)25}$
6.	$6\overline{)30}$	$6\overline{)18}$	$1\overline{)5}$	$5\overline{)40}$	$8\overline{)16}$	$7\overline{)21}$	$8\overline{)8}$	$6\overline{)48}$	$1\overline{)2}$	$7\overline{)28}$
7.	$6\overline{)42}$	$7\overline{)56}$	$7\overline{)7}$	$7\overline{)35}$	$7\overline{)0}$	$7\overline{)49}$	$6\overline{)54}$	$1\overline{)6}$	$7\overline{)42}$	$7\overline{)14}$
8.	$1\overline{)8}$	$7\overline{)63}$	$8\overline{)24}$	$9\overline{)72}$	$8\overline{)64}$	$9\overline{)54}$	$9\overline{)27}$	$8\overline{)56}$	$9\overline{)0}$	$8\overline{)72}$
9.	$1\overline{)9}$	$8\overline{)32}$	$9\overline{)9}$	$9\overline{)36}$	$9\overline{)45}$	$8\overline{)40}$	$9\overline{)63}$	$9\overline{)18}$	$9\overline{)81}$	$8\overline{)48}$

86 or more right? Go to page 94.

Less than 86 right? Go to page 91.

DIVISION FACTS TEST B (3 MINUTES)

1.	$1\overline{)9}$	$1\overline{)8}$	$8\overline{)32}$	$6\overline{)42}$	$7\overline{)63}$	$9\overline{)9}$	$6\overline{)30}$	$7\overline{)56}$	$8\overline{)24}$	$9\overline{)36}$
2.	$6\overline{)12}$	$6\overline{)18}$	$7\overline{)7}$	$9\overline{)72}$	$9\overline{)45}$	$5\overline{)15}$	$1\overline{)5}$	$7\overline{)35}$	$8\overline{)64}$	$8\overline{)40}$
3.	$5\overline{)10}$	$5\overline{)40}$	$7\overline{)0}$	$9\overline{)54}$	$9\overline{)63}$	$5\overline{)0}$	$8\overline{)16}$	$7\overline{)49}$	$9\overline{)27}$	$9\overline{)18}$
4.	$6\overline{)6}$	$7\overline{)21}$	$6\overline{)54}$	$8\overline{)56}$	$9\overline{)81}$	$5\overline{)20}$	$8\overline{)8}$	$1\overline{)7}$	$9\overline{)0}$	$8\overline{)48}$
5.	$6\overline{)36}$	$6\overline{)48}$	$7\overline{)42}$	$8\overline{)72}$	$5\overline{)30}$	$8\overline{)0}$	$7\overline{)14}$	$1\overline{)6}$	$7\overline{)28}$	$5\overline{)25}$

48 or more right? Go to page 94.

Less than 48 right? See your teacher.

DIVISION FACTS

To learn the division facts

- Use related multiplication facts.

Each division fact can be written in two ways:
$6 \div 2$ means the same as $2\overline{)6}$. Both equal 3.

EXAMPLE 1

Divide. $7\overline{)21}$
Think: What number times 7 is 21?
Since $3 \times 7 = 21$ is the related multiplication fact, 3 is that number.

So, $\begin{array}{r} \mathbf{3} \\ 7\overline{)21} \end{array}$.

EXAMPLE 2

Divide. $45 \div 9$
Think: What number times 9 is 45?
Since $5 \times 9 = 45$, 5 is that number. So, $45 \div 9 = \mathbf{5}$.

GETTING READY

Divide. Use related multiplication facts.

1. $8\overline{)32}$ **2.** $9\overline{)36}$ **3.** $6\overline{)42}$ **4.** $5\overline{)40}$

5. $24 \div 4$ **6.** $30 \div 6$ **7.** $64 \div 8$ **8.** $28 \div 7$

9. List all of the division facts you missed on Facts Test A and practice them until you know them.

EXERCISES

Divide.

1. $7\overline{)56}$ **2.** $4\overline{)36}$ **3.** $8\overline{)48}$ **4.** $2\overline{)18}$ **5.** $9\overline{)81}$

6. $30 \div 5$ **7.** $63 \div 9$ **8.** $27 \div 3$ **9.** $16 \div 2$ **10.** $49 \div 7$

Solve.

11. Anna Biehl ordered 48 boxes of pastels for her art class. The boxes came in cartons of 6 boxes. How many cartons were delivered?

12. The gym teacher divided a class of 36 students into 9 teams for gymnastics practice. How many students were on each team?

DIVIDING MULTIPLES

To divide multiples of 10, 100, or 1,000

- Use the division facts.
- Find the number of zeros needed by multiplying.

EXAMPLE 1

Divide. $\quad 2\overline{)60} \qquad 2\overline{)600}$

Think: $6 \div 2 = 3$.
To find how many zeros you need in the quotient, multiply.

$$\begin{aligned} \mathbf{3 \times 2} &= \mathbf{6} \\ \mathbf{30 \times 2} &= \mathbf{60} \\ \mathbf{300 \times 2} &= \mathbf{600} \\ \mathbf{3{,}000 \times 2} &= \mathbf{6{,}000} \end{aligned}$$

So, $2\overline{)60}$ with quotient **30** and $2\overline{)600}$ with quotient **300** ← **quotient**.

EXAMPLE 2

Divide. $\quad 450 \div 50 \qquad 45{,}000 \div 50$

Think: $45 \div 5 = 9$.
Multiply to find the number of zeros.

$$\begin{aligned} \mathbf{9 \times 50} &= \mathbf{450} \\ \mathbf{900 \times 50} &= \mathbf{45{,}000} \end{aligned}$$

So, $450 \div 50 =$ **9** and $45{,}000 \div 50 =$ **900.**

EXAMPLE 3

Divide. $\quad 40\overline{)2{,}000}$

Think: 4 does not divide 2, so use $20 \div 4 = 5$.
Multiply to find the number of zeros in the quotient.

$$\begin{aligned} \mathbf{5 \times 40} &= \mathbf{200} \\ \mathbf{50 \times 40} &= \mathbf{2{,}000} \end{aligned}$$

So, $40\overline{)2{,}000}$ with quotient **50**.

GETTING READY

Divide.

1. $6\overline{)480}$ **2.** $7\overline{)3{,}500}$ **3.** $800 \div 20$ **4.** $30{,}000 \div 50$

EXERCISES

Divide.

1. $8\overline{)80}$	**2.** $7\overline{)700}$	**3.** $6\overline{)6{,}000}$	**4.** $5\overline{)250}$
5. $240 \div 4$	**6.** $7\overline{)560}$	**7.** $9\overline{)360}$	**8.** $8\overline{)1{,}600}$
9. $1{,}800 \div 3$	**10.** $5\overline{)4{,}000}$	**11.** $5\overline{)3{,}000}$	**12.** $120 \div 6$
13. $30\overline{)2{,}400}$	**14.** $60\overline{)240}$	**15.** $70\overline{)350}$	**16.** $500 \div 50$
17. $70\overline{)2{,}100}$	**18.** $80\overline{)4{,}000}$	**19.** $50\overline{)2{,}000}$	**20.** $60\overline{)42{,}000}$
21. $80\overline{)7{,}200}$	**22.** $30\overline{)1{,}500}$	**23.** $3{,}600 \div 40$	**24.** $80\overline{)16{,}000}$
25. $800\overline{)8{,}000}$	**26.** $800\overline{)2{,}400}$	**27.** $800\overline{)4{,}000}$	**28.** $700\overline{)1{,}400}$
29. $700\overline{)2{,}800}$	**30.** $6{,}300 \div 700$	**31.** $900\overline{)7{,}200}$	**32.** $500\overline{)3{,}000}$
33. $600\overline{)4{,}200}$	**34.** $900\overline{)54{,}000}$	**35.** $3{,}200 \div 400$	**36.** $300\overline{)21{,}000}$

Solve.

37. Eddie O'Brien packed 180 books into cartons of 6 books each. How many cartons did he use?

Return to Facts Test B on page 90.

38. The bill for 200 school band uniforms was $6,000. How much did each uniform cost?

ON YOUR OWN

Solve the following puzzle. (Hint: Draw a diagram.)

Anna, Bart, Charlotte, Dick, and Emma work together. They sit at a round table for their business meetings. No two people sit next to each other if the first letters of their names are next to each other in the alphabet. Emma's brother sits on her right. Where is everyone sitting in relation to Charlotte?

DIAGNOSTIC TEST: FORM A (35 MINUTES)

Divide.

1. $7\overline{)541}$	**2.** $6\overline{)357}$	**3.** $8\overline{)968}$	**4.** $4\overline{)2,664}$
5. $3\overline{)1,435}$	**6.** $8\overline{)83,512}$	**7.** $34\overline{)950}$	**8.** $43\overline{)714}$
9. $92\overline{)3,084}$	**10.** $29\overline{)8,507}$	**11.** $62\overline{)7,739}$	**12.** $28\overline{)5,212}$
13. $15\overline{)24,090}$	**14.** $15\overline{)4,530}$	**15.** $26\overline{)52,260}$	**16.** $17\overline{)53,040}$
17. $34\overline{)17,238}$	**18.** $12\overline{)12,093}$	**19.** $21\overline{)11,042}$	**20.** $13\overline{)16,114}$

19 or more right? Go to page 100. *Less than 19 right? Go to page 95.*

DIAGNOSTIC TEST: FORM B (35 MINUTES)

Divide.

1. $9\overline{)497}$	**2.** $8\overline{)473}$	**3.** $6\overline{)492}$	**4.** $8\overline{)2,776}$
5. $7\overline{)3,657}$	**6.** $9\overline{)70,911}$	**7.** $28\overline{)348}$	**8.** $34\overline{)590}$
9. $17\overline{)7,004}$	**10.** $37\overline{)2,049}$	**11.** $34\overline{)6,295}$	**12.** $42\overline{)7,813}$
13. $15\overline{)39,090}$	**14.** $15\overline{)6,030}$	**15.** $13\overline{)91,299}$	**16.** $36\overline{)10,080}$
17. $14\overline{)91,099}$	**18.** $17\overline{)18,152}$	**19.** $31\overline{)11,042}$	**20.** $17\overline{)29,166}$

19 or more right? Go to page 100. *Less than 19 right? See your teacher.*

DIVIDING WITHOUT REMAINDERS

To divide by a one-digit number with no remainder

- Think of separating objects into groups.
- Estimate the quotient by multiplying.

EXAMPLE

Divide. $648 \div 4$

Estimate the quotient. $100 \times 4 = 400$ and $200 \times 4 = 800$.
Notice that 648 is between 400 and 800. So, the quotient is between 100 and 200.

Think: $6 \div 4 = ?$

$$\begin{array}{r} \mathbf{1} \\ \mathbf{4)648} \\ 1 \times 4 \longrightarrow \mathbf{-4} \\ \mathbf{24} \end{array}$$

Think: $24 \div 4 = ?$

$$\begin{array}{r} \mathbf{16} \\ \mathbf{4)648} \\ \mathbf{-4} \\ \mathbf{24} \\ 6 \times 4 \longrightarrow \mathbf{-24} \\ \mathbf{8} \end{array}$$

Think: $8 \div 4 = ?$

$$\begin{array}{r} \mathbf{162} \\ \mathbf{4)648} \\ \mathbf{-4} \\ \mathbf{24} \\ \mathbf{-24} \\ \mathbf{8} \\ 2 \times 4 \longrightarrow \mathbf{-8} \\ \mathbf{0} \end{array}$$

162 is between 100 and 200. So, $648 \div 4 =$ **162.**

GETTING READY

Divide.

1. $2\overline{)88}$ **2.** $5\overline{)95}$ **3.** $91 \div 7$ **4.** $684 \div 6$

EXERCISES

Divide.

1. $3\overline{)39}$ **2.** $6\overline{)72}$ **3.** $5\overline{)75}$ **4.** $9\overline{)99}$

5. $161 \div 7$ **6.** $54 \div 2$ **7.** $590 \div 5$ **8.** $192 \div 6$

Solve.

***9.** Hal has a total of 96 plants for the sides of a walk. How many plants does he have for one side?

***10.** Mrs. Berra teaches 170 students. She has 5 classes. About how many students does she have in each class?

DIVIDING WITH REMAINDERS

To divide by a one-digit number with a remainder

- Think of separating objects into groups.
- Estimate the quotient by multiplying.
- Subtract to find the remainder.

EXAMPLE 1

Divide. $38 \div 7$

Think: How many groups of 7 can be formed from 38?

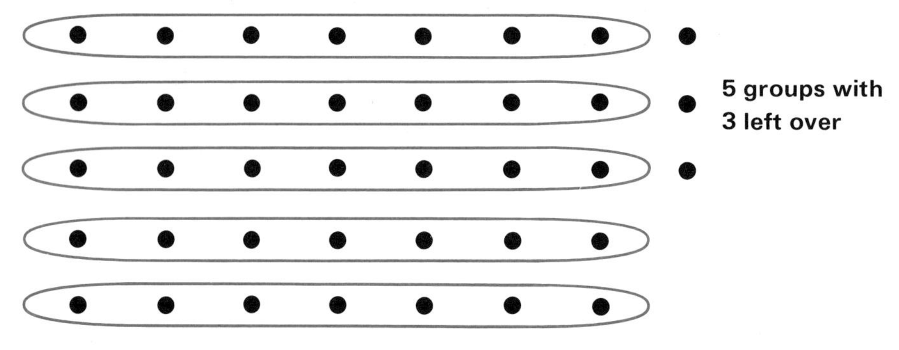

So, $38 \div 7 =$ **5 r 3.**

EXAMPLE 2

Divide and check. $9\overline{)79}$

Estimate the quotient. $8 \times 9 = 72$ and $9 \times 9 = 81$.
The quotient is between 8 and 9.

Subtract to find the remainder.

$$\begin{array}{rr} & \mathbf{8} \\ & \mathbf{9\overline{)79}} \\ 8 \times 9 \longrightarrow & \mathbf{-72} \\ \hline & \mathbf{7} \end{array}$$

So, $79 \div 9 =$ **8 r 7.**

> Parts in division problems:
>
> quotient → 6, remainder → 2: $6 \text{ r } 2$
>
> divisor → $3\overline{)20}$ ← dividend

Check:

$$\begin{array}{rl} \mathbf{8} & \longrightarrow \text{quotient} \\ \mathbf{\times 9} & \longrightarrow \text{divisor} \\ \hline \mathbf{72} & \\ \mathbf{+\ 7} & \longrightarrow \text{remainder} \\ \hline \mathbf{79} & \longrightarrow \text{dividend} \end{array}$$

So, 8 r 7 is correct.

> To check division, multiply the quotient by the divisor. Add the remainder, if there is one. The result equals the dividend.

EXAMPLE 3

Divide. $7\overline{)2,574}$

Estimate the quotient. $300 \times 7 = 2,100$ and $400 \times 7 = 2,800$.
The quotient is between 300 and 400.

Think: $25 \div 7 = ?$

$$\begin{array}{r} \mathbf{3} \\ 7\overline{)2,5\,7\,4} \\ 3 \times 7 \longrightarrow -2\,1 \\ \hline 4\,7 \end{array}$$

Think: $47 \div 7 = ?$

$$\begin{array}{r} \mathbf{3\,6} \\ 7\overline{)2,5\,7\,4} \\ -2\,1 \\ \hline 4\,7 \\ 6 \times 7 \longrightarrow -4\,2 \\ \hline 5\,4 \end{array}$$

Think: $54 \div 7 = ?$

$$\begin{array}{r} \mathbf{3\,6\,7} \\ 7\overline{)2,5\,7\,4} \\ -2\,1 \\ \hline 4\,7 \\ -4\,2 \\ \hline 5\,4 \\ 7 \times 7 \longrightarrow -4\,9 \\ \hline 5 \end{array}$$

So, $2,574 \div 7 =$ **367 r 5.**

GETTING READY

Divide and check.

1. $46 \div 8$ **2.** $7\overline{)65}$ **3.** $530 \div 4$ **4.** $9\overline{)4,537}$

EXERCISES

Divide.

1. $5\overline{)22}$ **2.** $9\overline{)15}$ **3.** $8\overline{)67}$ **4.** $5\overline{)49}$

5. $56 \div 9$ **6.** $7\overline{)69}$ **7.** $25 \div 8$ **8.** $47 \div 6$

9. $5\overline{)183}$ **10.** $3\overline{)256}$ **11.** $7\overline{)360}$ **12.** $9\overline{)478}$

13. $127 \div 4$ **14.** $793 \div 6$ **15.** $500 \div 8$ **16.** $235 \div 7$

17. $6\overline{)8,837}$ **18.** $2\overline{)3,617}$ **19.** $3\overline{)2,299}$ **20.** $4\overline{)2,643}$

***21.** $5\overline{)78,193}$ ***22.** $7\overline{)20,684}$ ***23.** $4\overline{)12,419}$ ***24.** $4\overline{)13,519}$

Solve.

25. The four children in the Roberts family found a box of old stamps in their attic. They decided to share them. The box contained 247 stamps. How many did each child get? How many were left over?

TWO-DIGIT DIVISORS

To divide a whole number by a two-digit number

- Use multiplication to estimate the quotient.
- Subtract to find the remainder, if any.

EXAMPLE 1

Divide. $48 \div 24$

Think: $2 \times 24 = 48$.

Subtract to find the remainder.

$$\begin{array}{r} 2 \\ 24\overline{)48} \\ -48 \\ \hline 0 \end{array}$$

So, $48 \div 24 =$ **2.**

EXAMPLE 2

Divide and check. $56{,}987 \div 35$

Estimate the quotient. $1{,}000 \times 35 = 35{,}000$ and $2{,}000 \times 35 = 70{,}000$.

So the quotient is between 1,000 and 2,000.

Think: $56 \div 35 = ?$

$$\begin{array}{r} 1 \\ 35\overline{)56{,}987} \\ -35 \\ \hline 219 \end{array}$$

Think: $219 \div 35 = ?$ Try $21 \div 3$.
$7 \times 35 = 245$, but $6 \times 35 = 210$.
So, 6 is correct.

$$\begin{array}{r} 1{,}6 \\ 35\overline{)56{,}987} \\ -35 \\ \hline 219 \\ -210 \\ \hline 98 \end{array}$$

Think: $98 \div 35 = ?$ Try $9 \div 3$.
$3 \times 35 = 105$, but $2 \times 35 = 70$.
So, 2 is correct.

$$\begin{array}{r} 1{,}62 \\ 35\overline{)56{,}987} \\ -35 \\ \hline 219 \\ -210 \\ \hline 98 \\ -70 \\ \hline 287 \end{array}$$

Think: 287 ÷ 35 = ? Try 28 ÷ 3.
9 × 35 = 315, but 8 × 35 = 280.
So, 8 is correct.

```
     1,6 2 8
35)5 6,9 8 7
  -3 5
   2 1 9
  -2 1 0
       9 8
      -7 0
       2 8 7
      -2 8 0
           7
```

Check:

```
  1,628
   ×35
   8140
  48840
  56980
+     7
 56,987
```

So, 56,987 ÷ 35 = **1,628 r 7.**

GETTING READY

Divide and check.

1. 58 ÷ 16 **2.** 23)146 **3.** 385 ÷ 14 **4.** 60)485

5. 3,751 ÷ 17 **6.** 31)2,638 **7.** 43)67,490 **8.** 51,627 ÷ 84

EXERCISES

Divide.

1. 76 ÷ 32 **2.** 15)64 **3.** 325 ÷ 13 **4.** 26)252

5. 42)950 **6.** 658 ÷ 13 **7.** 27)849 **8.** 14)3,168

9. 4,530 ÷ 15 **10.** 32)1,568 **11.** 1,324 ÷ 44 **12.** 15)3,913

13. 12)8,436 **14.** 4,448 ÷ 18 **15.** 21)5,523 **16.** 8,942 ÷ 34

17. 23)4,738 **18.** 51)1,989 **19.** 6,500 ÷ 18 **20.** 3,472 ÷ 62

21. 13)65,845 **22.** 28)61,740 **23.** 23)13,984 **24.** 27,738 ÷ 46

25. 67)12,663 **26.** 48)14,888 **27.** 11,442 ÷ 93 **28.** 98)25,186

Solve.

***29.** In his will, Mr. Buchanan wanted his savings account of $58,272 to be divided equally among his 24 grandchildren. How much was each grandchild to receive?

Return to Form B on page 94.

THREE-DIGIT DIVISORS

To divide a whole number by a three-digit number

- Use multiplication to estimate the quotient.
- Subtract to find the remainder, if any.

EXAMPLE

Divide. 235,974 ÷ 587

Estimate the quotient. 400 × 587 = 234,800 and 500 × 587 = 293,500.
The quotient is between 400 and 500.

```
       4                  4 0                  4 0 2
587)2 3 5,9 7 4    587)2 3 5,9 7 4    587)2 3 5,9 7 4
   -2 3 4 8           -2 3 4 8           -2 3 4 8
       1 1 7              1 1 7              1 1 7
                      -       0          -       0
                          1 1 7 4            1 1 7 4
                                           - 1 1 7 4
                                                   0
```

So, 235,974 ÷ 587 = **402.**

GETTING READY

Divide.

1. 7,196 ÷ 514 **2.** 29,766 ÷ 123 **3.** 97,036 ÷ 379

EXERCISES

Divide.

1. $209\overline{)7,315}$ **2.** $326\overline{)3,912}$ **3.** 79,570 ÷ 218

4. 103,075 ÷ 217 **5.** 267,425 ÷ 475 **6.** 56,088 ÷ 123

7. $659\overline{)153,689}$ **8.** $218\overline{)37,946}$ **9.** $927\overline{)316,023}$

10. Ms. Garcia worked 245 days last year. Her take-home pay for the year was \$23,275. How much did she earn per day?

11. Ms. Gelbard hired 28 new typists. Each typist receives the same salary. Their combined weekly salaries total \$11,060. What is the weekly salary for each typist?

DIVIDING MENTALLY

To divide mentally

- Use division facts to divide each digit or pair of digits.

EXAMPLE 1

Divide mentally. 8,462 ÷ 2

Divide each digit by 2.

$$2\overline{)8{,}462}\quad\text{quotient } \mathbf{4{,}231}$$

The answer is **4,231.**

EXAMPLE 2

Divide mentally. 49,637 ÷ 7

Put a 0 in the empty column over 6.

Use 49 ÷ 7, 63 ÷ 7, and 7 ÷ 7.

$$7\overline{)4\ 9{,}6\ 3\ 7}\quad\text{quotient } \mathbf{7{,}0\ 9\ 1}$$

The answer is **7,091.**

EXAMPLE 3

Fifteen scout troops shared $165 in banquet expenses. How much did each troop owe?

16 ÷ 15 = 1 with 1 left over.
15 ÷ 15 = 1.

$$15\overline{)165}\quad\text{quotient } \mathbf{11}$$

The answer is **$11.**

GETTING READY.

Divide mentally.

1. 844 ÷ 4 **2.** $6\overline{)246}$ **3.** $6\overline{)318}$ **4.** 42,000 ÷ 7

EXERCISES

Divide mentally.

1. 804 ÷ 4 **2.** $8\overline{)1{,}608}$ **3.** $3\overline{)3{,}219}$ **4.** 1,050 ÷ 2

5. $5\overline{)685}$ **6.** $3\overline{)597}$ **7.** 288 ÷ 9 **8.** $6\overline{)858}$

9. $5\overline{)795}$ **10.** 6,461 ÷ 7 **11.** $9\overline{)3{,}699}$ **12.** $15\overline{)4{,}560}$

FINDING AVERAGES

To find the average of a set of whole numbers

- Add the numbers to find the sum.
- Divide the sum by the number of items in the set.

EXAMPLE 1

Find the average height of 5 students whose heights are 64 in., 61 in., 67 in., 60 in., and 58 in.

> The average is sometimes called the **mean.**

To average, you **add** and **divide.**

Add the heights.

$$\begin{array}{r} \mathbf{64} \\ \mathbf{61} \\ \mathbf{67} \\ \mathbf{60} \\ \underline{\mathbf{+58}} \\ \mathbf{310} \end{array}$$

Divide by the number of students.

$$\begin{array}{r} \mathbf{62} \\ 5\overline{)\mathbf{310}} \\ \underline{\mathbf{30}} \\ \mathbf{10} \\ \underline{\mathbf{10}} \\ \mathbf{0} \end{array}$$

So, the average height is **62 in.**

EXAMPLE 2

In a city election, Johnson, Thomas, and Ramirez ran for the same office. The average number of votes cast per candidate was 379 votes. If Johnson received 580 votes and Thomas received 64 votes, how many votes did Ramirez receive?

Find the total votes cast.

$$\begin{array}{rl} \mathbf{379} & \longleftarrow \text{average} \\ \underline{\mathbf{\times 3}} & \longleftarrow \text{number of candidates} \\ \mathbf{1,137} & \longleftarrow \text{total votes cast} \end{array}$$

$$\begin{array}{lr} & \mathbf{1,137} \\ \text{Subtract Johnson's votes.} & \underline{\mathbf{-580}} \\ & \mathbf{557} \\ \text{Subtract Thomas's votes.} & \underline{\mathbf{-64}} \\ & \mathbf{493} \end{array}$$

So, Ramirez received **493** votes.

GETTING READY

1. Find the average of 48, 52, 46, and 54.
2. Find the average of 90, 110, 106, and 94.
3. Find the average of 89, 75, 91, 86, and 94.
4. The average of six numbers is 21. Five of the numbers are 23, 19, 17, 24, and 27. Find the sixth number.

EXERCISES

Solve.

1. Sara got the following scores on her science tests: 76, 78, 60, and 66. Her average must be at least 70 to get credit. Will she get credit?
2. Each of four runners ran 400 m in a relay race. Their times were 56 s, 54 s, 58 s, and 52 s. What was their average time?
3. Beatriz's average on four tests was 85. What was her score on the fourth test if the first three test scores were 81, 89, and 79?
4. The height of each basketball player is as follows: Jerry, 144 cm; Lee, 152 cm; Harrell, 140 cm; Ray, 142 cm; and Luis, 132 cm. What is the average height?
5. The article in the DeKalb School paper was torn. The number of students in grade 12 could not be read. Find the missing number.

Students in each class

Grade	Number
10	1,383
11	1,251
12	
Average per class:	1,269

CALCULATOR

Use the calculator to solve these problems.

1. Find the average: 23,000; 17,893; 4,992; 71,309; 10,387; 37,413; 85,000; and 46,086.
2. For his monthly expense account, a traveling salesman recorded the following trip distances in miles: 87, 153, 46, 188, 79, 61, 202, 35, 52, 91, and 73. What was his average mileage per trip?

ESTIMATING QUOTIENTS

To estimate answers to division problems

- Round each number to its largest place.
- Then divide.

EXAMPLE 1

Estimate. 3,185 ÷ 63

Round the divisor. 63 rounds to 60.

Round the dividend. 3,185 rounds to 3,000.

Now divide. **3,000 ÷ 60 = 50**

So, **50** is the estimate.

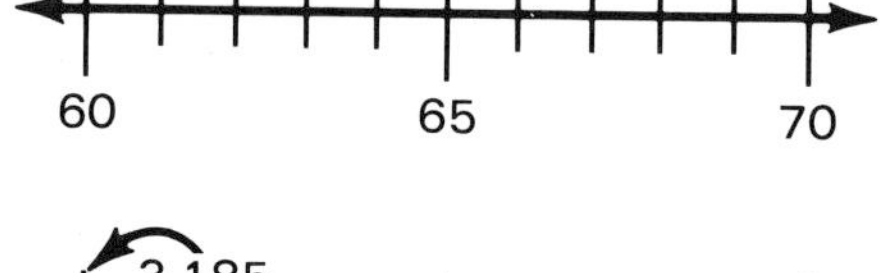

EXAMPLE 2

Estimate. 79,810 ÷ 423

Round the divisor. 423 rounds to 400.

Round the dividend. 79,810 rounds to 80,000.

Now divide. **80,000 ÷ 400 = 200**

So, **200** is the estimate.

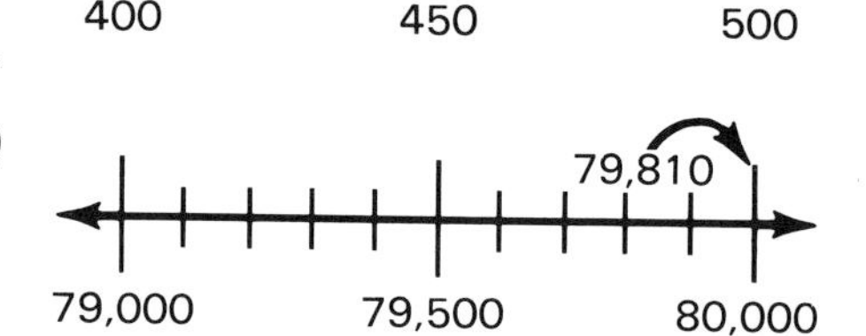

EXAMPLE 3

Three shirts cost \$21.95. Estimate how much one shirt costs.

Use 3 as the divisor.

Since 20 is not divisible by 3, the nearest dollar amount that is divisible by 3 is 21.

So, round \$21.95 to \$21, not \$20. \$21.95 ⟶ **\$21**

Now divide. **\$21 ÷ 3 = \$7**

So, one shirt costs about **\$7**.

GETTING READY

Estimate.

1. $8\overline{)96}$ **2.** $460 \div 95$ **3.** $215\overline{)3{,}930}$ **4.** $8\overline{)\$39.77}$

Solve.

5. Six pairs of socks sold for $7.12. Estimate the price per pair.

EXERCISES

Estimate.

1. $46\overline{)51{,}732}$ **2.** $12{,}172 \div 44$ **3.** $84\overline{)72{,}710}$ **4.** $84\overline{)48{,}970}$

5. $59\overline{)61{,}281}$ **6.** $51\overline{)68{,}001}$ **7.** $63{,}819 \div 56$ **8.** $44\overline{)40{,}261}$

9. $550 \div 98$ **10.** $24\overline{)40{,}133}$ **11.** $924\overline{)54{,}217}$ **12.** $94\overline{)560}$

13. $40\overline{)8{,}495}$ **14.** $25{,}016 \div 51$ **15.** $568\overline{)704}$ **16.** $4{,}714 \div 75$

17. $686 \div 33$ **18.** $64\overline{)296}$ **19.** $445\overline{)948}$ **20.** $51\overline{)7{,}099}$

21. $686 \div 43$ **22.** $95\overline{)460}$ **23.** $393 \div 115$ **24.** $102\overline{)35{,}876}$

Solve.

25. A bolt of cloth 20 m long costs $64.95. Estimate the price per meter.

26. Six baseballs cost $23.54. Estimate the cost of each ball.

27. Five tapes cost $49.75. About how much does one tape cost if they all cost the same?

Problem Solving Applications

READ • PLAN • SOLVE • CHECK

At Montfort High School students take physical education as a part of the requirements for graduation. Classes are offered in many sports. Montfort has varsity, junior varsity, and intramural teams. Some students are team managers, helping with records and keeping score at games.

AS YOU READ

- freestyle—a swimming stroke
- medley relay—a relay race in which each person uses a different stroke

Answer these questions about physical education at Montfort.

1. Mary's specialties are the 1,000-m *freestyle* and the *medley relay.* If the pool is 25 m in length, how many laps is the 1,000-m freestyle?

2. During second period 108 students are playing volleyball. There are 18 players on each court. How many volleyball courts are there?

3. Paul keeps records for the varsity track team. A 4-member relay team clocked 55 s, 56 s, 54 s, and 53 s on the 4 laps. What was the total running time for the relay?

4. In a basketball game, Madeline scored 22 out of the 77 points for Montfort High. The remaining points were scored equally by 5 of her teammates. How many points did each of the 5 score?

5. Terry jogs 12 laps around the track each morning before school. If 4 laps equal a mile, how many miles does he jog each morning?

*6. Chris averages 21 points a game. In the last 5 games he scored 16, 28, 13, 17, and 21 points. Was his average for these games above or below his playing average?

Career: Recreation Workers

READ • PLAN • SOLVE • CHECK

Municipal recreation departments, businesses, and athletic clubs provide a variety of recreational programs. These programs create employment for lifeguards, exercise instructors, parks maintenance workers, and others. Training for these jobs is often on-the-job experience.

AS YOU READ

- kick board—a board used to practice kicking in swimming

Solve.

1. James Peters is a swimming instructor for the Health Spa. He needs 60 *kick boards*, and there are 43 in storage. How many kick boards should he request?

2. Alma Frye owns Par 4 Driving Range. She charges $6 for a bucket of golf balls. If she took in $1,680 in a day for the golf balls, how many buckets were used?

3. A crew of 60 people maintains about 6,000 acres of park grounds throughout the city. How many acres of land does this average per crew member?

4. Paul Furman directs the Acme Warehouse Soccer League. If there are 11 players on each team and 94 have signed up, at most how many teams can Paul have?

***5.** Amy Johnson is a summer camp director. She ordered craft supplies for $265, playground balls for $120, and softball equipment for $185. Is this total within her budget of $500?

***6.** Arnold Bowes runs the Recreation Basketball tournament. The Blazers have scored 45, 54, and 63 points in their games so far. How many points must they score in the next game to have an average of 58 points?

Problem Solving Skill

To solve multistep word problems

- Choose the needed operations.
- Decide the order of operations.
- Compute the answer.

EXAMPLE 1

Mr. Hurango has 23 students in his math class. How much will it cost to supply each of his students with a book and workbook if the book costs \$5.75 and the workbook costs \$1.93?

Think: To find the cost for one student, **add** the book cost to the workbook cost. Then, **multiply** by the number of students to find the total cost.

Add first.

$$\begin{array}{rl} \mathbf{\$5.75} & \longleftarrow \text{cost of book} \\ \underline{\mathbf{+1.93}} & \longleftarrow \text{cost of workbook} \\ \mathbf{\$7.68} & \longleftarrow \text{cost per student} \end{array}$$

Then, multiply.

$$\begin{array}{rl} \mathbf{\$7.68} & \longleftarrow \text{cost per student} \\ \underline{\mathbf{\times 23}} & \longleftarrow \text{number of students} \\ \mathbf{23\ 04} & \\ \underline{\mathbf{153\ 60}} & \\ \mathbf{\$176.64} & \longleftarrow \text{cost for 23 students} \end{array}$$

So, the cost for 23 students is **\$176.64.**

EXAMPLE 2

The odometer of Jayne's car read 47,818 mi when she began her business trip. When she reached her destination, the odometer read 48,376 mi. Her car gets 18 mi to the gallon. How many gallons of gasoline did she use for the trip?

Subtract first to find the number of miles Jayne traveled.
Then, **divide** miles by 18 to find the number of gallons used.

Subtract. $48{,}376 - 47{,}818 = 558 \longleftarrow$ miles traveled
Divide. $558 \div 18 = 31$

So, she used **31 gal** of gas for the trip.

EXERCISES

Select the order of operations that can be used to solve each problem.

1. Mrs. Hundley bought 2 new suits at \$89.95 each and an extra pair of slacks for \$22. What was the cost of her new clothes?

   A add and divide **B** multiply and add
   C multiply and subtract **D** divide and subtract

2. Ms. Dana wants to buy an archery set. A bow will cost \$16, the arrows \$5, and a target \$4. If she saves \$4 a week, for how many weeks will she need to save in order to have enough money to buy the archery set?

   A add and multiply **B** subtract and divide
   C multiply and subtract **D** add and divide

Solve.

3. In the Summer Soccer League, there were 6 teams of 11 players each. Three additional players served as league substitutes. How many players were in the Soccer League?

4. Petunia plants at the Greenery Shop usually sell for 19¢ each. Mr. Ramos put them on sale at 6 for \$1.00. Mr. McBroom bought a dozen plants on sale. How much did he save?

5. Marilyn receives 5 bread deliveries a week at her diner. In each delivery she gets 10 packages of dinner rolls. Each package holds a dozen rolls. How many dinner rolls does she use each week?

6. Annette uses 3 boxes of frozen fries per day at her restaurant. Each box contains 2 bags of fries, and each bag yields 11 individual servings. How many individual servings of fries does Annette sell each day?

CHAPTER TEST 5

Divide.

1. 99 ÷ 3
2. $6\overline{)192}$
3. 26 ÷ 12
4. $15\overline{)330}$
5. 315 ÷ 21
6. $65\overline{)21{,}060}$
7. 20,646 ÷ 37
8. $29\overline{)20{,}880}$
9. 5,504 ÷ 18
10. $4\overline{)7{,}604}$
11. 96,481 ÷ 54
12. $56\overline{)39{,}450}$
13. 61,750 ÷ 26
14. $36\overline{)34{,}416}$
15. 46,212 ÷ 37
16. $30\overline{)21{,}000}$

Solve.

17. Joanna kept a record of her math test scores. Her scores were 88, 73, 62, 93, and 94. What was her average?

18. David bought 36 m^2 of carpeting for $324. How much did he pay per square meter?

19. On a summer job, Marie earned $213 in June, $492 in July, and $576 in August. Find her average monthly salary.

20. A bus can hold 68 passengers. How many buses are needed to transport 3,128 students?

21. If the total cost of three shirts is $28.95, estimate the cost of one shirt.

22. Juan had $5.40. He lost 2 coins. Now he has $5.20. Which 2 coins did he lose?

23. Ono has $20. She wants to buy 4 tubes of lip gloss at $3 each and 2 packages of blusher at $5 each. How much more money does she need?

24. The total attendance at the basketball tournament was 29,742. If 12 games were played, estimate the average number attending each game.

25. A passenger train needs one engine for each 5 cars. Each car can carry 60 passengers. How many engines will be needed for a train carrying 700 passengers?

COMPUTER LITERACY

Computers have become an important part of everyday life. Automobile tune-ups can be done with computer analyzers that diagnose engine problems. Computerized supermarket checkouts save time while keeping track of inventory. Home computers are used to do routine tasks such as tax returns and word processing. These are just a few of the many uses of computers.

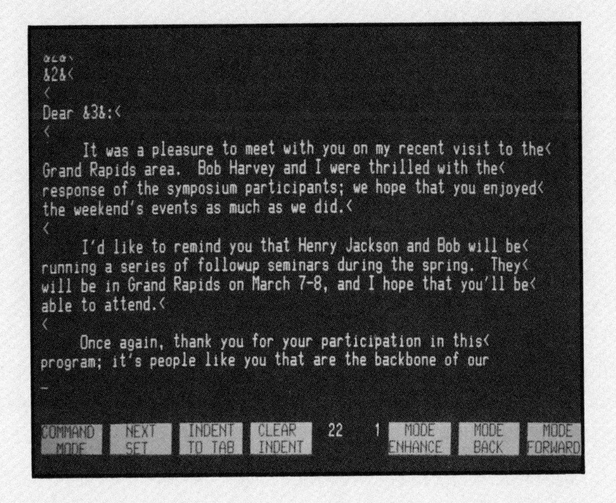

AS YOU READ

- program—an ordered list of instructions for a computer
- programmer—a person who writes computer programs
- BASIC—*B*eginner's *A*ll-Purpose *S*ymbolic *I*nstruction *C*ode, a computer language

A computer carries out a *program* written in a special language such as *BASIC*. Each BASIC instruction has a special form that must be used or the computer will reject it. An important instruction in BASIC is the PRINT statement. This instruction makes the computer write the answer to a calculation or, when quotation marks are used, print the characters between them. The RUN statement is used to tell the computer to follow the instructions and produce the output.

EXAMPLE

```
10  PRINT 3 + 5
20  PRINT "3 + 5"
30  PRINT "MY NAME IS MARK."

]RUN
8
3 + 5
MY NAME IS MARK.
```

Each line in a program is numbered. Often multiples of 10 are used so that more statements may be added to the program if necessary. The word PRINT is always followed by a space.

Another type of PRINT statement is called PRINT TAB. It allows the *programmer* to place characters anywhere on the output. PRINT TAB does this by printing in numbered columns like the TAB on a typewriter. The program below shows how to use PRINT TAB to write the letter P. The word PRINT without instructions produces a line of space in the output.

```
10  PRINT
20  PRINT   TAB( 5);"PPPPPP"
30  PRINT   TAB( 5);"PP"; TAB( 11)
    ;"P"
40  PRINT   TAB( 5);"PP"; TAB( 11)
    ;"P"
50  PRINT   TAB( 5);"PP"; TAB( 11)
    ;"P"
60  PRINT   TAB( 5);"PPPPPP"
70  PRINT   TAB( 5);"PP"
80  PRINT   TAB( 5);"PP"
90  PRINT   TAB( 5);"PP"
100  PRINT   TAB( 5);"PP"
110  END

]RUN

    PPPPPP
    PP    P
    PP    P
    PP    P
    PPPPPP
    PP
    PP
    PP
    PP
```

If you want to actually run the programs, consult your teacher.

EXERCISES

1. Write a program that will print the first names of the members of your family. Use PRINT.

2. Write a program that will print your first initial. (Hint: Put the letter on graph paper and number the columns first.) Use PRINT TAB.

3. Write a program that will print the word "RUN." Use PRINT TAB.

4. Develop a design of your own using the letter X. Write a program that will reproduce your design.

Addition and Subtraction of Decimals

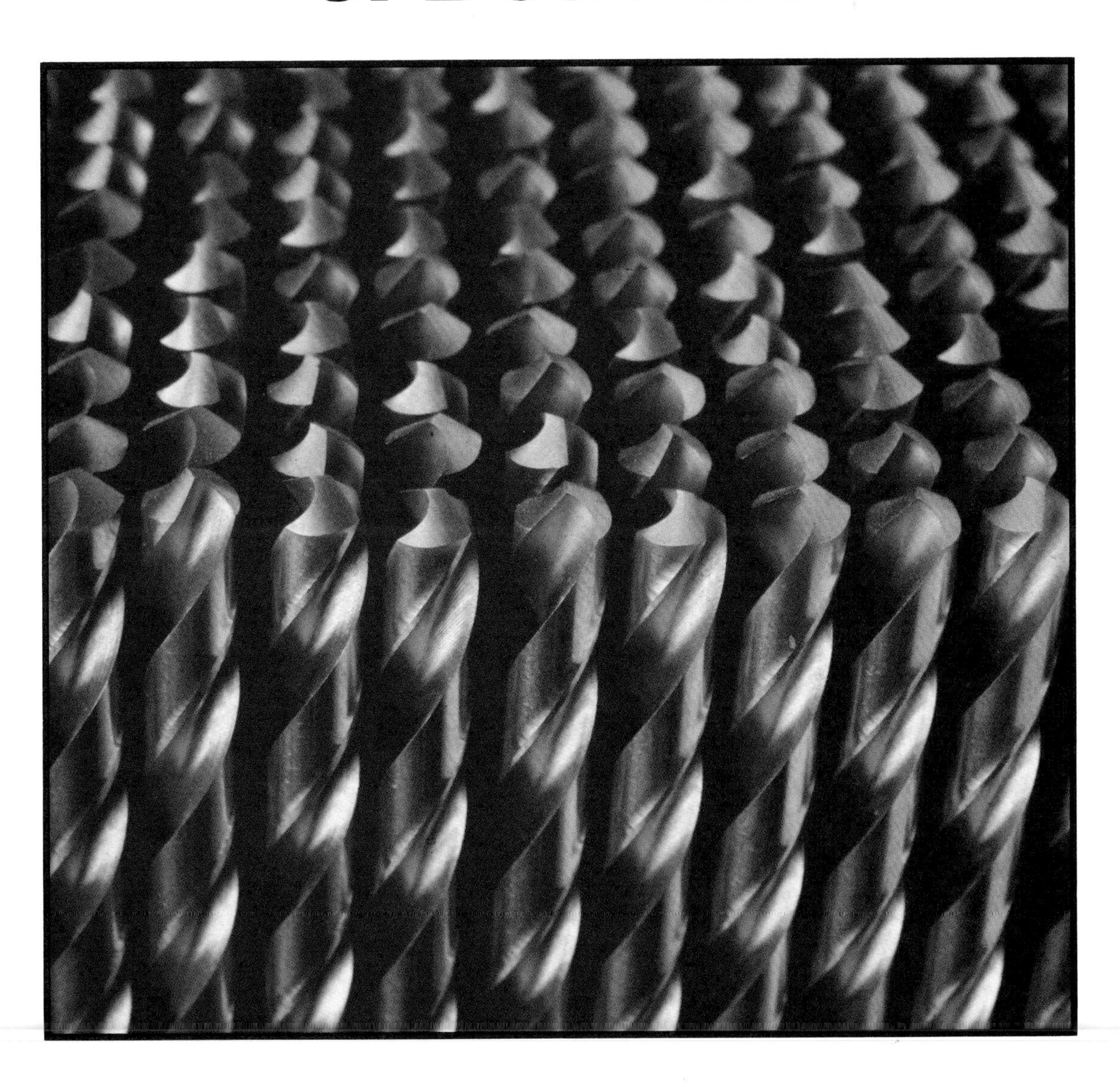

MAKING MATH COUNT

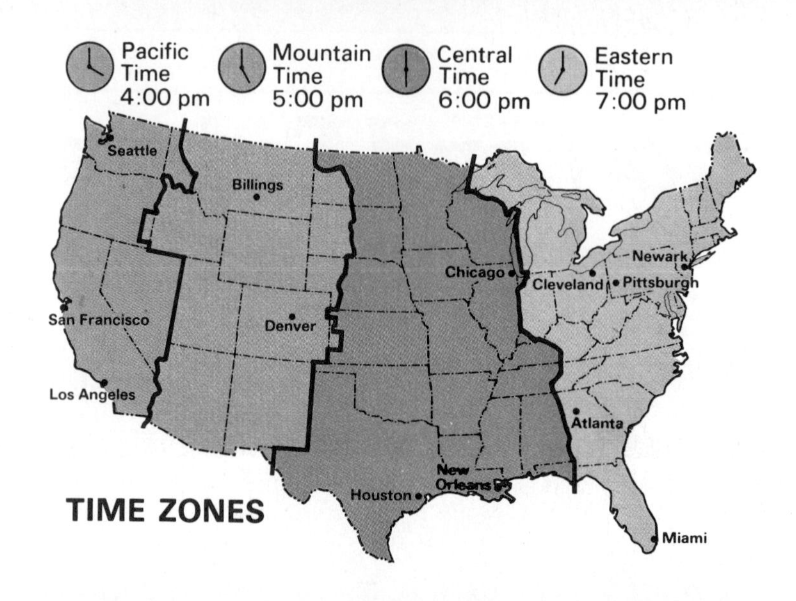

EXAMPLE

A plane leaves Chicago at 2:00 pm. It lands in Denver at 3:15 pm. How long is the flight?

A 2 h 15 min **B** 1 h 45 min
C 1 h 15 min **D** 1 h

First, find the time in Chicago when it is 3:15 pm in Denver.

Chicago is 1 h later ⟶ 4:15
Subtract the times. −2:00
2:15, or
2 h 15 min

So, **A** is the answer.

Choose the correct answer.

1. A plane leaves Los Angeles at 11:00 am and lands in Atlanta at 6:15 pm. How long is the flight?
A 4 h 15 min **B** 5 h 15 min
C 6 h 15 min **D** 7 h 15 min

2. A plane leaves Denver at 9:20 pm. It lands in Cleveland at 2:45 am. How long is the flight?
A 2 h 25 min **B** 3 h 25 min
C 4 h 25 min **D** 5 h 25 min

3. A plane leaves Houston at 11:00 am. The flight takes 2 h 30 min. What time will it be in Chicago when the plane lands?
A 1:30 pm **B** 10:30 pm
C 11:30 pm **D** 11:30 am

4. A flight leaves Atlanta for Los Angeles at 11:50 am. The flight takes 4 h 20 min. What time will it be in Los Angeles when it arrives?
A 7:10 pm **B** 4:10 pm
C 2:10 pm **D** 1:10 pm

DIAGNOSTIC TEST: FORM A (20 MINUTES)

Write a decimal.

1. two hundred forty-six and seven tenths

2. eighty-two and nine thousandths

Round to the nearest whole number.

3. 18.62 **4.** 129.1 **5.** 7.086

Add or subtract.

6. $0.3 + 2.04$

7. $4.1 + 7.8 + 19$

8. $29.2 + 3.07 + 589 + 7$

9. $64.09 + 0.85 + 0.017 + 10$

10. $32.6 - 18.3$ **11.** \$65.52 − \$5.40 **12.** \$329 − \$2.04 **13.** $41.6 - 19.87$

14. $\begin{array}{r} 83.47 \\ +\ 7.86 \\ \hline \end{array}$ **15.** $\begin{array}{r} 6.063 \\ +9.957 \\ \hline \end{array}$ **16.** $\begin{array}{r} \$478.68 \\ -198.32 \\ \hline \end{array}$ **17.** $\begin{array}{r} \$60.04 \\ -13.79 \\ \hline \end{array}$

15 or more right? Go to page 122. *Less than 15 right? Go to page 116.*

DIAGNOSTIC TEST: FORM B (20 MINUTES)

Write a decimal.

1. three hundred fifty-two and four tenths

2. thirty-one and twenty-four thousandths

Round to the nearest whole number.

3. 21.53 **4.** 378.2 **5.** 11.009

Add or subtract.

6. $6.71 + 0.4$

7. $3.6 + 5.4 + 26$

8. $465 + 8 + 31.6 + 2.03$

9. $55.61 + 0.42 + 19 + 0.022$

10. $27.9 - 15.4$ **11.** \$47.61 − \$6.50 **12.** \$456 − \$5.09 **13.** $42.5 - 23.97$

14. $\begin{array}{r} 49.85 \\ +\ 3.37 \\ \hline \end{array}$ **15.** $\begin{array}{r} 8.865 \\ +5.097 \\ \hline \end{array}$ **16.** $\begin{array}{r} \$358.89 \\ -288.71 \\ \hline \end{array}$ **17.** $\begin{array}{r} \$70.03 \\ -13.28 \\ \hline \end{array}$

15 or more right? Go to page 122. *Less than 15 right? See your teacher.*

READING AND WRITING DECIMALS

To read and write decimals

The decimal point separates the whole number part and the fractional part of a decimal number.

EXAMPLE 1

Read 145.862.

The decimal point is read *and*.

145.862 ⟵ 2 is in the thousandths place.

145.862

one hundred forty-five ↙ ↓ and ↘ eight hundred sixty-two thousandths

EXAMPLE 2

Write the decimal for two thousand, eighty-six and forty-two thousandths.

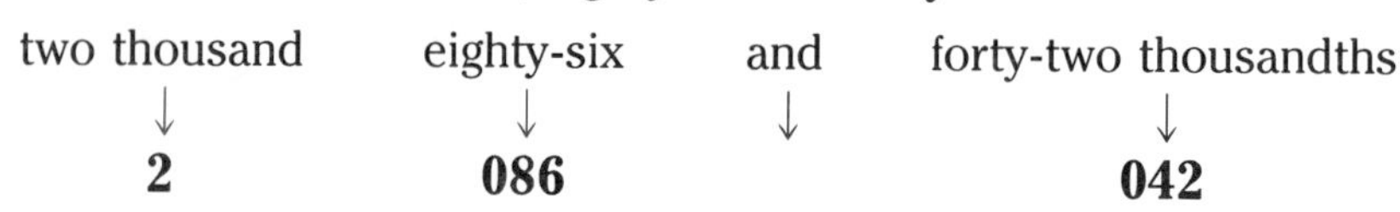

So, the decimal is **2,086.042.**

GETTING READY

Read.

1. 5.6 **2.** $134.06 **3.** 452.783 **4.** 19,205.98

Write a decimal.

5. two hundred one and eight hundredths

6. one thousand and thirty-two thousandths

EXERCISES

Read.

1. 3.9 **2.** $56.93 **3.** 600.05 **4.** 243,000.05

Write a decimal.

5. eighty-nine and six tenths

6. one hundred fifty-five and nine tenths

7. three hundred five thousand, six hundred and nineteen hundredths

8. seven hundred fifty thousand and three hundred seventy-eight thousandths

ROUNDING DECIMALS

To round a decimal to the nearest whole number

- Find the whole number closest to the decimal.

EXAMPLE 1

Round 12.6 to the nearest whole number.

Look at the number line.
12.6 is between 12 and 13.
Since 12.6 is closer to 13,
12.6 rounds up to **13.**

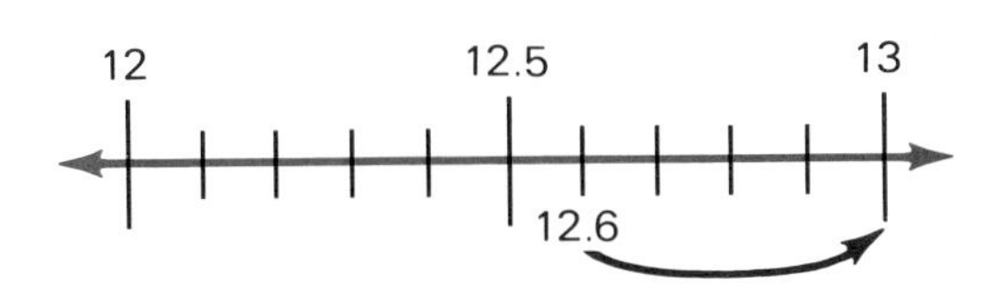

EXAMPLE 2

Round 147.348 to the nearest whole number.

147.348 is between 147 and 148.
The digit in the **tenths** place is 3.
Round down.

147.(3)48
↓
147

So, 147.348 rounds down to **147.**

> To round a decimal to the nearest whole number, look at the digit in the **tenths** place. If the digit is 5 or more, **round up.** If the digit is less than 5, **round down.**

GETTING READY

Round to the nearest whole number.

1. 21.8 **2.** 88.4 **3.** 34.19 **4.** 156.73

5. 7.684 **6.** 19.009 **7.** 0.25 **8.** 399.6104

EXERCISES

Round to the nearest whole number.

1. 8.1 **2.** 125.6 **3.** 100.84 **4.** 52.28

5. 76.195 **6.** 16.008 **7.** 0.06 **8.** 0.878

9. 0.499 **10.** 98.098 **11.** 607.5108 **12.** 1,099.8

ADDING DECIMALS

To place the decimal point in a sum by estimation
To add decimals

- Line up the decimal points.
- Add like whole numbers.
- Place the decimal point in the sum.

EXAMPLE 1

Place the decimal point to make the sum correct.

45.6 + 12.9 + 23.28 = 8178

Round each decimal to the nearest whole number.

45.6 ⟶ **46**
12.9 ⟶ **13**
+23.28 ⟶ **+23**
82

> To estimate sums of decimals, round each number to the nearest whole number. Then add to get an estimate.

The estimate is 82, which is close to 81.78. So, the sum is **81.78**.

EXAMPLE 2

Find the sum of 0.4, 18.3, and 106.7.

Line up the decimal points.

0.4
18.3
+106.7
125.4

Add.

Place the decimal point.

So, the sum is **125.4.**

To check, round each number to the nearest whole number. Then add.

0.4 ⟶ **0**
18.3 ⟶ **18**
+106.7 ⟶ **+107**
125

Compare 125 with your answer.

EXAMPLE 3

Find the sum of 18, 19.7, and 188.82.

Line up the decimal points.

18.00
19.70
+188.82
226.52

Add.

Place the decimal point.

> If necessary, use zeros as placeholders before adding decimals.

Check by rounding.

18 + 20 + 189 = 227

Compare with your answer.

227 is close to 226.52.

So, the sum is **226.52.**

GETTING READY

Place the decimal point to make the sum correct.

1. 8.3 + 6.5 + 19.7 = 345

2. 79.5 + 6.79 + 12.81 = 991

3. 676.4 + 452 + 7.6 = 1136

4. 9.258 + 1.369 + 0.03 = 10657

Add and check.

5. 6.46 + 88.07 + 326.01

6. 878.8 + 432.4 + 786

7. 18.6 + 9.23 + 2.647

8. 588.5 + 58.85 + 0.5885

EXERCISES

Place the decimal point to make the sum correct.

1. 57.9 + 1.6 + 9.7 = 692

2. 45.6 + 12.9 + 23.28 = 8178

3. 438.7 + 58.63 + 9.231 = 506561

4. 216 + 0.72 + 42.9 = 25962

Add and check.

5. 28.32 + 2.58 + 9.09

6. 3.7 + 23.7 + 9

7. 35.35 + 0.58 + 134

8. 8.02 + 6.7 + 1.986

9. 436.41 + 0.82 + 0.0037 + 31

10. 63.29 + 0.27 + 7 + 65.7601

11. $28.77 + $0.85 + $169.29 + $7

12. $32 + $0.06 + $21.59 + $322.79

ON YOUR OWN

Find a number that meets all of these conditions.

Divide it by 2; the remainder is 0.
Divide it by 7; the remainder is 0.
Divide it by 28; the remainder is 0.
Divide it by 98; the remainder is 0.
The sum of its digits is 16.

Is the number more than 150 and less than 200?

SUBTRACTING DECIMALS

To place the decimal point in a difference by estimation
To subtract decimals

- Line up the decimal points.
- Subtract like whole numbers.
- Place the decimal point in the difference.

EXAMPLE 1

Place the decimal point to make the difference correct.

743.19 − 292.8 = 45039

Round each decimal to the nearest whole number.

$$\begin{array}{r} 743.19 \\ -292.8 \\ \hline \end{array} \longrightarrow \begin{array}{r} 743 \\ -293 \\ \hline 450 \end{array}$$

The estimate is 450, which is close to 450.39.
So, the difference is **450.39.**

> To estimate differences of decimals, round each number to the nearest whole number. Then subtract to get an estimate.

EXAMPLE 2

Subtract 78.6 from 169.4.

Line up the decimal points.
Subtract.

$$\begin{array}{r} 169.4 \\ -\ \ 78.6 \\ \hline 90.8 \end{array}$$

Place the decimal point.

To check, round each decimal to the nearest whole number. Then subtract.

$$\begin{array}{r} 169.4 \\ -\ \ 78.6 \\ \hline \end{array} \longrightarrow \begin{array}{r} 169 \\ -\ \ 79 \\ \hline 90 \end{array}$$

Compare 90 with your answer.
So, the difference is **90.8.**

EXAMPLE 3

Earl's checkbook balance shows $500. His monthly bank statement shows that he needs to subtract $0.72. What is his actual balance?

$$\begin{array}{r} \$500.00 \\ -\ \ \ \ 0.72 \\ \hline \$499.28 \end{array}$$

So, the actual balance is $499.28.

> If necessary, use zeros as placeholders when subtracting decimals.

GETTING READY

Place the decimal point to make the difference correct.

1. 376.4 − 218.5 = 1579

2. 156.83 − 57.9 = 9893

3. 186.01 − 87.009 = 99001

4. 880 − 380.16 = 49984

Subtract and check.

5. 704.9 − 106.7

6. $943.12 − $74.08

7. 45.678 − 38.967

8. 5.926 − 0.947

9. 328.09 − 233

10. 234.57 − 38.6

EXERCISES

Place the decimal point to make the difference correct.

1. 284.36 − 152.12 = 13224

2. 158.1 − 62.35 = 9575

3. $800 − $83.75 = $71625

4. 5.93 − 0.947 = 4983

Subtract and check.

5. $60.06 − $37.56

6. 576.94 − 95.08

7. 42.4 − 8.96

8. 912.009 − 732.89

9. 0.973 − 0.810

10. 4,783.16 − 4,175.932

Return to Form B on page 115.

CALCULATOR

Use a calculator. Find the daily totals of the expense account below.

DATE	TRAVEL	MEALS	MOTEL	PHONE	OTHER	TOTALS
Feb. 1	$48.75	$13.65	$33.75	$4.55	$7.88	
2	33.55	12.98	32.50	3.55	7.89	
3	47.70	14.55	32.50	2.66	4.67	
4	54.25	18.50	29.50	1.88	7.55	

What is the total amount spent for the 4 days?

Problem Solving Applications

READ • PLAN • SOLVE • CHECK

The students in Mr. Bloch's metal shop classes are working on their projects. Some are making jewelry and art objects, while others are practicing welding and metal working.

AS YOU READ

- cast—to form in a mold
- etching—causing an impression in metal
- lathe—a machine for turning and shaping

Answer these questions about the students' projects.

1. Vera is *etching* a design on a metal plate. The metal is 0.125 in. thick. The etching process removes another 0.003 in. How thick is the plate now?

2. May is using a *lathe* to turn a steel rod 2.5 cm in diameter. If she decreases the diameter by 0.14 cm, what is the diameter now?

3. Peter is welding together 3 layers of metal. The thicknesses are as follows: 0.03 cm, 0.1 cm and 1.05 cm. What is the total thickness of the layers?

4. José is operating the cutting bar. Each time he makes a cut in the metal stock, 0.125 in. is wasted. If he makes 5 cuts, how much is wasted altogether?

5. Jack is making fine copper wire pins. Each pin takes 1.85 cm of wire. How much wire is needed for 3 pins?

***6.** Mabel made a mold to *cast* lead sinkers. The mold casts eight 0.5-oz sinkers at one time. How many ounces of lead are needed for 2 doz sinkers?

Career: Machinists

READ • PLAN • SOLVE • CHECK

Many industries, such as machinery, transportation, metals, textiles, and food processing, employ machinists. Most machinists learn their trade through apprenticeships lasting up to 4 years.

AS YOU READ

- dress—to prepare for use
- tapered—decreasing gradually
- threaded—having ridges like a bolt

Solve.

1. A machinist is *dressing* a grinding wheel with a diamond cutting tool. If 0.003 in. of metal is removed with each lathe setting, how many total inches of metal are removed after 6 settings?

2. A rough casting weighs 64.25 lb as it comes from the mold. When finished, the casting weighs 42.75 lb. What is the weight of the metal removed during the finishing process?

3. Tony Rice, a hand molder, uses 16.25 g of sand in each mold for one production run of castings. How many grams of sand are needed for 3 castings?

4. Anna and Tim are cutting and finishing gear wheels. The teeth on Anna's wheel are set 0.8 cm apart. The teeth on Tim's wheel are set 0.78 cm apart. How much farther apart are the teeth on Anna's wheel?

5. Ben Sterns, a finisher, is checking *tapered* rollers used in large flywheels. Each roller tapers from 3.875 in. in diameter to 3.5 in. What is the difference in the two diameters?

*6. Mindy Wright, a machinist, is making 26.5 cm *threaded* connecting rods. How many centimeters of metal stock are needed for 10 rods?

Problem Solving Skill

To read data from charts

- Notice the headings and types of information given.
- Decide what is being asked.

EXAMPLE 1

Use this chart to find the amount of vitamin C in 8 fl oz of grapeade.

RATINGS OF FRUIT DRINKS

PER 8 FL OZ

Product	Cost	Vitamin C	Calories
Orange Mix	7¢	12 mg	93
Orangeade	8¢	37 mg	90
Orange Punch	14¢	96 mg	113
Grape Mix	7¢	10 mg	90
Grapeade	12¢	87 mg	112
Grape Punch	14¢	86 mg	116

What types of information are given in the chart?
cost, amount of vitamin C, and calories in 8 fl oz of some fruit drinks.
What is being asked?
amount of vitamin C in 8 fl oz of grapeade
Find *Grapeade* under "Product." Then find the vitamin C column.
Read across to find the amount of vitamin C. So, the amount of vitamin C is **87 mg**.

Use the chart to answer the following.

1. What is the cost per 8-fl-oz serving of the orange punch?

2. How many calories are there in 8 fl oz of grape mix?

3. Which fruit drink contains the most vitamin C per 8-fl-oz serving?

4. Which fruit drink contains 112 cal per 8-fl-oz serving?

***5.** What is the ratio of vitamin C to calories for the grape mix?

***6.** What is the cost of 1 fl oz of the orange punch?

EXAMPLE 2

Use the chart below to find the charge for shipping a 7-oz package a distance of 250 mi.

DISTANCE FROM DISTRIBUTION CENTER

SHIPPING WEIGHT	Local Zone	Zones 1 and 2 (not over 150 miles)	Zone 3 (151 to 300 miles)	Zone 4 (301 to 600 miles)	Zone 5 (601 to 1000 miles)
1 oz to 8 oz	$0.69	$0.73	$0.73	$0.75	$0.80
9 oz to 15 oz	1.07	1.07	1.09	1.12	1.15
1 lb to 2 lb	1.38	1.38	1.42	1.47	1.53
2 lb 1 oz to 3 lb	1.47	1.47	1.52	1.61	1.70
3 lb 1 oz to 5 lb	1.57	1.57	1.65	1.78	1.91
5 lb 1 oz to 10 lb	1.80	1.80	1.93	2.14	2.45

What types of information are given in the chart?
shipping weight and shipping charge by zone

Find 7 oz under "Shipping Weight." 7 oz is between 1 oz and 8 oz.

Find the zone column that includes 250 mi. 250 mi is between 151 mi and 300 mi.

Read across to find the shipping charge for zone 3.

So, the charge is **$0.73**.

Find the shipping charges using the chart in Example 2.

1. 3 lb 7 oz, 400 mi

2. 9 oz, 300 mi

3. 2 lb, 301 mi

4. 5 lb 4 oz, 159 mi

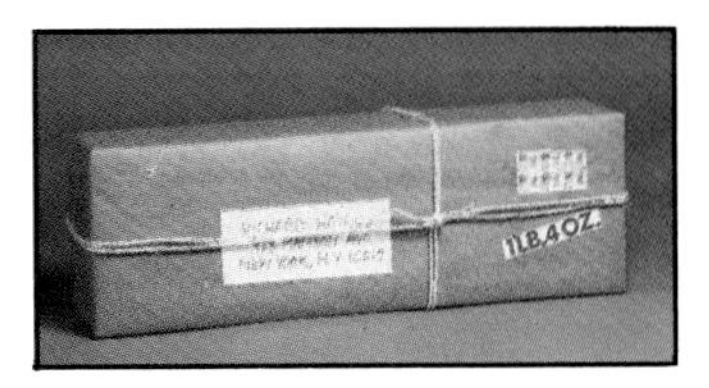

Solve.

5. Marie has two packages to ship to zone 5. One package weighs 1 lb 4 oz, and the other weighs 4 lb 10 oz. What is the total shipping charge for both packages?

6. Sammy has three packages for shipping. Each package weighs 6 oz. He is shipping one package locally and two packages to zone 3. What are the total shipping charges?

7. John lives 600 mi from the Distribution Center. How much will it cost to have two 10-lb packages sent to his home?

***8.** Arlene lives in zone 3. She places an order that weighs 3 lb 15 oz. The order is to be sent to her sister in zone 5. What will be the shipping charges?

CHAPTER TEST 6

Write a decimal.

1. two hundred twenty-six and four tenths
2. six thousand fifty and six thousandths

Round to the nearest whole number.

3. 36.82
4. 517.298

Place the decimal point to make the answer correct.

5. 18.9 + 40.7 + 59.24 = 11884
6. 187.6 − 19.25 = 16835

Add or subtract.

7. 52.46 + 10.23 + 9.11
8. 192.7 − 6.8
9. 6.17 + 2.8 + 3.9
10. 23 + 76.24 + 9.83
11. 5.02 − 2.9
12. 465.83 − 84.876
13. $76.46 + $0.33 + $0.43 + $63
14. $190 − $1.68

Solve.

15. Sean's dinner cost him $7.27 including tax. How much change did he receive from a $10 bill?
16. Rita bought a jogging suit on sale for $18.88 plus $1.51 in tax. How much change did she receive from $25?
17. Anne's monthly deductions from her paycheck are federal tax, $87.52; state tax, $18.12; and retirement, $77.50. Find her total monthly deductions.
18. Alex owed $900.85 for repairs to his roof. He made payments of $300 and $275.50. How much does he still owe?

Use the chart to answer **19** *and* **20.**

19. Ira has a package that weighs 15 oz. He must ship it 160 mi. What is the shipping charge?
20. Phyllis is shipping a package locally. The package weighs 1 lb 2 oz. What is the shipping charge?

DISTANCE FROM DISTRIBUTION CENTER

SHIPPING WEIGHT	Local Zone	Zones 1 and 2 (not over 150 miles)	Zone 3 (151 to 300 miles)
1 oz to 8 oz	$0.69	$0.73	$0.73
9 oz to 15 oz	1.07	1.07	1.09
1 lb to 2 lb	1.38	1.38	1.42

Multiplication of Decimals

MAKING MATH COUNT

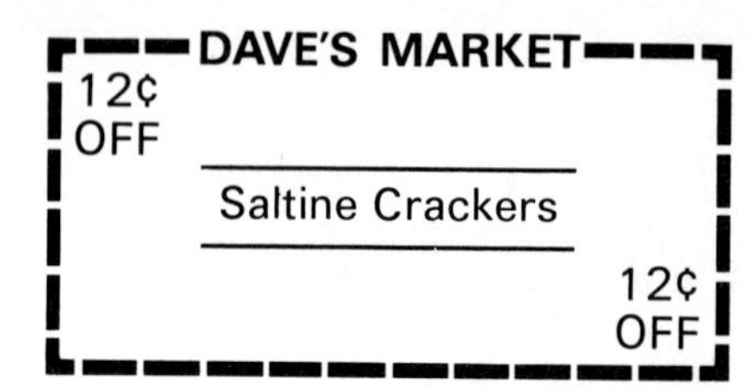

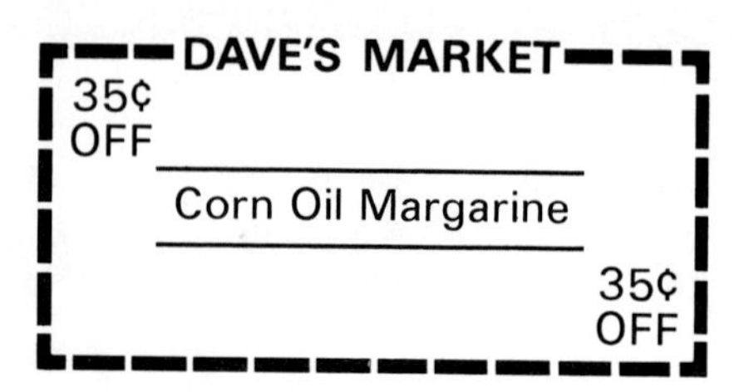

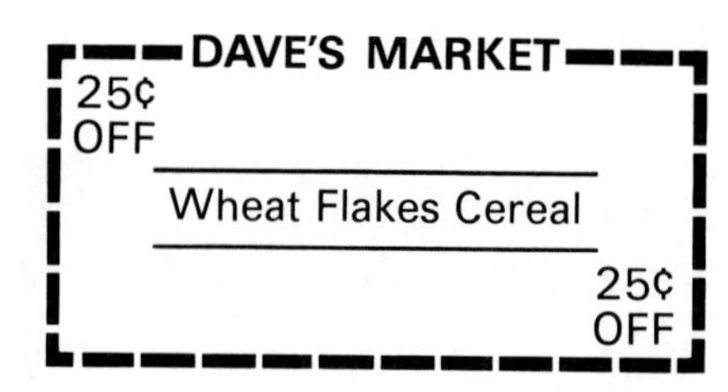

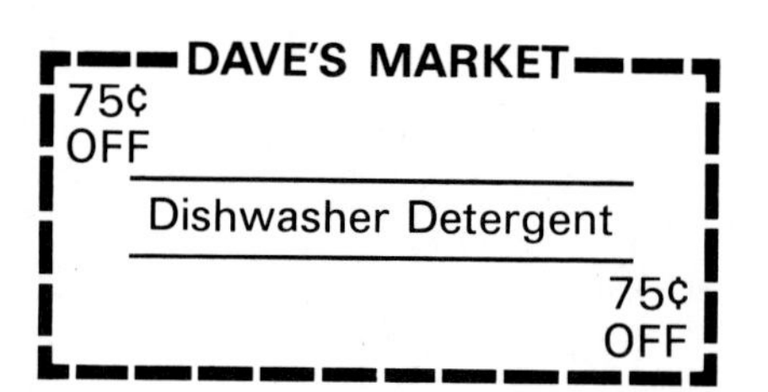

EXAMPLE

Mrs. Reese bought wheat flakes cereal for $1.89, corn oil margarine for $1.09, and saltines for $1.19. How much would she save by using the coupons above?

A $0.12 **B** $0.25
C $0.35 **D** $0.72

Add the amount of the coupon for each of the items purchased.

25¢ + 35¢ + 12¢ = 72¢

So, **D** is the answer.

Choose the correct answer.

1. Dishwasher detergent costs $2.79 and saltines cost $1.19. How much would be saved by using coupons?
A $0.12 **B** $0.75
C $0.87 **D** $1.19

2. Mr. Allen bought eggs for $0.99 and corn oil margarine for $1.09. What is the cost of the margarine with a coupon?
A $0.74 **B** $1.63
C $3.42 **D** $3.77

3. Dave's Market has double coupon day on Saturdays. How much would you save by using all of the coupons?
A $2.94 **B** $2.44
C $1.44 **D** $0.94

***4.** Ms. Bart bought a ham for $6.57, wheat flakes cereal for $1.89, and saltines for $1.19. How much would she spend if she used coupons on double coupon day?
A $9.28 **B** $8.91
C $9.40 **D** $9.65

DIAGNOSTIC TEST: FORM A (35 MINUTES)

Place the decimal point to make the product correct.

1. 4.75 × 38 = 18050

2. 73.3 × 87.6 = 642108

3. 45.7 × 10 = 4570

4. 825.96 × 100 = 8259600

5. 5.09 × 0.92 = 46828

6. 0.76 × 4,000 = 304000

Multiply.

7. 62.7 × 6.4

8. 22.3 × 1.3

9. 2.6 × 27

10. \$1.71 × 38

11. 8.08 × 7.98

12. 92.6 × 2.01

13. 1.4 × 1.5

14. 4.9 × 0.723

15. 0.26 × 0.79

16. 0.04 × 0.6

17. 1.824 × 100

18. 5.45 × 1,000

17 or more right? Go to page 136.

Less than 17 right? Go to page 130.

DIAGNOSTIC TEST: FORM B (35 MINUTES)

Place the decimal point to make the product correct.

1. 42.7 × 3.21 = 137067

2. 8.25 × 38 = 31350

3. 5.47 × 10 = 5470

4. 537.58 × 100 = 5375800

5. 4.12 × .88 = 36256

6. 0.98 × 2,000 = 196000

Multiply.

7. 27.6 × 4.6

8. 33.2 × 1.2

9. 6.2 × 39

10. \$1.17 × 26

11. 7.07 × 8.79

12. 62.9 × 1.03

13. 1.5 × 1.4

14. 9.4 × 0.237

15. 0.72 × 0.97

16. 0.06 × 0.04

17. 1.428 × 100

18. 4.54 × 1,000

17 or more right? Go to page 136.

Less than 17 right? See your teacher.

ESTIMATING PRODUCTS OF DECIMALS

To place the decimal point in a product by estimation

- Round each decimal number to its largest place.
- Multiply the whole numbers to get an estimate.
- Use the estimate to place the decimal point.

EXAMPLE 1

Place the decimal point to make the product correct.

$$\begin{array}{r} 3.44 \\ \times 2.5 \\ \hline 8600 \end{array}$$

Round each decimal.
Multiply the whole numbers.

$$\begin{array}{rcr} \mathbf{3.44} & \longrightarrow & \mathbf{3} \\ \mathbf{\times 2.5} & \longrightarrow & \mathbf{\times 3} \\ \hline & & \mathbf{9} \end{array}$$

The estimate is 9, which is close to 8.600.
So, the product is **8.600,** or **8.6.**

> To estimate products of decimals, round each number to its largest place.
>
> $$\begin{array}{rcr} 265.3 & \longrightarrow & 300 \\ \times 72.4 & \longrightarrow & \times 70 \end{array}$$
>
> Then multiply to get an estimate.

Notice that any number of zeros after the last digit in a decimal does not change its value. For example, 2.4 = 2.40 = 2.400.

EXAMPLE 2

Place the decimal point to make the product correct.

$$\begin{array}{r} 187.66 \\ \times 3.02 \\ \hline 5667332 \end{array}$$

Round each decimal and multiply.

$$\begin{array}{rcr} \mathbf{187.66} & \longrightarrow & \mathbf{200} \\ \mathbf{\times 3.02} & \longrightarrow & \mathbf{\times 3} \\ \hline & & \mathbf{600} \end{array} \longleftarrow \text{estimate}$$

The estimate is 600, which is close to 566.7332.
So, the product is **566.7332.**

EXAMPLE 3

Place the decimal point to make the product correct.

$$\begin{array}{r} 4.082 \\ \times 4.7 \\ \hline 191854 \end{array}$$

Round each decimal and multiply.

$$\begin{array}{rcr} \mathbf{4.082} & \longrightarrow & \mathbf{4} \\ \mathbf{\times 4.7} & \longrightarrow & \mathbf{\times 5} \\ \hline & & \mathbf{20} \end{array} \longleftarrow \text{estimate}$$

The estimate is 20, which is close to 19.1854.
So, the product is **19.1854.**

GETTING READY

Place the decimal point to make the product correct.

1. 9.01 × 7.06 = 636106

2. 6.66 × 1.07 = 71262

3. 4.75 × 3.8 = 18050

4. 18.35 × 7.24 = 1328540

5. 55.83 × 9.87 = 5510421

6. 3.947 × 6.93 = 2735271

EXERCISES

Place the decimal point to make the product correct.

1. 4.27 × 3.21 = 137067

2. 8.25 × 3.8 = 31350

3. 5.278 × 19 = 100282

4. 8.125 × 88 = 715000

5. 32 × 39.54 = 126528

6. 81 × 67.50 = 546750

7. 345 × 1.64 = 56580

8. 525 × 2.04 = 107100

9. 75.25 × 1.44 = 1083600

10. 81.45 × 3.66 = 2981070

11. 428.2 × 12.5 = 535250

12. 78.375 × 80 = 6270000

13. 476.5 × 9.97 = 4750705

14. 6.476 × 3.87 = 2506212

15. $\begin{array}{r} 674 \\ \times 58.3 \\ \hline 392942 \end{array}$

16. $\begin{array}{r} 208.7 \\ \times 4.07 \\ \hline 849409 \end{array}$

17. $\begin{array}{r} 869.3 \\ \times 2.01 \\ \hline 1747293 \end{array}$

18. $\begin{array}{r} 5.73 \\ \times 2.08 \\ \hline 119184 \end{array}$

19. $\begin{array}{r} 0.93 \\ \times 69.5 \\ \hline 64635 \end{array}$

20. $\begin{array}{r} 4.376 \\ \times 8.9 \\ \hline 389464 \end{array}$

21. $\begin{array}{r} 8{,}001 \\ \times 1.9 \\ \hline 152019 \end{array}$

22. $\begin{array}{r} 7{,}111 \\ \times 5.09 \\ \hline 3619499 \end{array}$

***23.** If 6 cans of dog food cost $2.55, how much will 12 cans cost at the same rate? Place the decimal point:
$2.55 × 2 = 510

★SPECIAL★
12 Cans of BARKZ
Now on Sale!

MULTIPLYING DECIMALS

To multiply two decimals

- Multiply as whole numbers.
- Place the decimal point in the product.

EXAMPLE 1

Multiply. 9.8 × 7.24

Multiply as whole numbers.

$$\begin{array}{r@{\quad}l} \mathbf{7.24} & \longleftarrow \text{ 2 places} \\ \underline{\times \mathbf{9.8}} & \longleftarrow \text{ 1 place} \\ \mathbf{5\,792} & \\ \underline{\mathbf{65\,16}} & \\ \mathbf{70.952} & \longleftarrow \text{ 3 places} \end{array}$$

> Add the number of decimal places in each factor to find the number of decimal places in the product.

Place the decimal point.
Check by estimating the product.
Round each decimal to the nearest whole number.
7 × 10 = 70 ⟵ estimate
The estimate is 70, which is close to 70.952.
So, the product is **70.952.**

EXAMPLE 2

Multiply. $6.05 × 1.3

$$\begin{array}{r} \mathbf{\$6.05} \\ \underline{\times \mathbf{1.3}} \\ \mathbf{1\,815} \\ \underline{\mathbf{6\,05}} \\ \mathbf{\$7.865} \end{array}$$

So, the product is **$7.865,** or **$7.87.**

Check by estimating the product.

$$\begin{array}{r@{\quad}c@{\quad}r} \mathbf{\$6.05} & \longrightarrow & \mathbf{\$6} \\ \underline{\times \mathbf{1.3}} & \longrightarrow & \underline{\times \mathbf{1}} \\ & & \mathbf{\$6} \end{array}$$

Compare with your answer.

GETTING READY

Multiply and check.

1. $\begin{array}{r} 7.99 \\ \underline{\times 6.2} \end{array}$ **2.** $\begin{array}{r} 108.75 \\ \underline{\times 12.3} \end{array}$ **3.** $\begin{array}{r} \$15.56 \\ \underline{\times 1.2} \end{array}$ **4.** $\begin{array}{r} 42.6 \\ \underline{\times 0.98} \end{array}$

Find the cost.

5. 4 gal of milk at $1.59 a gallon

6. 7 pens at $0.59 each

EXERCISES

Multiply.

1. 2.07 $\times$ 1.5 **2.** \$130 $\times$ 2.3 **3.** 1.47 $\times$ 24 **4.** 1.86 $\times$ 21

5. 62.4 $\times$ 8.3 **6.** 1.03 $\times$ 5.5 **7.** \$2.07 $\times$ 0.99 **8.** 0.92 $\times$ 6.3

9. \$6.09 $\times$ 3.4 **10.** 2.46 $\times$ 1.3 **11.** 82.8 $\times$ 0.9 **12.** 475 $\times$ 7.5

13. 1.9 $\times$ 0.87 **14.** 23.1 $\times$ 29.8 **15.** \$5.80 $\times$ 1.53 **16.** 39.6 $\times$ 20.5

17. 3.08 $\times$ 2.9 **18.** \$953 $\times$ 0.88 **19.** 35.2 $\times$ 1.9 **20.** 6.2 $\times$ 4.7

21. \$60.20 $\times$ 62.5 **22.** 0.191 $\times$ 0.07 **23.** 0.93 $\times$ 0.52 **24.** 0.414 $\times$ 94

25. 62.1 $\times$ 0.7 **26.** 5.12 $\times$ 0.32 **27.** \$10.60 $\times$ 2.8 **28.** 858 $\times$ 0.06

29. 0.019 $\times$ 2.8 **30.** 4.271 $\times$ 3.07 **31.** 4.271 $\times$ 30.7 **32.** 4.271 $\times$ 307

Find the cost.

33. 18 yd of cloth at \$4.95 a yard

34. 27 ft of moulding at \$0.98 a foot

35. 3 boxes of soap at \$5.98 each

36. 7 cartons of juice at \$2.69 each

37. 4 tires at \$69.99 each

38. 4 batteries at \$1.39 each

ON YOUR OWN

What is the least number of airplanes needed to fly in this formation:
2 planes in front of a plane,
2 planes behind a plane, and
1 plane between 2 planes?

MULTIPLYING BY 10; 100; OR 1,000

To multiply a decimal by 10, 100, or 1,000

- Move the decimal point one place to the right for each zero in 10; 100; or 1,000.
- Use zeros as place holders when necessary.

EXAMPLE 1

Multiply. 34.56 × 10

Move the decimal point 1 place to the right. **34.56**

So, 34.56 × 10 = **345.6.**

EXAMPLE 2

Multiply. 38.507 × 100

Move the decimal point 2 places to the right. **38.507**

So, 38.507 × 100 = **3,850.7.**

EXAMPLE 3

Multiply. 0.3 × 1,000

Move the decimal point 3 places to the right and add 2 zeros. **0.300**

So, 0.3 × 1,000 = **300.**

GETTING READY

Multiply.

1. 7.2 × 10   7.2 × 100   7.2 × 1,000

2. 0.65 × 10   0.65 × 100   0.65 × 1,000

3. $5.37 × 10 $5.37 × 100 $5.37 × 1,000

4. 6.008 × 10   6.008 × 100   6.008 × 1,000

5. 4.57 × 10

6. 3.46 × 10

7. 65.31 × 100

8. 8.543 × 100

9. 0.42 × 1,000

10. 5.4 × 1,000

11. 0.675 × 100

12. 5.135 × 1,000

EXERCISES

Multiply.

1. 0.123×10 0.123×100 $0.123 \times 1{,}000$

2. 3.456×10 3.456×100 $3.456 \times 1{,}000$

3. 0.03×10 0.03×100 $0.03 \times 1{,}000$

4. 0.007×10 0.007×100 $0.007 \times 1{,}000$

5. 43.246×10 43.246×100 $43.246 \times 1{,}000$

6. 15.005×10 15.005×100 $15.005 \times 1{,}000$

7. $\$1.83 \times 10$ $\$1.83 \times 100$ $\$1.83 \times 1{,}000$

8. $6{,}004 \times 10$ $6{,}004 \times 100$ $6{,}004 \times 1{,}000$

9. 5.47×10

10. 4.36×10

11. 56.13×100

12. 5.387×100

13. $0.27 \times 1{,}000$

14. $4.7 \times 1{,}000$

15. 0.705×100

16. $1{,}355 \times 1{,}000$

17. 612.8×10

18. 82.596×100

19. $9.2518 \times 1{,}000$

20. $0.7 \times 1{,}000$

21. Find the cost of 100 pens that sell for $2.29 each.

22. Find the cost of 10 tapes that sell for $4.98 each.

Return to Form B on page 129.

CALCULATOR

For each number, find the number you would multiply by to give you 1 on the calculator.

1. 0.001   **2.** 0.125   **3.** 0.0625   **4.** 0.8

5. 5   **6.** 50   **7.** 500   **8.** 4,000

9. 0.4   **10.** 0.025   **11.** 800   **12.** 200

For each number, find the number you would add to give you 1 on the calculator.

13. 0.1   **14.** 0.01   **15.** 0.001   **16.** 0.0001

17. 0.234   **18.** 0.567   **19.** 0.891   **20.** 0.8998

ROUNDING PRODUCTS

To round a product to the nearest tenth or hundredth

- Check the digit to the right of the tenths or hundredths place.
- If the digit is 5 or greater, round up.
- If the digit is less than 5, round down.

EXAMPLE 1

Multiply. Round the product to the nearest tenth.
8.407 × 5.4

3 is in the **tenths** place.	**45.3978**	⟵ the product of 8.407 × 5.4
The digit to the right is 9.	**45.3(9)78**	
Round up.	**45.4**	

So, 45.3978 rounded to the nearest tenth is **45.4.**

EXAMPLE 2

Multiply. Round the product to the nearest hundredth.
7.113 × 0.46

7 is in the **hundredths** place.	**3.27198**	⟵ the product of 7.113 × 0.46
The digit to the right is 1.	**3.27(1)98**	
Round down.	**3.27**	

So, 3.27198 rounded to the nearest hundredth is **3.27.**

GETTING READY

Multiply. Round the product to the nearest tenth and hundredth.

1. 23.89 × 8.03 **2.** 0.67 × 0.75 **3.** 6,211 × 0.009

EXERCISES

Multiply. Round the product to the nearest tenth.

1. 3.4 × 1.7 **2.** 20.5 × 0.39 **3.** 1.784 × 10.9

4. 7.19 × 0.3 **5.** 14.3 × 7.6 **6.** 1,543 × 0.005

Multiply. Round the product to the nearest hundredth.

7. 16.94 × 7.05 **8.** 0.43 × 0.98 **9.** 8,200 × 0.004

Problem Solving Skill

To solve word problems by making a reasonable estimate

Exact answers are not always needed in problem solving situations. Sometimes a reasonable estimate of the answer is all that is needed.

EXAMPLE

One gallon covers about 450 ft^2. Would 5 gal of paint be enough to cover 975 ft^2 of wall space with two coats of paint?

Round. $\quad$ **975 ft^2 ⟶ 1,000 ft^2**

For two coats of paint, multiply by 2. $\quad$ **1,000 ft^2 × 2 = 2,000 ft^2**

Round. $\quad$ **450 ft^2 ⟶ 500**

Then, number of gallons needed is $\frac{2,000}{500} = 4$ gal.

Yes, 5 gal of paint should be enough.

EXERCISES

Solve by estimating.

1. Cathy reads 21 pages of a novel in 40 min. Can she finish the remaining 384 pages in 10 h by reading at this rate?

2. The full dinner prices on a restaurant menu range from $5.95 to $12.95. You have three $20 bills. Will you be able to pay for 4 dinners including a 15% tip?

3. Mrs. MacAdam saves $89 a month to pay a $1,200 a year tax on her property. Will she have enough to pay the tax?

4. Helen has chartered a bus for $80 for a class trip. Will $3.25 from each of the 30 students in the class be enough to pay for the charter?

5. A pound of baking potatoes serves 4 people. Is a 5-pound bag of baking potatoes enough to serve 8 couples?

6. A half-gallon of paint is enough to paint a ceiling measuring 180 ft^2. Is a half-gallon of paint enough to paint a ceiling 15 ft by 15 ft?

7. The road distance from Pittsburgh to San Francisco is 2,642 mi. By averaging 50 mph, could 3 drivers make it non-stop in 2 days?

8. An automobile averages 29 mpg. The gas tank holds 14 gal. Would you be able to take a 350-mi trip on a tankful of gas?

Problem Solving Applications

READ • PLAN • SOLVE • CHECK

The students in Mr. Manzo's auto shop class learn how to maintain and repair cars. They also learn safe driving habits and money-saving techniques. Each student is encouraged to practice regular auto maintenance and energy conservation.

AS YOU READ

- antifreeze—a substance added to lower the freezing point of water
- tread life—the time that the ridges on a tire last

Answer these questions about the auto shop class.

1. A car is driven about 15,000 mi a year. If the car gets 20 mpg, how many gallons of gas are used in a year?

2. The *tread life* of the new tires on Barbara's car is 35,000 mi. If she drives 10,000 mi a year, how many years will her tires last?

3. Mr. Manzo said driving with 12 lb less air in the tires will decrease the tread life by a factor of 0.4. If Barbara does this, what will the tread life of the tires be then?

4. Mr. Manzo said that running the car air conditioner increases the cost of gasoline about $0.017 a mile. If a car is driven 3,750 mi in the summer, how much would be added to the gasoline cost?

5. Frank is draining and flushing out the radiator of his car. Mr. Manzo recommends a half-water, half-*antifreeze* mixture. If Frank's radiator holds 3.5 gal, how much antifreeze should he add?

6. Sara is repairing rust areas on her car's fenders. One part hardening compound is mixed with 10 parts patching compound. She mixes 0.5 oz of hardening compound with how many ounces of patching compound?

Career: Automobile Mechanics

READ • PLAN • SOLVE • CHECK

Automobile mechanics work for auto dealers, repair shops, gasoline service stations, taxicab companies, and auto leasing companies. Training is gained on the job, often through apprenticeships lasting 2 to 4 years.

AS YOU READ

- discount—a reduction in the usual cost
- lease—a contract to rent an item for a certain amount of time and money

Solve.

1. Tom Allen owns RCO Gas Depot. His station pumps about 4,730 gal of gas in an 8-h day. About how many gallons of gas are pumped each hour?

2. Andrea Roper spent $350 for a set of socket wrenches, $29.95 for work pants, and $19.95 for a work shirt. How much did she spend in all?

3. Ace Towing Service charges $25 for towing distances of 5 mi or less. A charge of $1.50 per mile is added to distances beyond 5 mi. If a car is towed 18 mi, what is the charge?

4. Mary Tolliver is an auto body repair mechanic. She charges $325 for fender replacement and labor. She buys the fender unit for $175. How much does she charge for labor?

5. Richie Dent works as a gasoline station service attendant. He gets a *discount* of 10¢ on a gallon of gas. If gas costs 133.9¢ a gallon, what would he pay for a monthly purchase of 50 gal of gas?

6. Roberta Sims is a front-end mechanic at an auto-*leasing* and used-car dealership. She works on both the leased cars and used cars. Last week, 0.75 of her 40-h total was spent on the leased cars. How many hours did she spend on the used cars?

CHAPTER TEST 7

Place the decimal point to make the product correct.

1. $107.85 \times 37 = 399045$
2. $66.32 \times 4.8 = 318336$
3. $6.875 \times 1.64 = 1127500$
4. $9.062 \times 160 = 1449920$

Multiply.

5. 65.7×100
6. $4.29 \times 1{,}000$
7. 8.675×10
8. $0.5 \times 1{,}000$
9. 2.3×100
10. $\$1.08 \times 10$
11. 4.8×2.7
12. 2.37×38
13. 687.9×2.25
14. 4.375×2.8
15. $\$4.40 \times 0.95$
16. 8.675×1.04

Multiply. Round the product to the nearest tenth.

17. $\begin{array}{r} 16.72 \\ \times 2.03 \\ \hline \end{array}$
18. $\begin{array}{r} 51.7 \\ \times 5.16 \\ \hline \end{array}$
19. $\begin{array}{r} 34.56 \\ \times 227 \\ \hline \end{array}$

Multiply. Round the product to the nearest hundredth.

20. $\begin{array}{r} 452.9 \\ \times 3.67 \\ \hline \end{array}$
21. $\begin{array}{r} 9.78 \\ \times 0.6 \\ \hline \end{array}$
22. $\begin{array}{r} 6.009 \\ \times 88.5 \\ \hline \end{array}$

Solve.

23. Mark purchased 7 cassette tapes for the school tape library. Each tape cost $6.95. What was the total cost?
24. Anne purchased 16 new chairs for the school lounge. Each chair cost $27.88. Find the total cost.

Solve by estimating.

25. If one gallon of fruit drink serves 18 campers, will 7 gal be enough to serve 150?

*26. Bev is planning a swim meet. The average length of an event is $3\frac{1}{2}$ min. If there are 58 events, will the meet take less than 4 h?

Division of Decimals

DIAGNOSTIC TEST: FORM A (40 MINUTES)

Place the decimal point to make the quotient correct.

1. 173.90 ÷ 18.5 = 94 **2.** 1,080.975 ÷ 8.75 = 12354

Divide.

3. 7)0.35 **4.** 3)6.9 **5.** 5)6.5 **6.** 3)0.9

7. 9)0.459 **8.** 4)1.248 **9.** 2)46.88 **10.** 40)8.44

11. 0.5)16.5 **12.** 0.3)243 **13.** 0.08)560 **14.** 3.14)6.28

15. 12.5)10 **16.** 0.22)1012 **17.** 0.84)29.988 **18.** 1.9)4.826

19. 24 ÷ 100 **20.** 3.6 ÷ 1,000 **21.** 56 ÷ 0.008 **22.** 72 ÷ 1.2

20 or more right? Go to page 150. *Less than 20 right? Go to page 143.*

DIAGNOSTIC TEST: FORM B (40 MINUTES)

Place the decimal point to make the quotient correct.

1. 146.38 ÷ 56.3 = 26 **2.** 1,195.125 ÷ 12.5 = 9561

Divide.

3. 8)0.56 **4.** 4)9.2 **5.** 5)8.5 **6.** 4)0.8

7. 7)0.1477 **8.** 6)1.872 **9.** 2)68.82 **10.** 14)52.36

11. 0.5)15.5 **12.** 0.4)268 **13.** 0.07)560 **14.** 4.23)8.46

15. 12.5)40 **16.** 0.33)1485 **17.** 0.67)3.015 **18.** 3.7)8.658

19. 42 ÷ 100 **20.** 6.4 ÷ 1,000 **21.** 72 ÷ 0.008 **22.** 56 ÷ 1.4

20 or more right? Go to page 150. *Less than 20 right? See your teacher.*

ESTIMATING QUOTIENTS OF DECIMALS

To place the decimal point in a quotient by estimation

- Round each decimal number to its largest place.
- Divide the whole numbers to get an estimate.
- Use the estimate to place the decimal point.

EXAMPLE

Place the decimal point to make the quotient correct.

109.65 ÷ 21.5 = 51

Round each decimal. 109.65 ⟶ **100**
21.5 ⟶ **20**

Divide the whole numbers to get an estimate. $20\overline{)100}$ with quotient **5**

The estimate is 5, which is close to 5.1.
So, the quotient is **5.1.**

> To estimate quotients of decimals, round each number to its largest place.
>
> $18.2\overline{)78.26} \longrightarrow 20\overline{)80}$
>
> Then divide to get an estimate.

GETTING READY

Place the decimal point to make the quotient correct.

1. 23.16 ÷ 1.2 = 193

2. 297.54 ÷ 51.3 = 580

3. 31.5798 ÷ 4.38 = 721

4. 110.162 ÷ 1.235 = 892

EXERCISES

Place the decimal point to make the quotient correct.

1. 30.02 ÷ 1.9 = 158

2. 61.2 ÷ 3.4 = 18

3. 342.16 ÷ 2.35 = 1456

4. 15,030.54 ÷ 158 = 9513

5. 333.79944 ÷ 12.89 = 25896

6. 1,374.8181 ÷ 86.63 = 1587

7. 947,139.6 ÷ 652.3 = 1452

8. 46,534.202 ÷ 9,571 = 4862

9. 429 ÷ 1.3 = 330

10. 113 ÷ 2.5 = 452

11. 432.276 ÷ 16.3 = 2652

12. 3,598.17 ÷ 4.38 = 8215

***13.** 880.65 ÷ 0.9 = 9785

***14.** 22.218 ÷ 84 = 2645

DIVIDING BY A WHOLE NUMBER

To divide a decimal by a whole number

- Place the decimal point in the quotient above the decimal point in the dividend.
- Divide like whole numbers.

EXAMPLE 1

Divide. $18\overline{)63.36}$

Place decimal point in quotient. $18\overline{)63.36}$

Divide.

$$\begin{array}{r} 3.52 \\ 18\overline{)63.36} \\ \underline{54} \\ 9\,3 \\ \underline{9\,0} \\ 36 \\ \underline{36} \end{array}$$

Check by estimating. Round each number to its largest place.

$$18\overline{)63.36} \longrightarrow \begin{array}{r} 3 \\ 20\overline{)60} \end{array}$$

The estimate is 3, which is close to 3.52.
So, **3.52** is the quotient.

When dividing a decimal by a whole number, the decimal point in the quotient is placed directly above the decimal point in the dividend.

$$\begin{array}{c} \downarrow \\ \begin{array}{r} 0.4 \\ 4\overline{)1.6} \end{array} \\ \uparrow \end{array}$$

EXAMPLE 2

Divide. $38\overline{)77.9}$

Place decimal point in quotient. $38\overline{)77.9}$

Divide.

$$\begin{array}{r} 2.05 \\ 38\overline{)77.90} \\ \underline{76} \\ 1\,90 \\ \underline{1\,90} \end{array} \longleftarrow \text{Add one zero.}$$

So, **2.05** is the quotient.

GETTING READY

Divide.

1. $7\overline{)14.7}$
2. $8\overline{)8.72}$
3. $24\overline{)55.2}$
4. $499.14 \div 59$
5. $52.65 \div 81$
6. $29.7 \div 66$

EXERCISES

Divide.

1. $8\overline{)5.6}$
2. $2\overline{)0.06}$
3. $9\overline{)18.63}$
4. $6\overline{)6.18}$
5. $8\overline{)1.7}$
6. $4\overline{)13.9}$
7. $17\overline{)52.02}$
8. $28\overline{)60.76}$
9. $74\overline{)\$224.96}$
10. $87\overline{)\$438.48}$
11. $15\overline{)30.060}$
12. $92\overline{)100.28}$
13. $12.288 \div 64$
14. $39.425 \div 95$
15. $\$16.79 \div 73$
16. $1{,}034.8 \div 26$
17. $110 \div 88$

*18. $0.00854 \div 305$

Solve.

19. Anne bought 13 hockey tickets for a total of $87.75. What was the cost of each ticket?
20. Frances drove 16,510 mi last year. Find the average number of miles driven per week.
21. Albert bought 3 shirts. He paid a total of $53.64. What was the average cost of one shirt?
22. David spent $570.72 on clothes last year. What was his average monthly expenditure for clothes?

ON YOUR OWN

How many squares can you see in each figure?

1 by 1 2 by 2 3 by 3 4 by 4

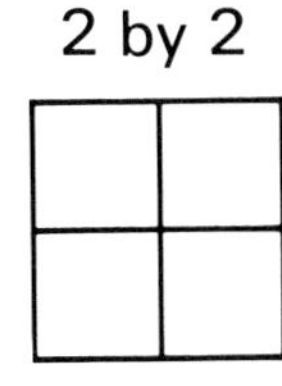

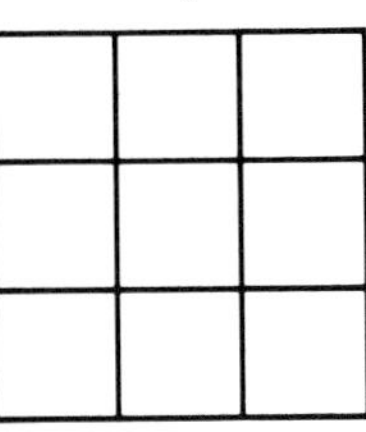

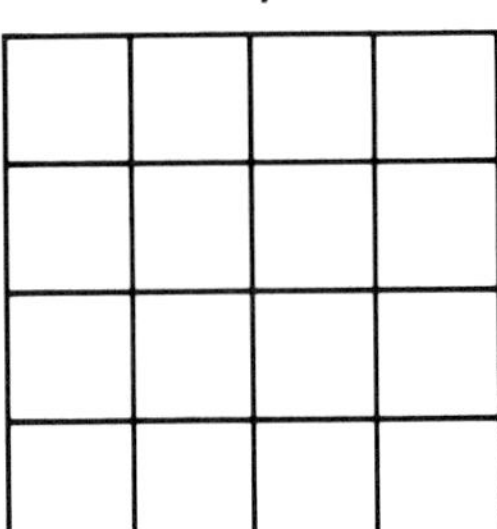

Can you predict the number of squares in a 5 by 5 square?

DIVIDING DECIMALS

To divide a decimal by a decimal

- Change the divisor to a whole number by moving the decimal point to the right.
- Move the decimal point in the dividend the same number of places.
- Place the decimal point in the quotient.
- Then divide.

EXAMPLE 1

Divide. $0.24\overline{)1.728}$

Move the decimal point 2 places to the right. $\mathbf{0.24\overline{)1.728}}$

Place decimal point in quotient. $24\overline{)172.8}$

Divide.

$$\begin{array}{r} \mathbf{7.2} \\ \mathbf{24\overline{)172.8}} \\ \underline{\mathbf{168}} \\ \mathbf{4\ 8} \\ \underline{\mathbf{4\ 8}} \end{array}$$

So, the quotient is **7.2.**

To divide by hundredths, move the decimal point 2 places to the right in both the divisor and dividend. This is the same as multiplying each number by 100.

EXAMPLE 2

Divide. $0.025\overline{)37.5}$

Move the decimal point 3 places to the right. $\mathbf{0.025\overline{)37.500}}$

Place decimal point in quotient. $25\overline{)37500.}$

Divide.

$$\begin{array}{r} \mathbf{1500.} \\ \mathbf{25\overline{)37500.}} \\ \underline{\mathbf{25}} \\ \mathbf{125} \\ \underline{\mathbf{125}} \end{array}$$

So, the quotient is **1,500.**

To divide by thousandths, move the decimal point 3 places to the right in both the divisor and dividend. This is the same as multiplying each number by 1,000.

GETTING READY

Divide.

1. $0.12\overline{)69.12}$ **2.** $2.4\overline{)9.12}$ **3.** $0.064\overline{)2.048}$

4. $50.4 \div 0.14$ **5.** $2.22 \div 0.6$ **6.** $99.33 \div 0.231$

EXERCISES

Divide.

1. $0.07\overline{)5.67}$ **2.** $0.09\overline{)9.09}$ **3.** $0.7\overline{)7.49}$ **4.** $0.4\overline{)81.212}$

5. $0.08\overline{)2.4}$ **6.** $0.06\overline{)42}$ **7.** $0.05\overline{)25.5}$ **8.** $0.8\overline{)72}$

9. $0.06\overline{)9.666}$ **10.** $0.08\overline{)87.2}$ **11.** $0.29\overline{)84.1}$ **12.** $0.46\overline{)149.50}$

13. $3.3\overline{)9.9}$ **14.** $2.5\overline{)164.5}$ **15.** $3.6\overline{)91.08}$ **16.** $2.6\overline{)6.864}$

17. $1.8\overline{)11.412}$ **18.** $1.3\overline{)0.91}$ **19.** $0.15\overline{)0.798}$ **20.** $0.78\overline{)1.794}$

21. $45 \div 0.009$ **22.** $39 \div 0.003$ **23.** $47 \div 23.5$

24. $66 \div 8.25$ **25.** $36.088 \div 0.052$ **26.** $1.6524 \div 0.068$

27. $1.0635 \div 0.709$ **28.** $6.9871 \div 0.653$ **29.** $8.8505 \div 0.571$

***30.** How many 8¢ pencils can be bought for $56?

***31.** How many 3¢ nails can be purchased with $6.33?

CALCULATOR

Use your calculator.

1. An assembly-line worker checked 278.25 circuits in 5.25 h. What was the average number checked per hour?

2. The sophomore class needs to earn $3,075 by selling cookbooks at $3.75 each. How many must be sold?

3. Mr. Janus spent $806.52 for fuel oil in the past 3 months. If the oil cost $1.43 a gallon, how many gallons did he buy?

4. Ms. Knight pays $0.068 per unit for electricity. If her bill was $37.74 for April, how many units did she use that month?

5. If the Jordans pay $87.35 per month on their furniture, how many payments are left if they still owe $1,572.30?

6. The freshman class sold tee shirts for $2.75 each. How many tee shirts did they sell if the income was $1,375?

DIVIDING BY 10; 100; OR 1,000

To divide a decimal by 10, 100, or 1,000

- Move the decimal point one place to the left for each zero in 10, 100, or 1,000.
- Add zeros when necessary.

EXAMPLE 1

Divide. 78.63 ÷ 10

Move the decimal point 1 place to the left. **78.63**

So, 78.63 ÷ 10 = **7.863.**

EXAMPLE 2

Divide. 5.7 ÷ 100

Move the decimal point 2 places to the left and add 1 zero. **05.7**

So, 5.7 ÷ 100 = **0.057.**

EXAMPLE 3

Divide. 7,864.6 ÷ 1,000

Move the decimal point 3 places to the left. **7864.6**

So, 7,864.6 ÷ 1,000 = **7.8646.**

GETTING READY

Divide.

1. $10\overline{)6.78}$ $100\overline{)6.78}$ $1,000\overline{)6.78}$

2. $10\overline{)7.654}$ $100\overline{)7.654}$ $1,000\overline{)7.654}$

3. $10\overline{)0.465}$ $100\overline{)0.465}$ $1,000\overline{)0.465}$

4. 87 ÷ 1,000

5. 68.5 ÷ 10

6. 9.543 ÷ 10

7. 64.86 ÷ 100

8. 456 ÷ 1,000

9. 538.22 ÷ 10

10. 54.6 ÷ 100

11. 956 ÷ 100

EXERCISES

Divide.

1. 927 ÷ 100
2. 482 ÷ 100
3. 0.735 ÷ 10
4. 8.06 ÷ 10
5. 90 ÷ 1,000
6. 87 ÷ 1,000
7. 0.78 ÷ 100
8. 45.8 ÷ 100
9. 4.95 ÷ 10
10. 78.642 ÷ 10
11. 975.6 ÷ 100
12. 987.56 ÷ 100
13. 0.776 ÷ 1,000
14. 6.56 ÷ 1,000
15. 2.376 ÷ 10
16. 7,455 ÷ 10
17. 4,729 ÷ 100
18. 50,000 ÷ 100
19. 5,000 ÷ 1,000
20. 8,428 ÷ 1,000
21. 800 ÷ 10
22. 7,575 ÷ 10
23. 5,986 ÷ 100
24. 6,787 ÷ 100
25. 888 ÷ 10
26. 5,466 ÷ 10
27. 6,752 ÷ 1,000

Return to Form B on page 142.

ON YOUR OWN

Refer to the chart for Problems 1–5.

1. How many sets of footwear are described in all?
2. How many black shoes and striped socks are there?
3. How many brown shoes and white socks are there?
4. How many tan socks are there?
5. How many black shoes and white socks are there?

SHOES	SOCKS: Stripe	White	Tan
Black	22	37	5
Brown	12	54	9

Solve.

6. The Rapid Package Company issued a statement about improperly wrapped packages after studying customers' mailings of 10,000 units. They stated that 4 out of every 100 packages were returned for improper wrapping. About how many packages did they detect in their survey that were improperly wrapped?

ROUNDING QUOTIENTS

To round a quotient to the nearest tenth or hundredth

- Check the digit to the right of the tenths or hundredths place.
- If the digit is 5 or greater, round up.
- If the digit is less than 5, round down.

EXAMPLE 1

Divide. Round the quotient to the nearest tenth.

$23\overline{)56.4}$

Divide to 2 decimal places (**hundredths**). $\mathbf{23\overline{)56.40}}$ with quotient **2.45**

4 is in the tenths place. **2.$\underline{4}$5** The digit to the right is 5. **2.4(5)**

Round up. **2.45** ⟶ **2.5**

So, the quotient to the nearest tenth is **2.5.**

EXAMPLE 2

Divide. Round the quotient to the nearest hundredth.

$0.037\overline{)0.47}$

Divide to 3 decimal places (**thousandths**). $\mathbf{0.037\overline{)0.470\ 000}}$ with quotient **12.702**

0 is in the hundredths place. **12.7$\underline{0}$2** The digit to the right is 2. **12.70(2)**

Round down. **12.702** ⟶ **12.70**

So, the quotient to the nearest hundredth is **12.70.**

GETTING READY

Divide. Round the quotient to the nearest tenth and hundredth.

1. $0.04\overline{)9.7862}$ **2.** $2.18\overline{)145.6}$ **3.** $309\overline{)745}$

EXERCISES

Divide. Round the quotient to the nearest tenth.

1. $0.02\overline{)0.6741}$ **2.** $2.16\overline{)5.272}$ **3.** $10.2\overline{)96.57}$

4. $26\overline{)85.8}$ **5.** $33.5\overline{)467.6}$ **6.** $18.33\overline{)0.1646}$

Divide. Round the quotient to the nearest hundredth.

7. $3.4\overline{)11.123}$ **8.** $2.84\overline{)800.7}$ **9.** $112\overline{)346}$

MAKING MATH COUNT

SALES TAX CHART FOR $4\frac{5}{8}\%$ SALES TAX

Sale	Tax	Sale	Tax	Sale	Tax	Sale	Tax	Sale	Tax
$ .00– .10	$.00	2.71–2.91	.13	5.52–5.72	.26	8.33– 8.54	.39	11.14–11.35	.52
.11– .32	.01	2.92–3.13	.14	5.73–5.94	.27	8.55– 8.75	.40	11.36–11.56	.53
.33– .54	.02	3.14–3.35	.15	5.95–6.16	.28	8.76– 8.97	.41	11.57–11.78	.54
.55– .75	.03	3.36–3.56	.16	6.17–6.37	.29	8.98– 9.18	.42	11.79–11.99	.55
.76– .97	.04	3.57–3.78	.17	6.38–6.59	.30	9.19– 9.40	.43	12.00–12.21	.56
.98–1.18	.05	3.79–3.99	.18	6.60–6.81	.31	9.41– 9.62	.44	12.22–12.43	.57
1.19–1.40	.06	4.00–4.21	.19	6.82–7.02	.32	9.63– 9.83	.45	12.44–12.64	.58
1.41–1.62	.07	4.22–4.43	.20	7.03–7.24	.33	9.84–10.05	.46	12.65–12.86	.59
1.63–1.83	.08	4.44–4.64	.21	7.25–7.45	.34	10.06–10.27	.47	12.87–13.08	.60
1.84–2.05	.09	4.65–4.86	.22	7.46–7.67	.35	10.28–10.48	.48	13.09–13.29	.61
2.06–2.27	.10	4.87–5.08	.23	7.68–7.89	.36	10.49–10.70	.49	13.30–13.51	.62
2.28–2.48	.11	5.09–5.29	.24	7.90–8.10	.37	10.71–10.91	.50	13.52–13.72	.63
2.49–2.70	.12	5.30–5.51	.25	8.11–8.32	.38	10.92–11.13	.51	13.73–13.94	.64

EXAMPLE

Ms. Barnes bought a shirt for half price. If the sale price was $13.50, how much tax did she pay?

A $0.61 **B** $0.62
C $0.63 **D** $0.64

Look at the chart. Find $13.50 in the column labeled "Sale." $13.50 is between $13.30 and $13.51. Read the column next to it labeled "Tax." The tax is $0.62. So, **B** is the correct answer.

Choose the correct answer.

1. John bought a TV game cartridge for $13.89. What is the sales tax?

A $0.62 **B** $0.63
C $0.64 **D** $0.65

2. Lucy bought a beach towel for $10.98. What is the sales tax?

A $0.51 **B** $0.52
C $0.53 **D** $0.54

3. Ernie bought a notebook for $3.98 and a package of pencils for $1.79. What was the total sales tax for both items?

A $0.24 **B** $0.25
C $0.26 **D** $0.27

***4.** Marge bought a pattern for $3.75 and thread for $0.89. What was the total cost including tax?

A $4.64 **B** $4.85
C $0.20 **D** $0.21

Problem Solving Skill

To find unit price
To determine the better buy

The cost of one item is called the **unit price**. Unit prices let you compare the costs of different sizes or quantities of the same item. In general, the size or quantity with the lowest price per unit is the better buy.

EXAMPLE 1

Find the unit price. box of 8 pencils for $2.10

Divide the total price by the number of items.

```
   $0.2625, or 26.25¢
8)$2.1000
   1 6
     50
     48
      20
      16
       40
       40
```

Divide to 4 decimal places because the unit price is less than $1.

> To find the unit price, divide the cost by the number of items or units. Unit prices of less than $1 are rounded to the nearest tenth of a cent.

Round to the nearest tenth of a cent. **26.25¢ ⟶ 26.3¢**
So, the unit price is **26.3¢.**

EXAMPLE 2

Which is the better buy? cheese: 16 slices for $1.37
12 slices for $0.98

Divide the total price by the number of items for each case.

```
    $0.085        $0.081
16)$1.370     12)$0.980
    1 28            96
      90            20
      80            12
```

Be sure to calculate enough decimal places for a price comparison.

Compare. $0.081 is less than $0.085.
So, **12 slices for $0.98** is the better buy.
Note that the largest size is not always the best buy.

> To determine the better buy, compare the unit prices of the items.

EXERCISES

Find the unit price.

1. jelly: 32 oz for $1.59
2. bread: 16 oz for $0.95
3. rice: 26 oz for $0.45
4. tomato sauce: 26 oz for $1.39
5. tuna (chunk): 7 oz for $1.09
6. tuna (solid): 7 oz for $1.69
7. macaroni: 32 oz for $1.19
8. potato flakes: 5.5 oz for $0.79
9. flour: 32 oz for $1.49
10. green beans: 15.5 oz for $0.34
11. baked beans: 12 oz for $0.27
12. applesauce: 29 oz for $0.59
13. bread: 12 oz for $0.54
14. bread: 18 oz for $0.69
15. salmon: 10 oz for $1.98
16. floor wax: 45 oz for $2.75
17. Brand A: 32 oz for $2.45
    Brand B: 48 oz for $3.55
    Brand C: 84 oz for $7.59
18. Brand X: 12 oz for $1.69
    Brand Y: 10.5 oz for $1.45
    Brand Z: 14.5 oz for $1.79

Which is the better buy?

19. 8 oz for $1.20
    6 oz for $0.75
20. 140 g for $2.10
    225 g for $2.93
21. 32 oz for $0.89
    64 oz for $1.98
22. 144 pens for $64.80
    75 pens for $32.25
23. 325 g for $4.55
    125 g for $2.00
24. 324 oz for $38.88
    120 oz for $14.40

Problem Solving Applications

READ • PLAN • SOLVE • CHECK

The Flint County Youth Employment Service encourages students to register for jobs in the community. They also sponsor student initiative projects for employment. One group of students specializes in pet care and grooming.

AS YOU READ

- grip—a handle
- remnant—a leftover piece of goods

Answer these questions about the student initiative project.

1. Robin is a pet sitter. She charges \$25 for a weekend at the owner's home or \$35 at her home. Last year she had jobs for 20 weekends. What was her total income if 3 jobs were at her home?

2. Don delivers pets to local veterinarians. His expenses were \$3.50 a week for a newspaper ad and \$20 a week for gasoline. How much were his expenses for the month? (1 mo = 4 wk)

3. Mrs. Wright and her son Richard have a dog-grooming business in their home. Richard is paid \$6.50 out of the normal \$10 grooming fee. If Mrs. Wright took in \$150, how much did Richard get?

***4.** Chain-link leashes are made by attaching a clip and leather *grip* to a length of chain. John has a chain 6.25 m long. How many 1.25 m lengths can he cut?

5. Some students make scratching posts that sell for \$7.50. Scrap lumber, nails, and carpet *remnants* cost about \$3.75 for each post. About how much profit do they make on 20 posts?

6. Paula makes leather collars for pets. She makes 3 different sizes. She charges \$2 for the small size, \$3 for medium, and \$4 for large. How much money would Paula earn if she sells 7 of each size?

Career: Animal-Care Workers

READ • PLAN • SOLVE • CHECK

Veterinarians are just one type of career in animal care and welfare. Other careers are animal technicians, attendants, trainers, and breeders. The training required for these careers ranges from a college degree plus veterinary school to on-the-job experience.

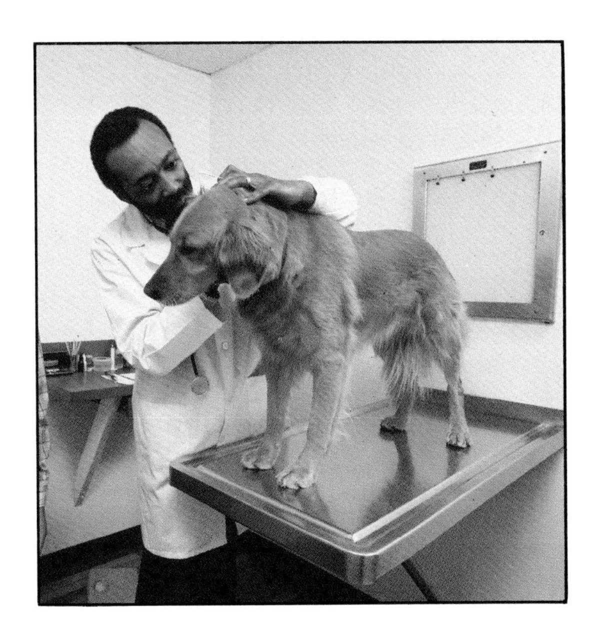

AS YOU READ

- medication—a medicine
- tips—money given in appreciation for a service

Solve.

1. Arthur Strider breeds Siamese cats. Each kitten is worth \$350. If the female had 4 kittens, how much are they worth?

2. Harvey Moore's cat has a temperature of 104.6°F. How many degrees above a cat's normal temperature of 101°F is this?

3. Amy Langston works in an animal-grooming salon. Including *tips*, she earned \$131.25 for a 35-h work week. How much did Amy make per hour?

4. Mary Walsh, a veterinarian, checks about 30 animals in an 8-h day. At this rate, how many minutes are spent on each animal?

5. May Adler, veterinary attendant, weighs pets by weighing herself and the pet. Then she subtracts her own weight. May weighs 125.75 lb. She and a cat weigh 134.25 lb. How much does the cat weigh?

6. Marie Todd is an animal technician. She is preparing *medication* for a dog. The dog is to be given 4 tablets a day for one week and 2 tablets a day for the next 3 wk. How many tablets are to be given in all?

CHAPTER TEST 8

Place the decimal point to make the quotient correct.

1. $82.72 \div 35.2 = 235$

2. $912.093 \div 2.57 = 3549$

Divide.

3. $3\overline{)2.619}$

4. $12\overline{)\$6.72}$

5. $0.4\overline{)4.48}$

6. $1.35\overline{)40.5}$

7. $0.009\overline{)63}$

8. $4.4\overline{)88.88}$

9. $43\overline{)0.903}$

10. $61.2\overline{)336.6}$

11. $0.34\overline{)2.2848}$

12. $45 \div 100$

13. $53.4 \div 1{,}000$

14. $\$3.60 \div 10$

Divide. Round the quotient to the nearest tenth.

15. $0.06\overline{)0.7836}$

16. $32.6\overline{)324.8}$

Divide. Round the quotient to the nearest hundredth.

17. $4.8\overline{)6.237}$

18. $84.8\overline{)365}$

Find the unit price.

19. shampoo: 16 oz for $1.35

Which is the better buy?

20. 20 oz for $2.70
12 oz for $1.70

Solve.

21. Andy read 421 pages of a novel in 8 h. Find his hourly reading rate.

22. Mr. Jones processed 342 orders in 12 h. Find his average rate per hour for processing.

***23.** Mildred's typing rate was 438 words in 7 min. At that rate, how many words could she type in 56 min?

***24.** Ms. Swift counseled 87 students in a 30-h week. At that rate, how many students could she counsel in 4 wk?

Measurement

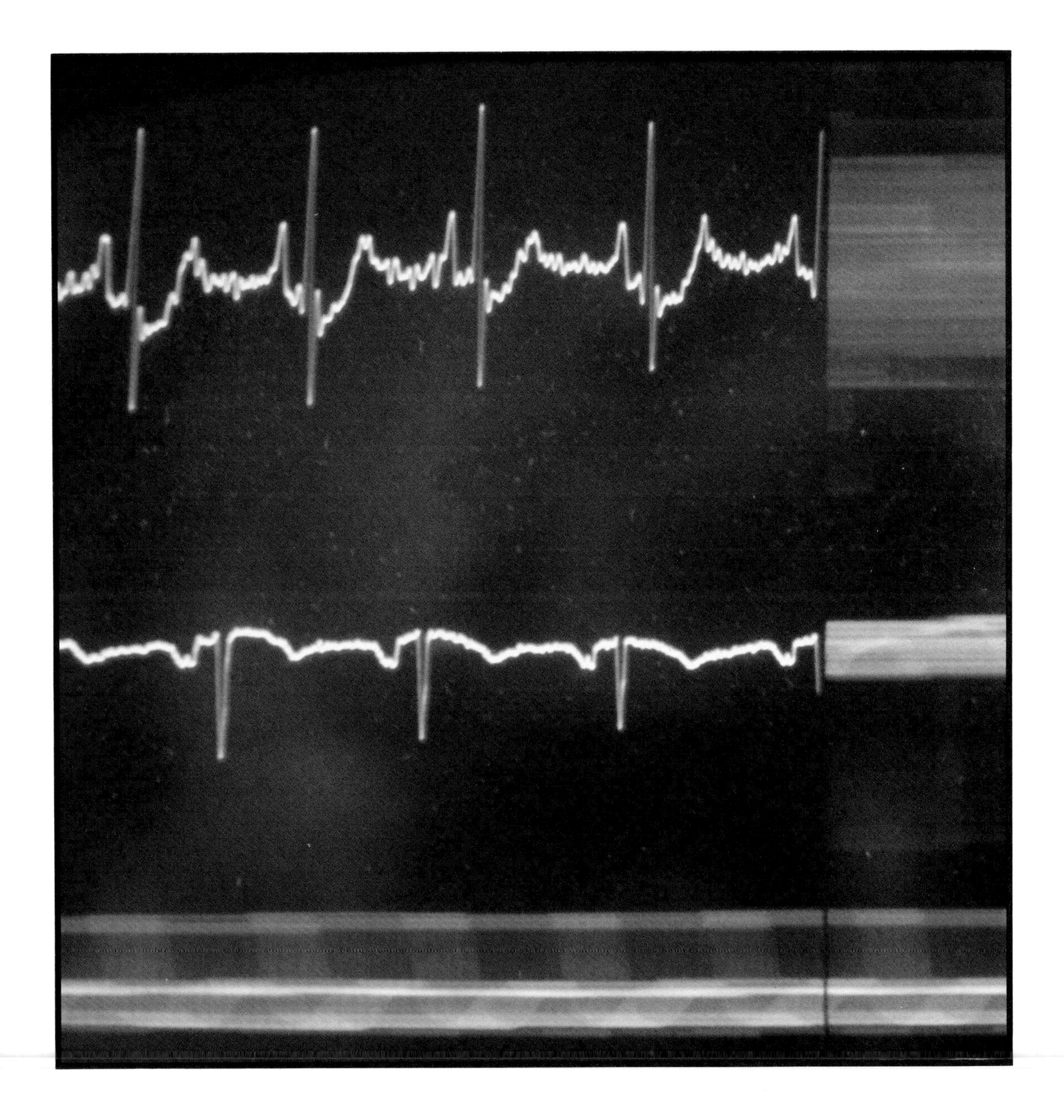

MAKING MATH COUNT

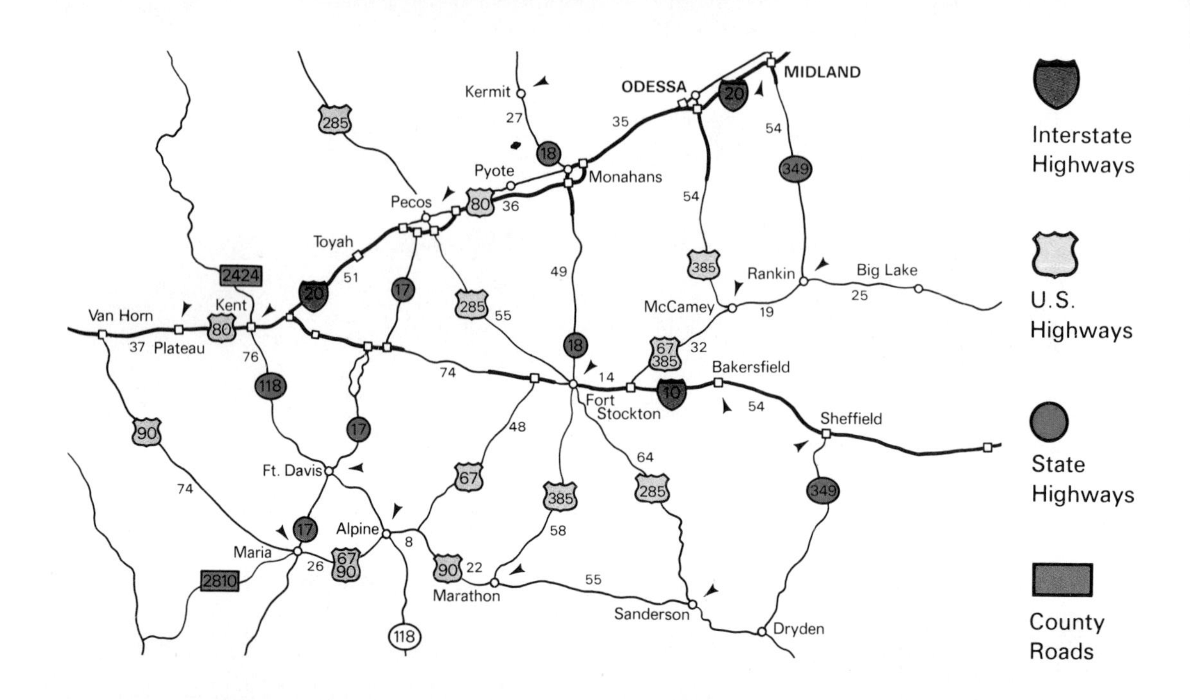

EXAMPLE

Adam lives in Fort Stockton and plans to drive to Pecos. How far is it to Pecos?

A 285 mi **B** 55 mi
C 49 mi **D** 18 mi

Look at the map. U. S. highway 285 connects Fort Stockton and Pecos. The arrows mark a distance of 55 mi.
So, **B** is the correct answer.

Choose the correct answer.

1. What is the shortest route north from Fort Stockton to Kermit?

A (18) **B** (285)
C (10) **D** (20)

2. Ms. Davidson is going from Kent to Marathon. What route should she take?

A (10) and (67) **B** (118) and (17)
C (90) and (67) **D** (118) and (90)

3. Traveling from Kent to Sanderson by (10), at which place would you exit?

A Plateau **B** Sheffield
C Fort Stockton **D** Bakersfield

4. How far is it from Fort Stockton to Rankin by way of McCamey?

A 19 mi **B** 36 mi
C 51 mi **D** 65 mi

ESTIMATING LENGTH

To choose the best estimate of metric length

The most commonly used metric units of length are millimeter (mm), centimeter (cm), meter (m), and kilometer (km). This chart gives you an approximate size for each unit.

Unit	Approximate Size
millimeter (mm)	thickness of a dime
centimeter (cm)	width of your smallest finger
meter (m)	width of a door
kilometer (km)	length of 10 football fields

EXAMPLE

Choose the best estimate.

Length of fountain pen

12 m 12 cm 12 mm

The length of a fountain pen is about 12 times the width of your smallest finger. So, the best estimate is **12 cm.**

GETTING READY

Choose the best estimate.

1. length of a paper clip

3 cm 3 m 3 mm

2. thickness of an aspirin tablet

4 m 4 cm 4 mm

EXERCISES

Choose the best estimate.

1. distance walked in 2 h

8 m 8 mm 8 km

2. width of a car tire

15 mm 15 cm 15 m

3. thickness of a toothpick

1 mm 1 cm 1 m

4. distance from Newtown to Alburgh

60 cm 60 mm 60 km

5. length of a pickup truck

4 m 4 cm 4 km

6. length of a sheet of paper

28 mm 28 cm 28 m

MEASURING LENGTH

To measure lengths in centimeters and millimeters

EXAMPLE 1

Describe the length of this ruler.

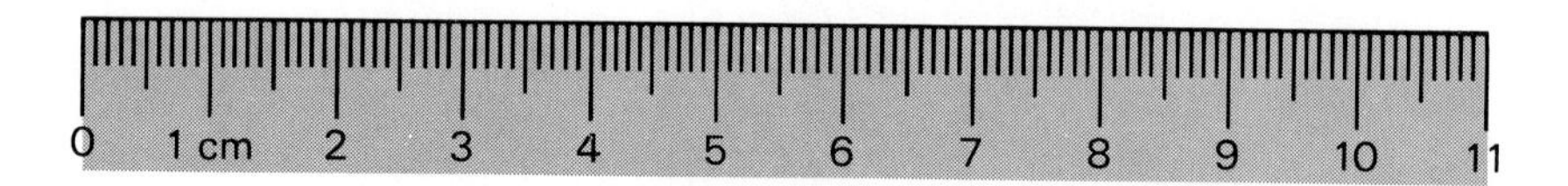

The ruler is 11 cm long. Since 1 cm = 10 mm and 11 × 10 = 110, the ruler is also **110 mm** long.

EXAMPLE 2

Find the length of the nail.

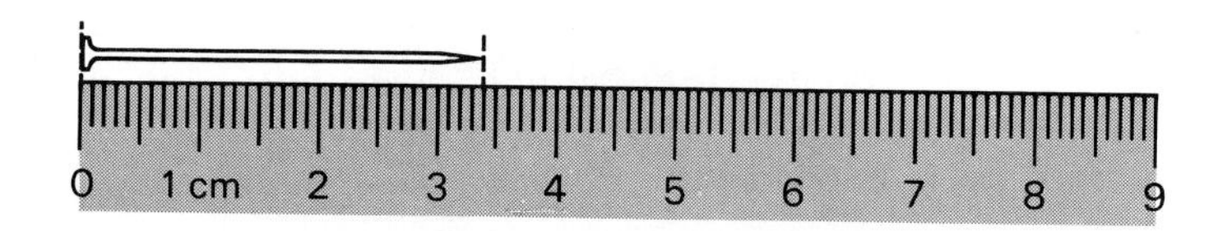

The length of the nail is **34 mm** or **3.4 cm.**

> To measure an object, place the ruler so that the endpoint or zero point of the ruler and one endpoint of the object are lined up.

GETTING READY

Measure each segment. Give the length in millimeters and in centimeters.

1. ____

2. ____

How long is $\overline{AB}$? $\overline{AC}$? $\overline{AD}$?

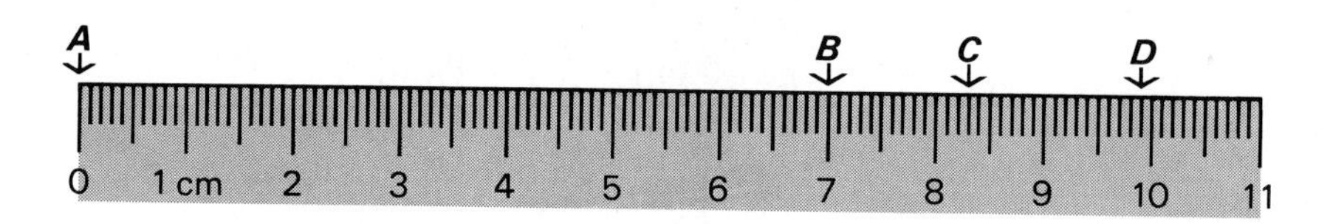

3. AB = ? **4.** AC = ? **5.** AD = ?

EXERCISES

Measure each segment. Give the length in millimeters and in centimeters.

1. ____________________________________

2. ____

3. ____

4. ____

5. ____

6. ______________________________

7. ____________________________

8. __________________________

9. ________________________________

10. ____

Use a metric ruler to draw segments of the following lengths.

11. 16 mm **12.** 35 mm **13.** 6.4 cm

Measure each indicated segment. Give the length in millimeters and in centimeters.

14. $\overline{AB}$
15. $\overline{BC}$
16. $\overline{CD}$
17. $\overline{DE}$
18. $\overline{EA}$
19. $\overline{AC}$
20. $\overline{AD}$
21. $\overline{BE}$
22. $\overline{BD}$
23. $\overline{CE}$

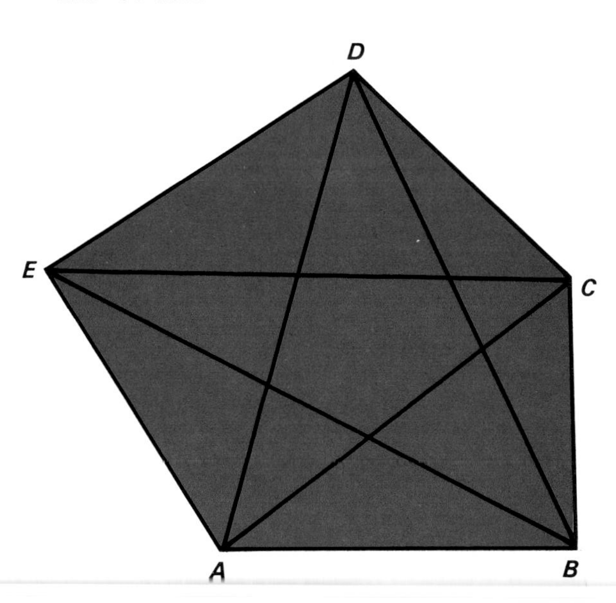

PERIMETER

To find the perimeter in centimeters

- Measure the length of each side with a ruler.
- Find the sum of the lengths of the sides.

The perimeter of a plane figure is the distance around it or the sum of the lengths of the sides.

EXAMPLE

Find the perimeter of each figure below. Measure the length of each side in centimeters.

2 cm + 5 cm + 2 cm + 5 cm = 14 cm

So, the perimeter is **14 cm.**

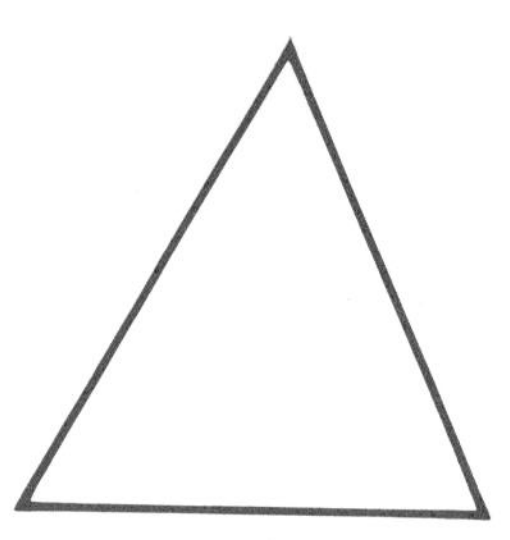

3.2 cm + 3.6 cm + 3.4 cm = 10.2 cm

So, the perimeter is **10.2 cm.**

GETTING READY

Find the perimeter of each figure. Measure the length of each side in centimeters.

1.

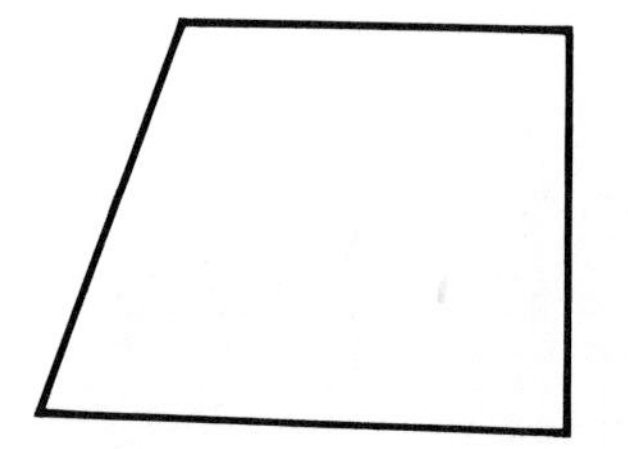

2.

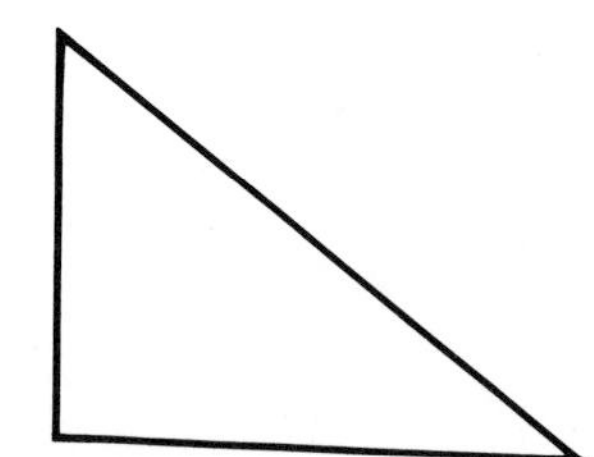

3.

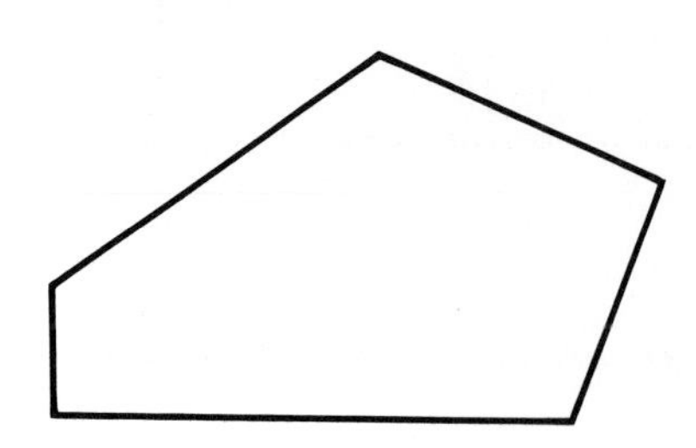

4.

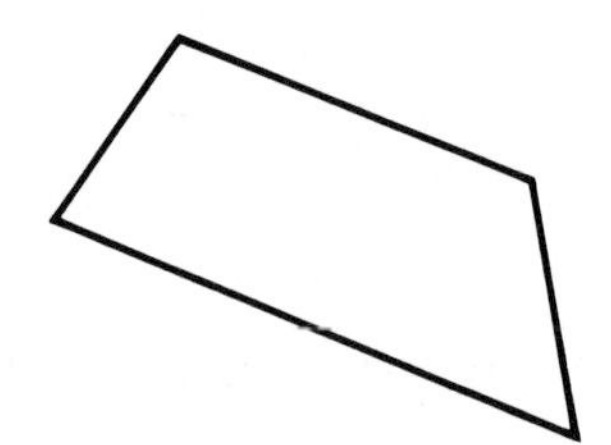

EXERCISES

Find the perimeter of each figure. Measure the length of each side in centimeters.

1.

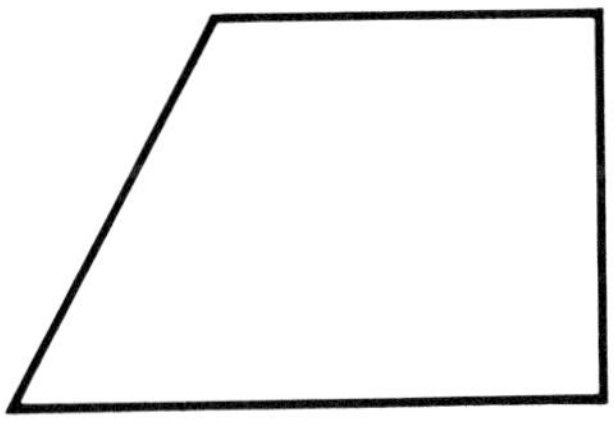

2.

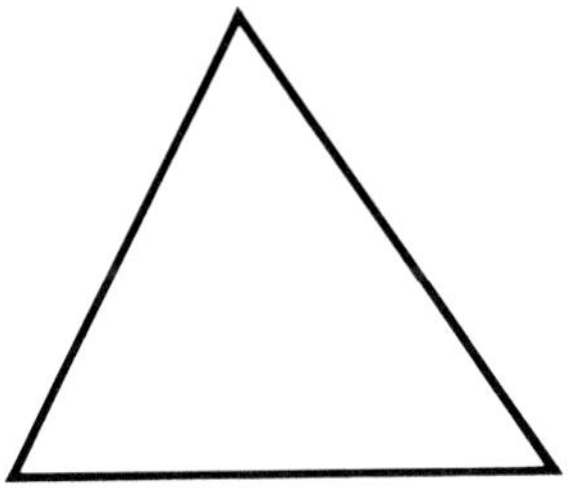

3.

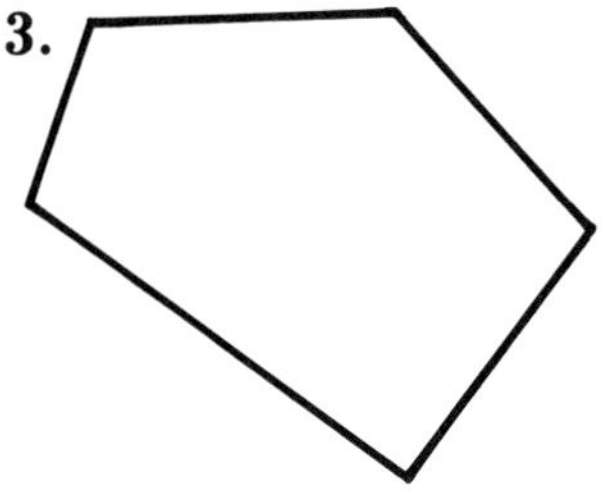

4.

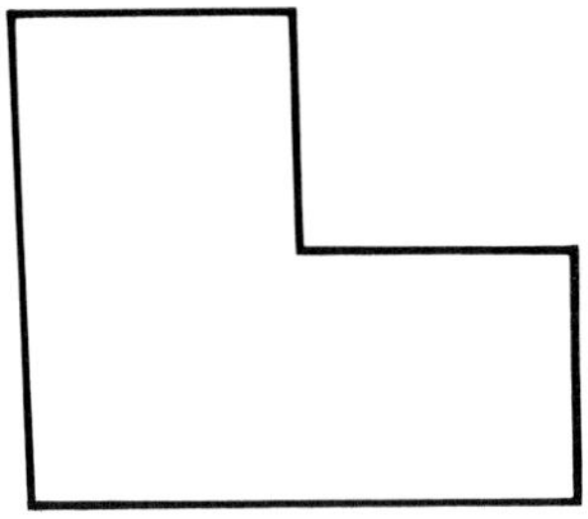

5.

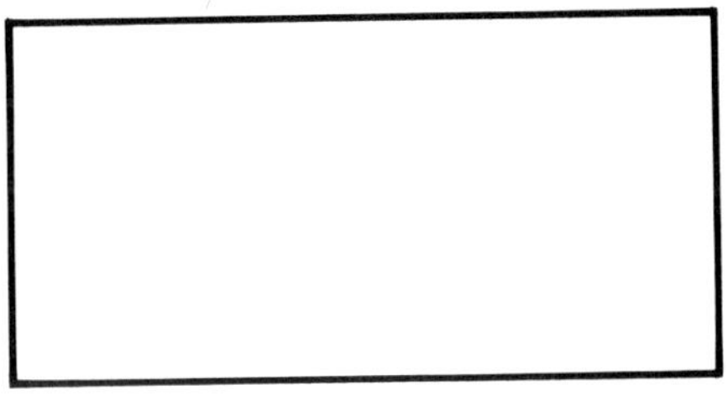

6.

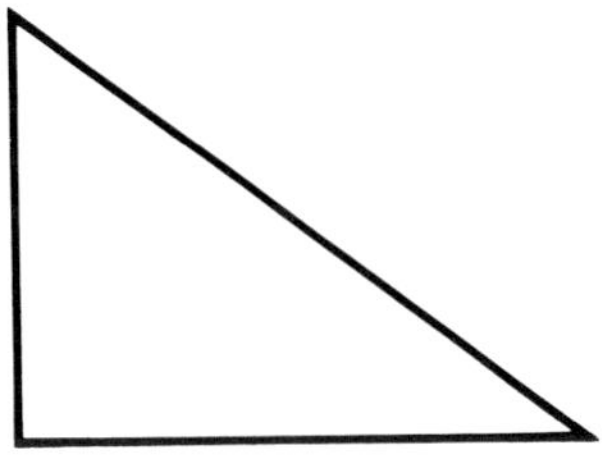

7.

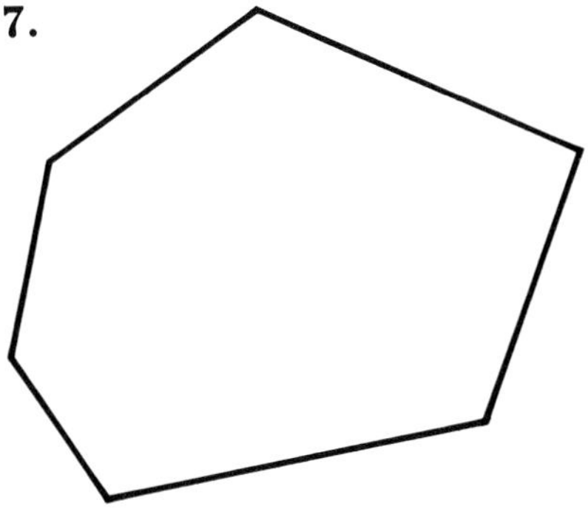

8.

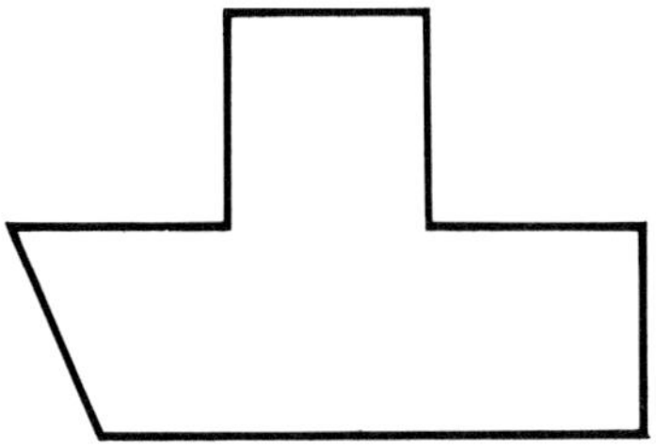

***9.** What happens to the perimeter of a square when the lengths of the sides are doubled?

AREA

To find the area of a region in square centimeters

EXAMPLE 1

Find the area of a rectangle 8 cm long and 4 cm wide.

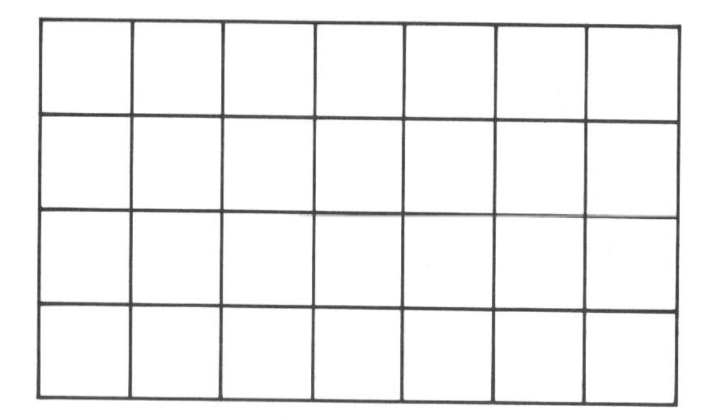

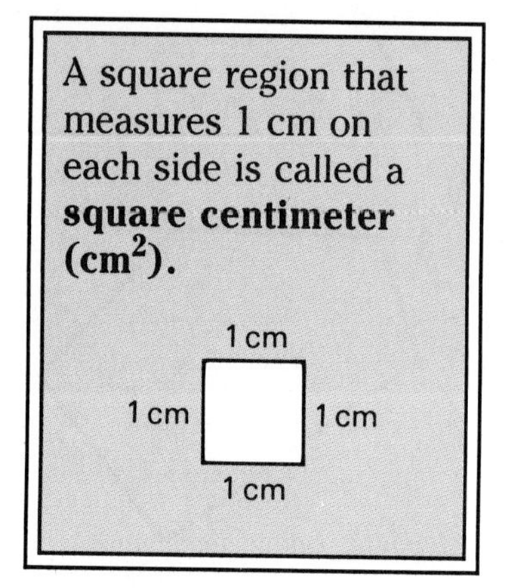
A square region that measures 1 cm on each side is called a **square centimeter (cm^2).**

There are 4 rows of square centimeters.
There are 8 squares in each row.
How many squares in all? $\mathbf{4 \times 8 = 32}$
Check by counting the squares.
So, the area of the rectangle is $\mathbf{32\ cm^2}$.

EXAMPLE 2

Find the area of the shaded region.

Count the number of square centimeters in each row.
Half of the area of a square is the same as $0.5\ cm^2$.

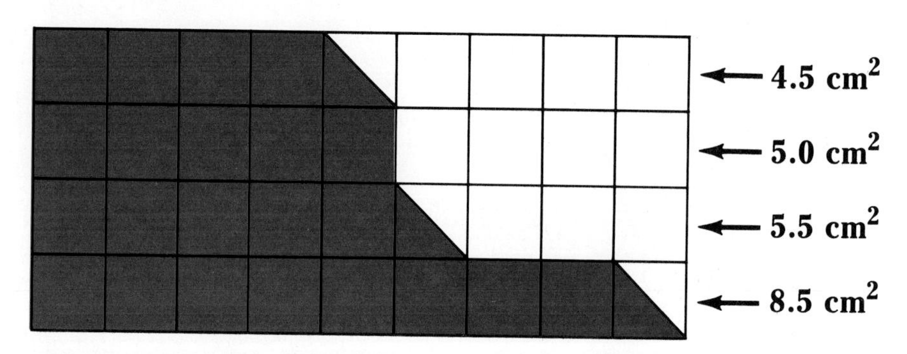

Add to find the area. $\mathbf{4.5 + 5.0 + 5.5 + 8.5 = 23.5}$
So, the area of the region is $\mathbf{23.5\ cm^2}$.

GETTING READY

Find the area of the shaded region in square centimeters.

1. 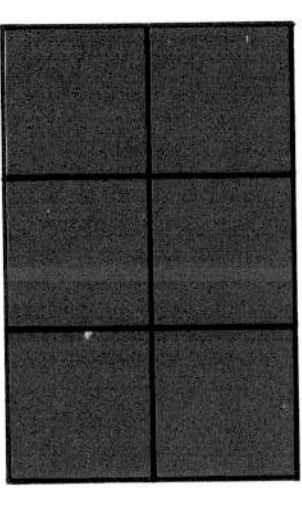

2. 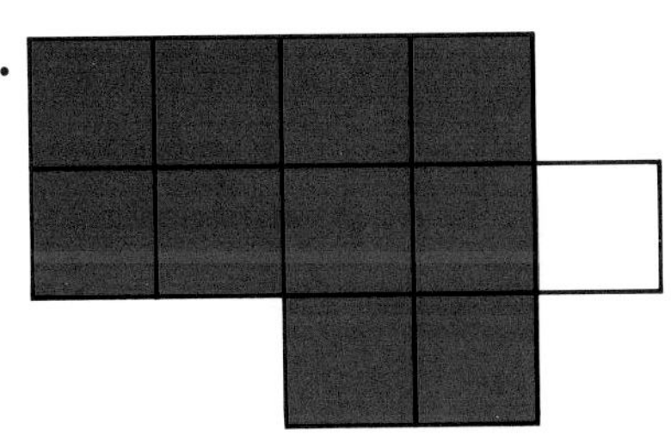

3. 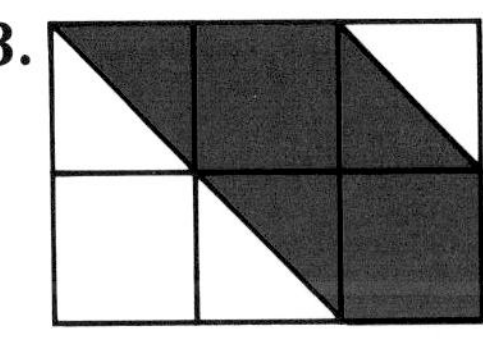

EXERCISES

Find the area of the shaded region in square centimeters.

1. 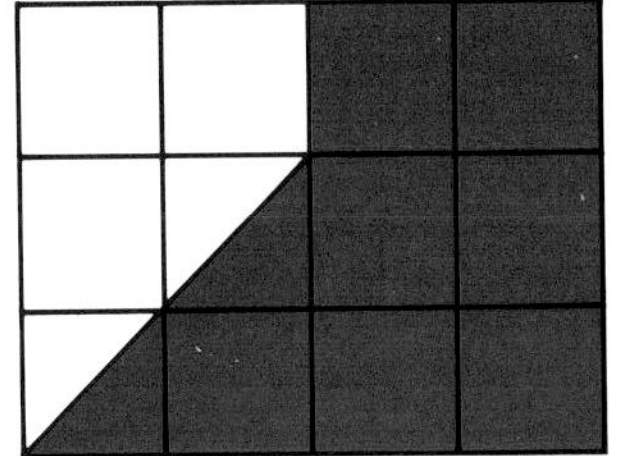

2. 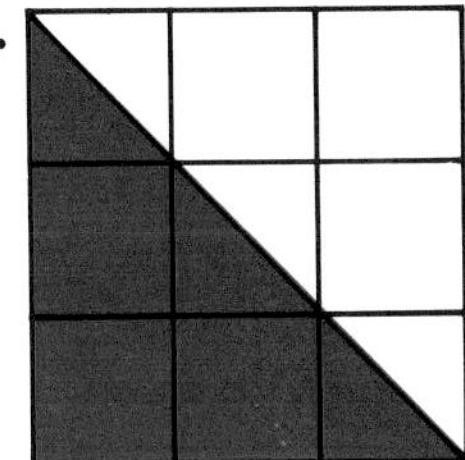

3. 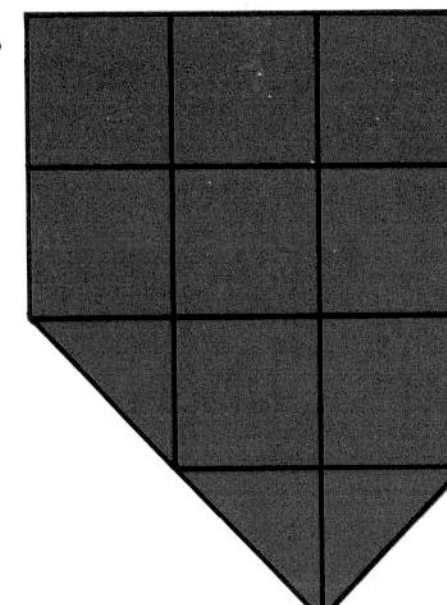

4. 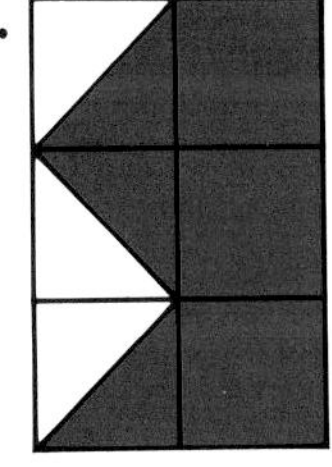

5. 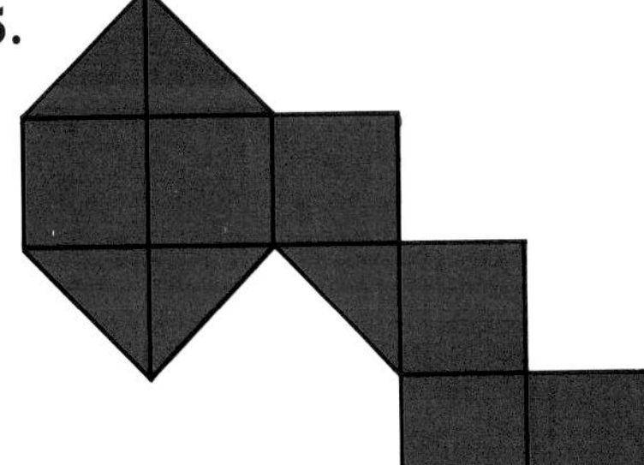

6. 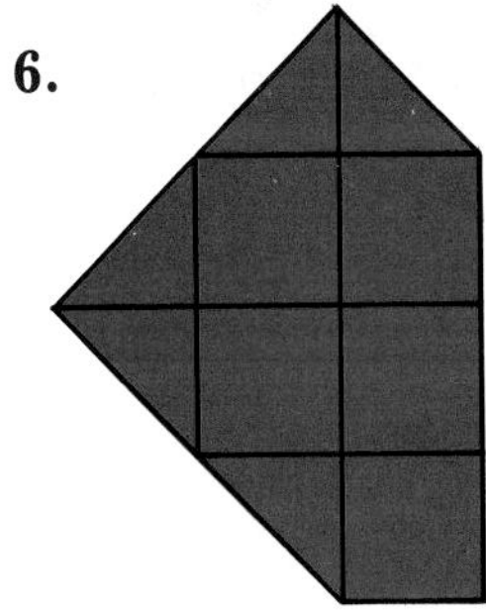

7. 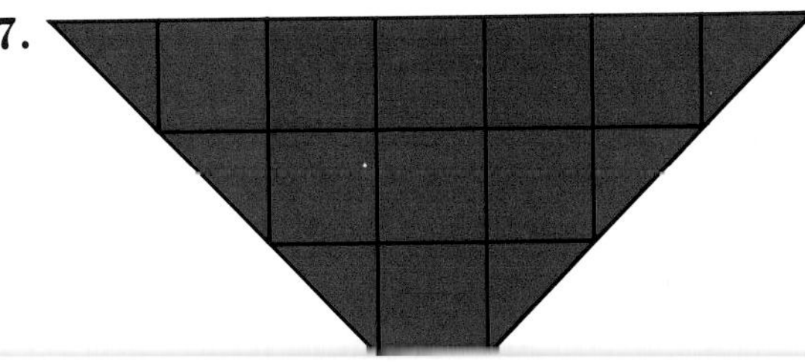

8. 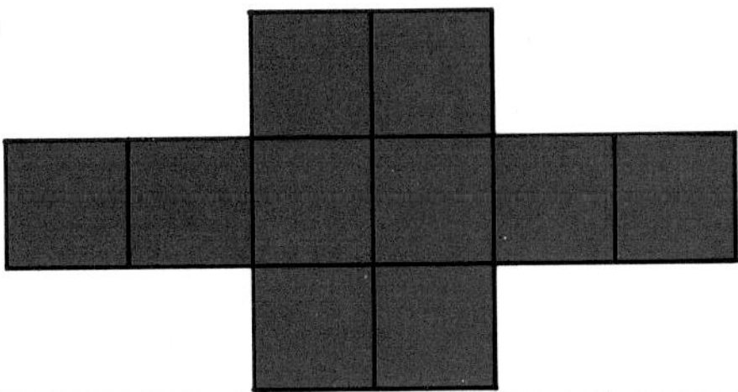

CHANGING BETWEEN UNITS OF LENGTH

To change between metric units of length

- Multiply when changing from larger units to smaller units.
- Divide when changing from smaller units to larger units.

The chart below shows you some of the relationships between metric units of length.

1 cm = 10 mm
1 m = 100 cm = 1,000 mm
1 km = 1,000 m

EXAMPLE 1

Complete. 6 cm = ___?___ mm
Look at the chart. A centimeter is larger than a millimeter.
Think: Larger unit to smaller unit, multiply.
1 cm = 10 mm
Multiply by 10. **6 × 10 = 60**
So, 6 cm = **60 mm.**

When changing from larger units of measurement to smaller units, **multiply.**

EXAMPLE 2

Complete. 4,500 m = ___?___ km
Look at the chart. A meter is smaller than a kilometer.
Think: Smaller unit to larger unit, divide.
1 km = 1,000 m
Divide by 1,000. **4,500 ÷ 1,000 = 4.5**
So, 4,500 m = **4.5 km.**

When changing from smaller units of measurement to larger units, **divide.**

GETTING READY

Complete.

1. 8 m = ___?___ cm

2. 240 mm = ___?___ cm

3. 80 cm = ___?___ m

4. 5.42 km = ___?___ m

5. 82.4 m = ___?___ mm

6. 8,328 cm = ___?___ m

EXERCISES

Complete.

1. 2 km = ___?___ m

2. 5 km = ___?___ m

3. 7.2 km = ___?___ m

4. 5.93 km = ___?___ m

5. 3,000 m = ___?___ mm

6. 4,670 km = ___?___ m

7. 2,890 km = ___?___ m

8. 9,000 m = ___?___ km

9. 750 cm = ___?___ m

10. 230 cm = ___?___ m

11. 110 cm = ___?___ m

12. 455 cm = ___?___ m

13. 0.35 m = ___?___ cm

14. 0.93 m = ___?___ cm

15. 450 m = ___?___ cm

16. 8.54 m = ___?___ cm

17. 17 mm = ___?___ cm

18. 247 mm = ___?___ cm

19. 3.65 mm = ___?___ cm

20. 462 mm = ___?___ m

21. 975 km = ___?___ m

22. 56.4 km = ___?___ m

***23.** 56,800 km = ___?___ mm

***24.** 0.00625 mm = ___?___ km

CALCULATOR

Find the answer.

1. The number of seconds in a day is found by multiplying $60 \times 60 \times 24$.

2. The number of seconds in a year is found by multiplying $60 \times 60 \times 24 \times 365.25$.

3. The approximate number of hours that you have lived since your birth is found by calculating $(24 \times 365.25 \times \text{your age in years}) + (24 \times \text{days beyond your birthday})$.

***4.** Can you find the number of seconds that you have lived since your birth?

ESTIMATING CAPACITY

To choose the better estimate of metric capacity

The most commonly used metric units of capacity are milliliter (mL) and liter (L). This chart gives you an approximate amount for each unit.

Unit	Approximate Amount
liter (L) milliliter (mL)	4 drinking glasses 10 drops of water

EXAMPLE

Choose the better estimate.

a regular soup can

300 mL or 300 L

One liter of soup would not fit into a soup can.
So, the better estimate is **300 mL.**

GETTING READY

Choose the better estimate.

1. a "family-size" beverage container

2 L or 2 mL

2. a cup of cocoa

250 mL or 250 L

EXERCISES

Choose the better estimate.

1. a pail of water

19 mL or 19 L

2. a thermos bottle of juice

473 mL or 473 L

3. a pan of soup

946 mL or 946 L

4. a tankful of gas in a car

60 L or 60 mL

5. a bathtub full of water

250 L or 250 mL

6. a swimming pool full of water

12,000 mL or 12,000 L

7. a can of paint

4,000 mL or 4,000 L

8. a jar of peanut butter

250 L or 250 mL

CHANGING BETWEEN UNITS OF CAPACITY

To change between metric units of capacity

- Multiply when changing from larger units to smaller units.
- Divide when changing from smaller units to larger units.

1 L = 1,000 mL

EXAMPLE 1

Complete. 5 L = ___?___ mL
Think: Larger unit to smaller unit, multiply.
Multiply by 1,000.
5 × 1,000 = 5,000
So, 5 L = **5,000 mL.**

When changing from larger to smaller units of measurement, **multiply.**

EXAMPLE 2

Complete. 250 mL = ___?___ L
Think: Smaller unit to larger unit, divide.
Divide by 1,000.
250 ÷ 1,000 = 0.250
So, 250 mL = **0.25 L.**

When changing from smaller units of measurement to larger units, **divide.**

GETTING READY

Complete.

1. 16 L = ___?___ mL **2.** 9,867 mL = ___?___ L **3.** 8.72 mL = ___?___ L

EXERCISES

Complete.

1. 5 L = ___?___ mL **2.** 9 L = ___?___ mL **3.** 12 L = ___?___ mL

4. 197 mL = ___?___ L **5.** 5,240 mL = ___?___ L **6.** 7,290 mL = ___?___ L

7. 9.5 L = ___?___ mL **8.** 12.5 L = ___?___ mL **9.** 0.49 L = ___?___ mL

10. 1.46 mL = ___?___ L **11.** 99 mL = ___?___ L **12.** 650 L = ___?___ mL

13. 2.09 mL = ___?___ L **14.** 3.75 L = ___?___ mL **15.** 5.50 mL = ___?___ L

Problem Solving Skill

To compute distances traveled from odometer readings

- Subtract the starting reading from the final reading.
- Remember that the last digit shows tenths.

An odometer measures the distance traveled in miles or kilometers. This odometer shows that the car has been driven 12,576.3 mi.

1 2 5 7 6 3

EXAMPLE

Mr. Kuh's odometer read 1 2 5 7 6 3 at the beginning of a trip. After the trip it read 1 5 3 8 1 4 . How many miles did Mr. Kuh drive on the trip?

Subtract the starting reading from the final reading.

$$\begin{array}{r} 15{,}381.4 \\ -\,12{,}576.3 \\ \hline 2{,}805.1 \end{array}$$

So, he drove the car **2,805.1 mi.**

EXERCISES

Solve.

1. A rental car's odometer reads 4 2 1 8 8 3 mi when it goes out and 4 2 7 2 0 1 when it is returned. How far has the car been driven?

2. Ms. Atkins takes her car on a business trip. The odometer reads 0 9 9 3 4 6 mi when she leaves and reads 1 0 5 6 1 8 when she returns. How long was her trip?

3. The directions say "Randolph Lake is 10.5 mi from the turnoff." If the odometer in your car reads 1 2 3 9 2 7 mi at the turnoff, what should it read when you reach the lake?

4. A racing circuit measures 7.6 km. If a racing car's odometer reads 3 2 7 3 2 2 km at the beginning of a 50-lap race, what will it read at the finish?

ESTIMATING MASS

To choose the best estimate of metric mass

The most commonly used metric units of mass are milligram (mg), gram (g), and kilogram (kg). This chart gives you an approximate size for each unit.

Unit	Approximate Size
milligram (mg)	a grain of sand
gram (g)	a paper clip
kilogram (kg)	a textbook

EXAMPLE

Choose the best estimate.

two thumbtacks

2 mg 2 g 2 kg

The mass of one thumbtack is about the same as the mass of one paper clip. So, the best estimate is **2 g.**

GETTING READY

Choose the best estimate.

1. an aspirin

500 mg 500 g 500 kg

2. a box of soap powder

1 mg 1 g 1 kg

EXERCISES

Choose the best estimate.

1. a pair of running shoes

1 mg 1 g 1 kg

2. a pencil

5 mg 5 g 5 kg

3. a toothpick

85 mg 85 g 85 kg

4. a playing card

500 mg 500 g 500 kg

5. a bag of cement

25 mg 25 g 25 kg

6. a wheelbarrow of sand

75 mg 75 g 75 kg

7. a cough drop

2 mg 2 g 2 kg

8. a bowling ball

7 mg 7 g 7 kg

CHANGING BETWEEN UNITS OF MASS

To change between metric units of mass

- Multiply when changing from larger units to smaller units.
- Divide when changing from smaller units to larger units.

The chart below shows you some of the relationships between metric units of mass.

1 g = 1,000 mg
1 kg = 1,000 g

EXAMPLE 1

Complete. 500 kg = ___?___ g

Look at the chart. A kilogram is larger than a gram.
Think: Larger unit to smaller unit, multiply.
1 kg = 1,000 g
Multiply by 1,000. **500 × 1,000 = 500,000**
So, 500 kg = **500,000 g.**

When changing from larger units of measurement to smaller units, **multiply.**

EXAMPLE 2

Complete. 156 mg = ___?___ g

Look at the chart. A milligram is smaller than a gram.
Think: Smaller unit to larger unit, divide.
1 g = 1,000 mg
Divide by 1,000. **156 ÷ 1,000 = 0.156**
So, 156 mg = **0.156 g.**

When changing from smaller units of measurement to larger units, **divide.**

GETTING READY

Complete.

1. 72 g = ___?___ mg

2. 52 kg = ___?___ g

3. 876,000 mg = ___?___ g

4. 4,000 g = ___?___ kg

5. 0.65 kg = ___?___ g

6. 1.58 g = ___?___ kg

7. 1,832 mg = ___?___ g

8. 228 mg = ___?___ g

EXERCISES

Complete.

1. 24 g = ? mg
2. 2 g = ? mg
3. 1,000 g = ? kg
4. 800 g = ? kg
5. 3 kg = ? g
6. 62 kg = ? g
7. 100 g = ? kg
8. 10 g = ? kg
9. 743,000 mg = ? g
10. 794 mg = ? g
11. 1 g = ? kg
12. 3,460 g = ? kg
13. 1.48 kg = ? g
14. 0.001 kg = ? g
15. 1,742 g = ? kg
16. 28 mg = ? g
17. 0.31 kg = ? g
18. 679 mg = ? g
19. 1,346 mg = ? g
20. 734 g = ? kg
21. 473 mg = ? g
22. 3.4 kg = ? g
23. 3.74 g = ? kg
24. 17.6 g = ? mg

*25. 4.872 mg = ? kg

*26. 0.004 kg = ? mg

ON YOUR OWN

One milliliter of water has a mass of exactly 1 g.

1. About what is the mass of one drop of water?
2. What is the mass of 1 L of water?
3. What is the mass of water in a 75,000-L swimming pool?
4. How many liters of water could you carry?

TEMPERATURE

To recognize temperature readings in degrees Celsius
To compute the amount of temperature change

Temperature is measured in degrees Celsius (°C) in the metric system. The thermometer below shows you some Celsius readings.

EXAMPLE 1

Is a room temperature of 35°C too cold, too hot, or comfortable? Normal room temperature is 20°C. 35°C is 15° warmer.
So, a room temperature of 35°C is **too hot.**

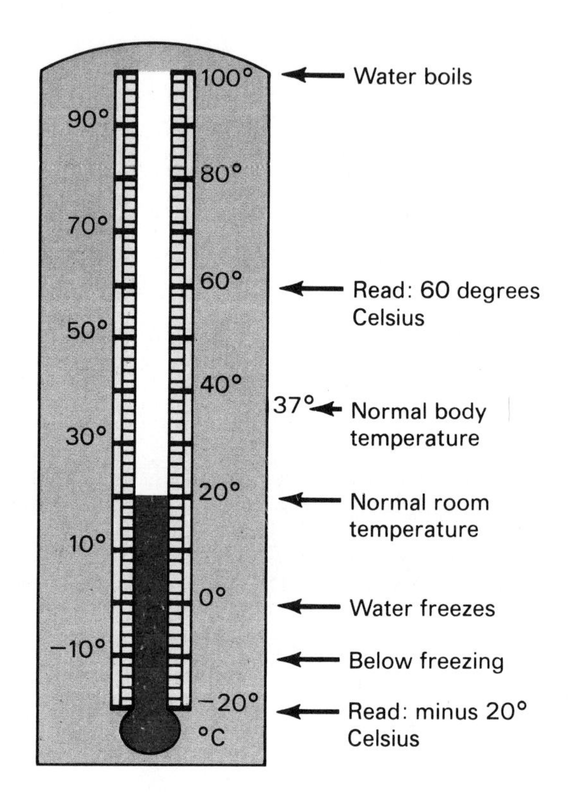

EXAMPLE 2

Find the temperature change from 15°C to −15°C.

15°C to 0°C is a 15° drop.
0°C to −15°C is a 15° drop.
15° + 15° = 30°

So, the temperature **dropped 30°.**

EXAMPLE 3

Find the temperature change from −23°C to 30°C.

−23°C to 0°C is a 23° rise.
0°C to 30°C is a 30° rise.
23° + 30° = 53°

So, the temperature **rose 53°.**

GETTING READY

Is each temperature too hot, too cold, or comfortable?

1. −5°C **2.** 50°C **3.** 25 °C **4.** 0°C

Which temperature is more likely?

5. a lake

15°C or 75°C

6. an apple in the refrigerator

10°C or 25°C

7. Find the temperature change from −15°C to 40°C.

EXERCISES

Is each temperature too hot, too cold, or comfortable?

1. 18°C
2. 62°C
3. −30°C
4. 45°C

Which temperature is more likely?

5. an ice cube

0°C or 20°C

6. a tub of bath water

90°C or 50°C

7. a snowstorm

15°C or −1°C

8. a sauna

45°C or 15°C

Solve.

9. At dawn the temperature was 5°C. At 3:00 pm, it was −15°C. Find the temperature change.

10. At 3:00 am the temperature was −17°C. At 5:00 pm, it was 11°C. Find the temperature change.

11. Find the temperature change from −13°C to −7°C.

12. Find the temperature change from 16°C to −19°C.

13. Find the temperature change from −5°C to 9°C.

14. Find the temperature change from 34°C to 11°C.

***15.** If the temperature rose 30° from −14°C, what is the new temperature reading?

***16.** If the temperature dropped 45° from 17°C, what is the new temperature reading?

ON YOUR OWN

Measure to the nearest sixteenth of an inch.

1. ____

2. ______________________________

3. Draw a line segment $3\frac{5}{8}$ in. long.

4. Draw a line segment $2\frac{11}{16}$ in. long.

CUSTOMARY MEASURES

To change between customary units of measurement

- Multiply when changing from larger units to smaller units.
- Divide when changing from smaller units to larger units.

The charts below show you some of the relationships between customary units of measurement.

Units of Length			
1 foot (ft)	= 12 inches (in.)		
1 yard (yd)	= 3 ft	= 36 in.	
1 mile (mi)	= 5,280 ft	= 1,760 yd	

Units of Capacity	
1 cup (c)	= 8 fluid ounces (fl oz)
1 pint (pt)	= 2 c
1 quart (qt)	= 2 pt
1 gallon (gal)	= 4 qt

Units of Weight	
1 pound (lb)	= 16 ounces (oz)
1 ton (T)	= 2,000 lb

EXAMPLE 1

Complete. 7 qt = __?__ pt

Look at the chart. A quart is larger than a pint.
Think: Larger unit to smaller unit, multiply.
1 qt = 2 pt
Multiply by 2. **7 × 2 = 14**
So, 7 qt = **14 pt.**

When changing from larger units of measurement to smaller units, **multiply.**

EXAMPLE 2

Complete. 108 in. = __?__ yd

Look at the chart. An inch is smaller than a yard.
Think: Smaller unit to larger unit, divide.
1 yd = 36 in.
Divide by 36. **108 ÷ 36 = 3**
So, 108 in. = **3 yd.**

When changing from smaller units of measurement to larger units, **divide.**

EXAMPLE 3

Complete. 64 oz = ___?___ lb

Look at the chart. An ounce is smaller than a pound.

Think: Smaller unit to larger unit, divide.

1 lb = 16 oz

Divide by 16. **64 ÷ 16 = 4**

So, 64 oz = **4 lb.**

GETTING READY

Complete.

1. 2 mi = ___?___ yd

2. 27 ft = ___?___ yd

3. 6 c = ___?___ fl oz

4. 16 pt = ___?___ qt

5. 12 lb = ___?___ oz

6. 2 T = ___?___ lb

EXERCISES

Complete.

1. 3 mi = ___?___ yd

2. 36 yd = ___?___ ft

3. 48 in. = ___?___ ft

4. 10,560 ft = ___?___ mi

5. 12 ft = ___?___ yd

6. 5,280 yd = ___?___ ft

7. 40 fl oz = ___?___ c

8. 6 qt = ___?___ pt

9. 8 gal = ___?___ qt

10. 4 pt = ___?___ qt

11. 2 c = ___?___ fl oz

12. 5 qt = ___?___ pt

13. 4 c = ___?___ pt

14. 7 pt = ___?___ c

15. 15 lb = ___?___ oz

16. 8 T = ___?___ lb

17. 5 lb = ___?___ oz

18. 24 oz = ___?___ lb

19. 5,000 lb = ___?___ T

20. 9 T = ___?___ lb

***21.** 0.598 mi = ___?___ in.

***22.** 484 gal = ___?___ fl oz

Problem Solving Applications

READ • PLAN • SOLVE • CHECK

Mrs. Lowry sponsors the Hospital Service Club. The students learn about jobs in the health care field by serving as volunteers. Club members volunteer their time in hospitals, nursing homes, and senior citizen centers.

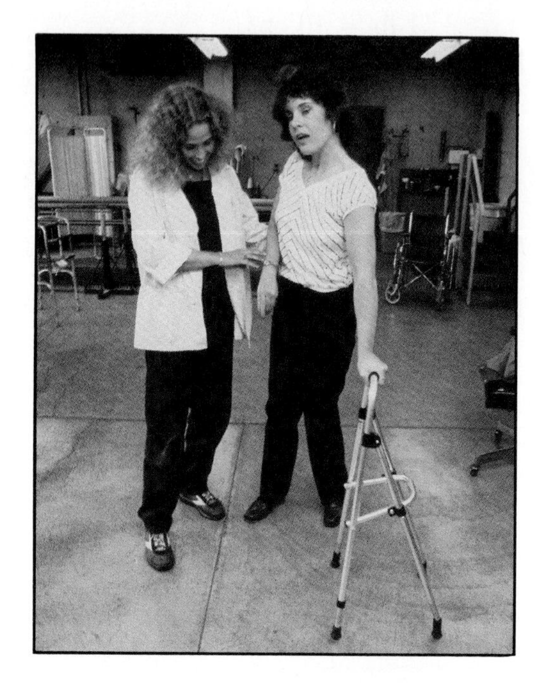

AS YOU READ

- physical therapy—treatment by exercise and massage
- whirlpool bath—a bath in which the water always moves in a circle

Answer these questions about the Hospital Service Club.

1. Carrie supervises patients who swim for *physical therapy*. If the pool is 25 m long, how far does a patient swim in 4 laps?

2. The temperature in the *whirlpool bath* is usually 37°C. If it dropped 3°, then rose 6°, what is the temperature now?

3. Stephanie volunteers at a senior citizen center. She is helping with a quilting project. How many squares measuring 10 cm × 10 cm can be cut from cloth measuring 1 m × 1 m?

4. Some students are finishing repairs at a nursing home. A door measures 1 m × 2 m. How many meters of weather stripping are needed for the door's top and sides?

*5. Ida is helping to prepare refreshments for a party for the patients in a hospital. If each punch cup holds 0.15 L, how many servings are there if she prepares 6 L of punch?

*6. Peter is planting flowers along the entrance walk to the nursing home. The walk is 3 m long. If the plants are placed 30 cm apart, how many plants does he need for both sides of the walk?

Career: Health Care Workers

READ • PLAN • SOLVE • CHECK

Health care workers are employed in hospitals, nursing homes, convalescent homes, and other long-term care facilities. Large hospitals employ nursing aides, orderlies, attendants, practical nurses, and technicians. Training varies from on-the-job experience to up to 4-year courses for certain types of technicians.

AS YOU READ

- pediatric—related to the care and treatment of children
- plasma—the fluid part of the blood

Solve.

1. Rocco Tanner is an emergency medical technician for the fire department. Last year he logged 11,250 mi for 250 d of work. On the average, how many miles does he travel a day?

2. Janet Parker is a *pediatric* aide. She weighs newborn infants daily. An infant weighed 3.15 kg at birth and 3.1 kg the next day. How many grams did the infant lose?

3. About half of the blood is *plasma.* If Adam Frank uses a blood sample of 50 mL, about how many milliliters of the sample is plasma?

4. Bill Arthur is an orderly. Part of his job is to move and set up heavy equipment. If the equipment weighs 0.125 T, how many pounds is this?

*5. Ernie Payne is a ward clerk in City Hospital. Each day he computes the daily amounts of medications needed by the patients. A certain patient needs 0.1 g of medication 3 times a day. How many milligrams is this a day?

6. Mary Alcott is a nursing aide at the Fleming Nursing Home. Part of her job is to serve meals to 82 patients. Each lunch cart has 4 shelves, 6 trays to a shelf. If the kitchen sends 3 filled carts, are there enough lunches for the patients?

CHAPTER TEST 9

Choose the best estimate.

1. width of a car

   2 mm   2m   2 km

2. length of a new pencil

   160 mm   160 cm   160 m

3. capacity of a can of paint

   4,000 mL   4,000 L

4. mass of an aspirin

   50 mg   500 g   500 kg

Measure each segment. Give the length in millimeters and in centimeters.

5. ____________________

6. ____

Complete.

7. 2,890 km = _?_ m
8. 455 cm = _?_ m
9. 72 in. = _?_ ft
10. 17 mm = _?_ cm
11. 12.5 L = _?_ mL
12. 7,000 lb = _?_ T
13. 32 fl oz = _?_ c
14. 148 g = _?_ kg
15. 15,840 ft = _?_ mi
16. 54 oz = _?_ lb
17. 9 pt = _?_ qt
18. 5 mi = _?_ yd
19. Find the perimeter. Measure the length of each side in centimeters.

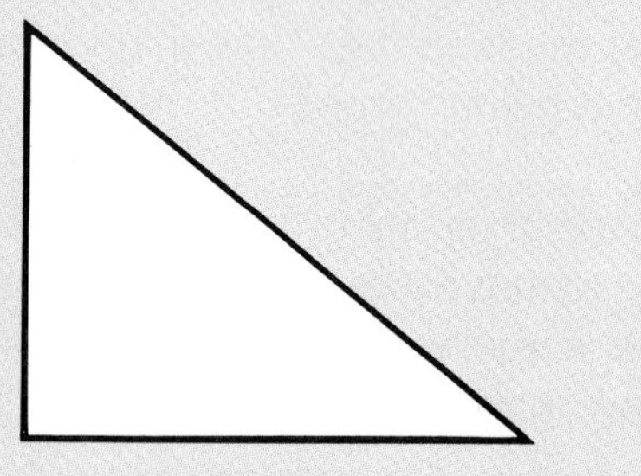

20. Find the area of the shaded region in cm^2.

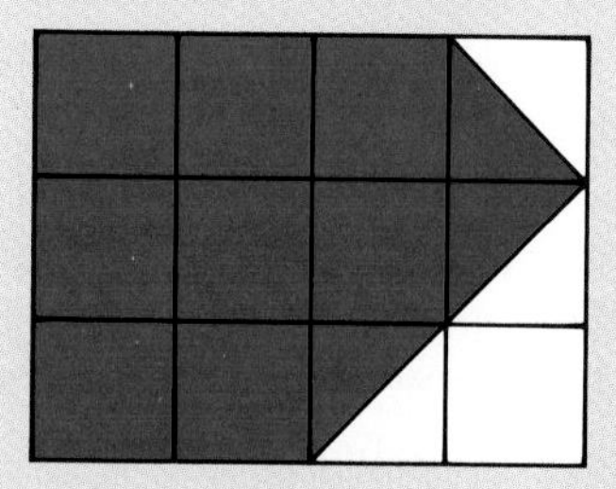

21. Is a room temperature of 22°C too hot, too cold, or comfortable?
22. Find the temperature change from 15°C to −7°C.
23. The distance between two rally checkpoints is 139.6 mi. If a driver's odometer reads 6 2 1 4 2 9 at the first checkpoint, what should it read at the second?

Cumulative Review 1-9

Add.

1. $\begin{array}{r} 279 \\ 483 \\ 670 \\ 405 \\ +976 \\ \hline \end{array}$ **2.** $\begin{array}{r} 168 \\ 72 \\ 549 \\ 83 \\ 147 \\ +295 \\ \hline \end{array}$ **3.** $\begin{array}{r} 650 \\ 832 \\ 2{,}086 \\ 73 \\ +\ 409 \\ \hline \end{array}$ **4.** $\begin{array}{r} 3{,}568 \\ 721 \\ 1{,}075 \\ 309 \\ 6{,}275 \\ +\ \ 82 \\ \hline \end{array}$

Subtract.

5. $\begin{array}{r} 954 \\ -672 \\ \hline \end{array}$ **6.** $\begin{array}{r} 8{,}093 \\ -3{,}758 \\ \hline \end{array}$ **7.** $\begin{array}{r} 3{,}000 \\ -\ \ 786 \\ \hline \end{array}$ **8.** $\begin{array}{r} 35{,}852 \\ -\ 4{,}178 \\ \hline \end{array}$

Multiply.

9. $\begin{array}{r} 87 \\ \times 42 \\ \hline \end{array}$ **10.** $\begin{array}{r} 576 \\ \times 93 \\ \hline \end{array}$ **11.** $\begin{array}{r} 3{,}060 \\ \times 240 \\ \hline \end{array}$ **12.** $\begin{array}{r} 4{,}753 \\ \times 605 \\ \hline \end{array}$

Divide.

13. $9\overline{)867}$ **14.** $3\overline{)4{,}062}$ **15.** $43\overline{)3{,}679}$ **16.** $46\overline{)19{,}694}$

Add.

17. $76.84 + 5.93 + 1.35$ **18.** $1.25 + 5.6 + 0.78$ **19.** $89.47 + 8.6 + 0.95$

Subtract.

20. $94.7 - 12.8$ **21.** $127.43 - 12.8$ **22.** $3.42 - 1.786$

Multiply.

23. 2.3×6.1 **24.** 1.742×3.05 **25.** 45.8×6.27

26. 7.4×10 **27.** 0.6×100 **28.** $4.98 \times 1{,}000$

Divide.

29. $7\overline{)42.56}$ **30.** $3.4\overline{)27.302}$ **31.** $0.006\overline{)42}$

32. Round 2,688 to the nearest hundred. **33.** Round 94.73 to the nearest tenth.

Complete.

34. 245 cm = ____?____ m

35. 12 mm = ____?____ cm

36. 5.5 L = ____?____ mL

37. 48 in. = ____?____ ft

38. 24 fl oz = ____?____ c

39. 136 kg = ____?____ g

40. Find the perimeter. Measure the length of each side in centimeters.

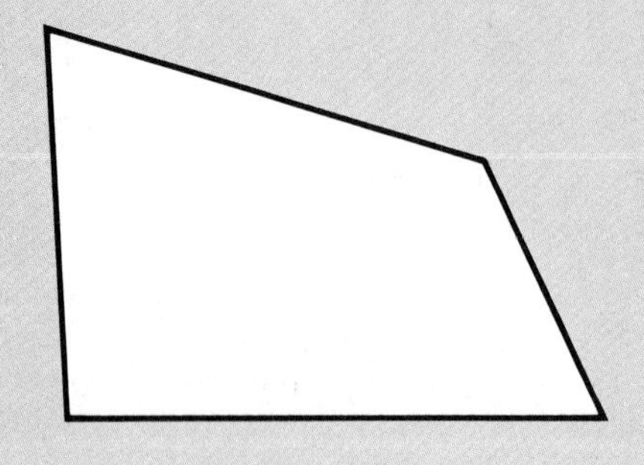

41. Find the area of the shaded region in square centimeters.

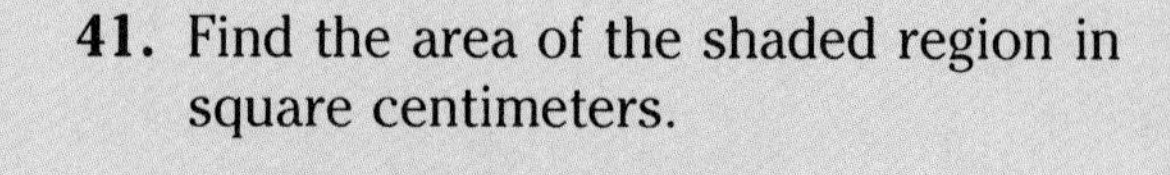

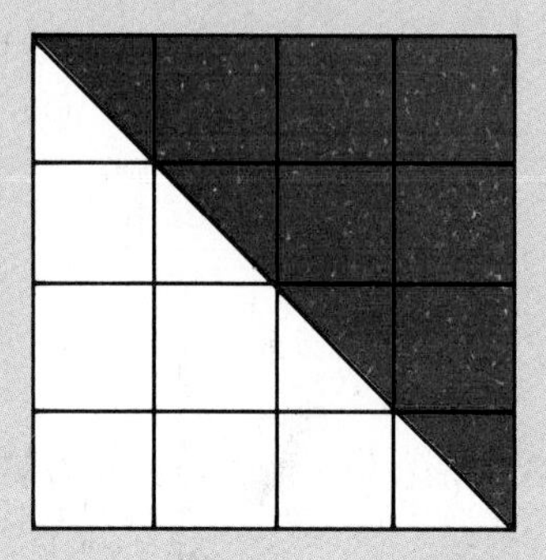

Read the problem and answer the questions.

Amy worked 19 h over the weekend. She earned \$86.64. How much did she earn each hour?

42. How much did Amy earn?

43. How many hours did she work?

44. What are you to find?

45. Which operation should you use?

What information is missing to solve this problem?

46. Erica paid \$9.98 for some shirts. How much was each shirt?

What is the extra information in this problem?

47. Tom bought orange juice for \$1.89, rolls for \$0.97 and hot dogs for \$2.49. How much did he pay for the rolls and hot dogs?

Solve.

48. Jake has a square garden. Each side is 18 m. How many meters of fencing will he need to enclose the garden?

49. Margaret bought 8 lb of apples for \$3.12. Mary bought 6 lb for \$2.46. Who got the better buy?

50. Ms. Martin purchased 64 yd^2 of carpet for \$648.55 including a sales tax of \$36.71. How much did she pay per square yard without tax?

51. A gallon of juice will serve 18 people. Will 5 gal be enough to serve 120 people? Solve by estimating.

The Meaning of Fractions

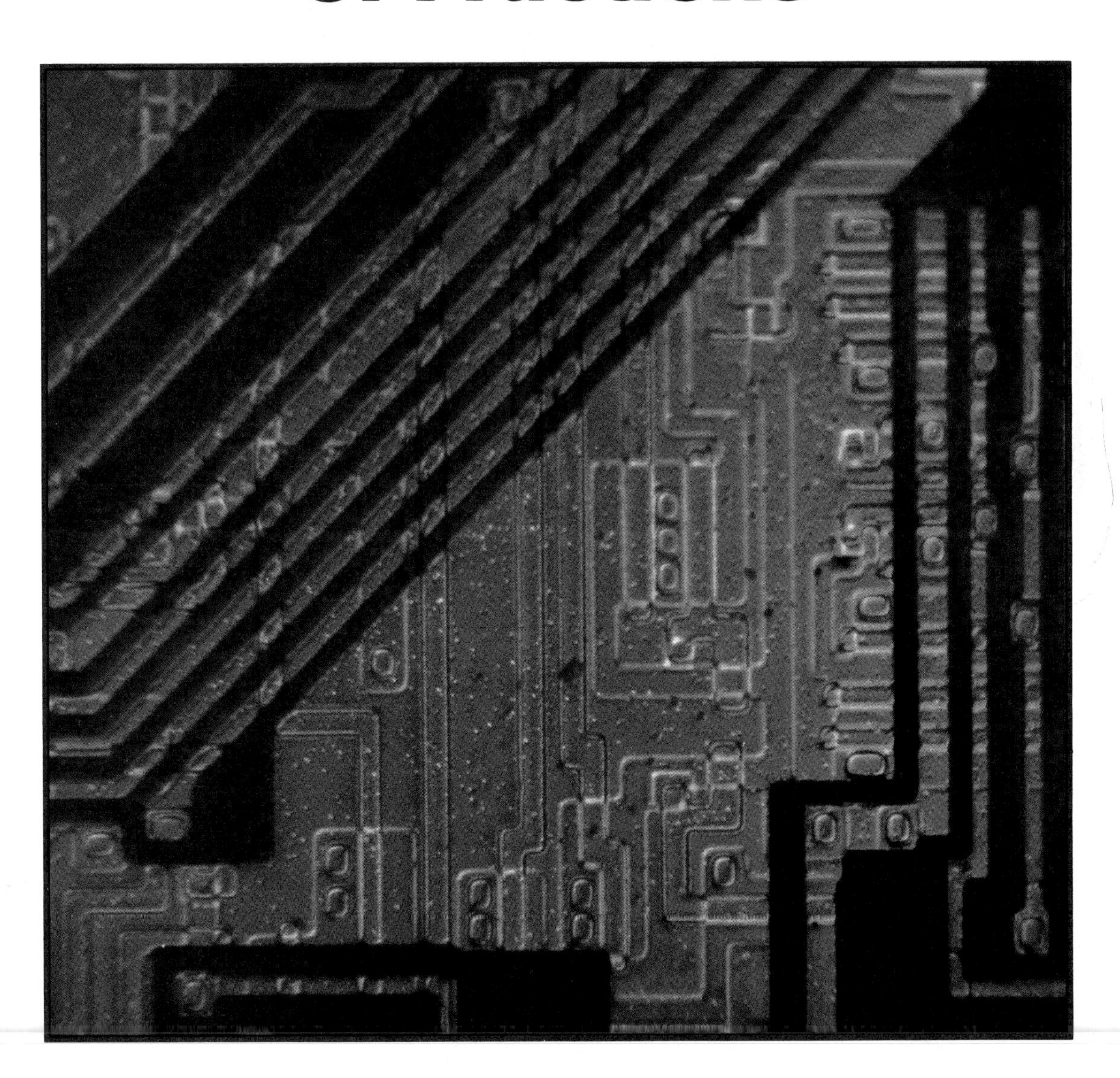

DIAGNOSTIC TEST: FORM A (30 MINUTES)

Write a fraction for the part that is not shaded.

1. 

2. 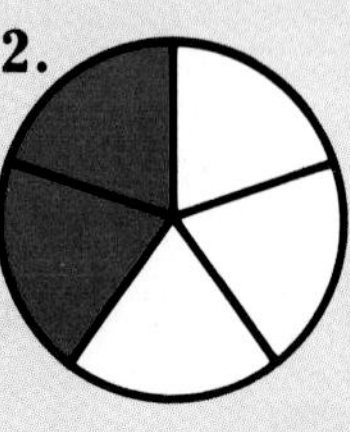

3. 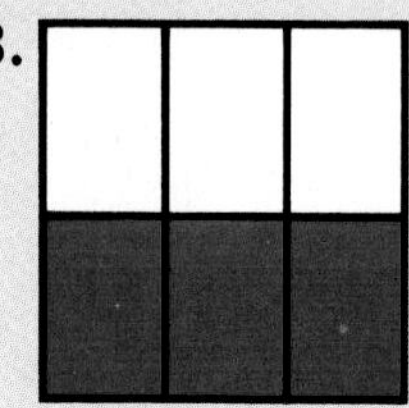

Write a fraction for the part of the group that is circled.

4. 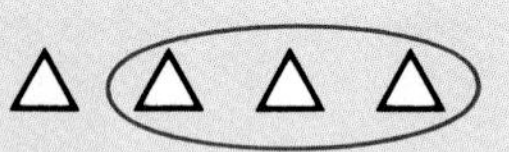

5. 

Write 2 fractions for the part that is shaded.

6. 

7. 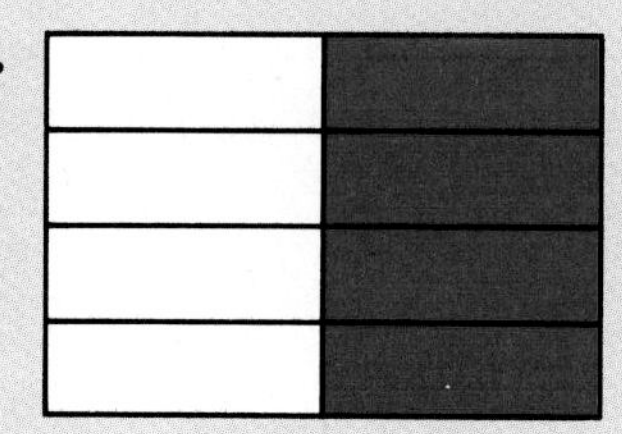

Write a fraction for the part of the cup that is filled.

8. 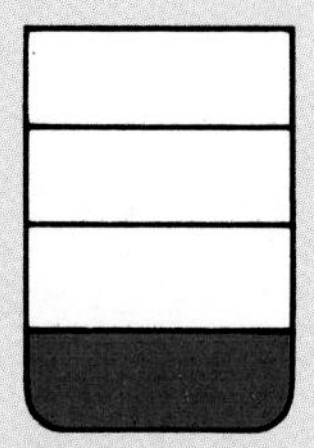

9. 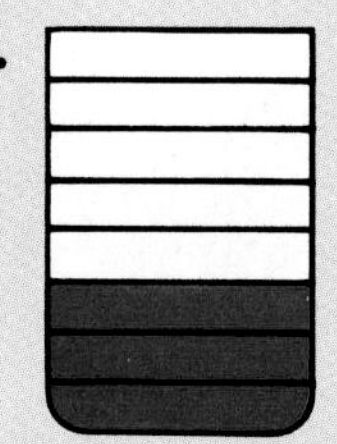

Draw a number line diagram. Tell whether the sentence is true.

10. $\frac{1}{2} = \frac{8}{16}$

11. $\frac{3}{4} = \frac{6}{8}$

12. $\frac{2}{3} = \frac{5}{6}$

13. $\frac{20}{25} = \frac{4}{5}$

14. $\frac{1}{2} = \frac{2}{6}$

15. $\frac{8}{10} = \frac{3}{4}$

14 or more right? Go to page 190.

Less than 14 right? Go to page 186.

DIAGNOSTIC TEST: FORM B (30 MINUTES)

Write a fraction for the part that is not shaded.

1. 

2. 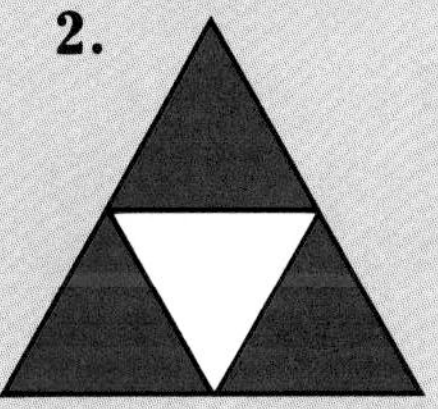

3. 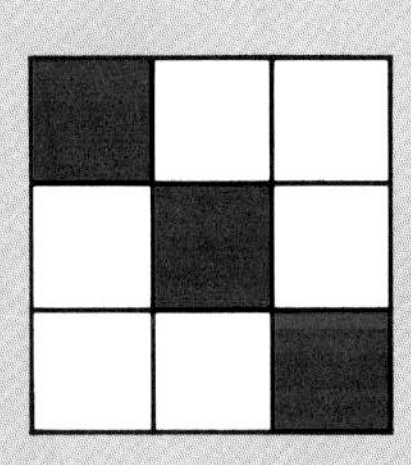

Write a fraction for the part of the group that is circled.

4. 

5. 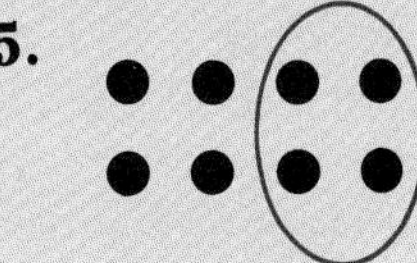

Write 2 fractions for the part that is shaded.

6. 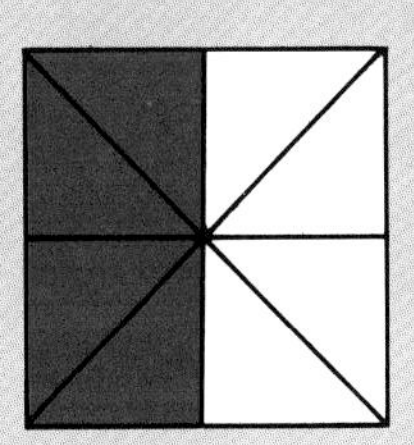

7. 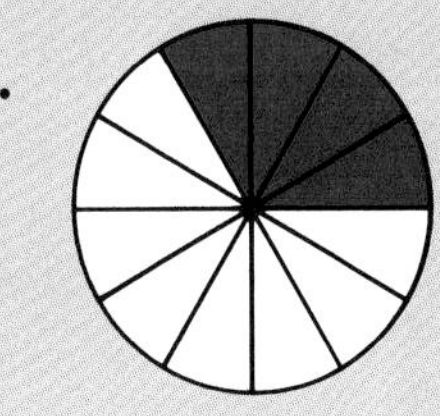

Write a fraction for the part of a pound that the meat weighs.

8. 

9. 

Draw a number line diagram. Tell whether the sentence is true.

10. $\frac{1}{3} = \frac{3}{12}$

11. $\frac{6}{10} = \frac{4}{5}$

12. $\frac{7}{5} = \frac{21}{15}$

13. $\frac{3}{4} = \frac{15}{20}$

14. $\frac{7}{21} = \frac{1}{4}$

15. $\frac{10}{25} = \frac{2}{5}$

14 or more right? Go to page 190.

Less than 14 right? See your teacher.

MEANING OF FRACTIONS

To recognize a fraction as part of a whole or part of a group

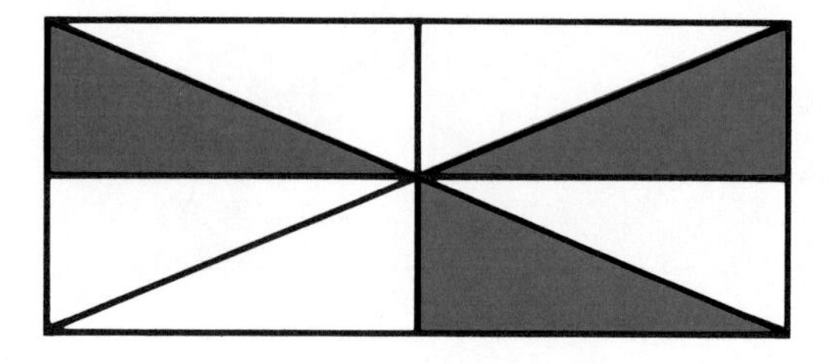

$\frac{5}{8}$ of the rectangle is not shaded.

$\frac{3}{7}$ of the books are math books.

EXAMPLE 1

Write a fraction for the part that is shaded.

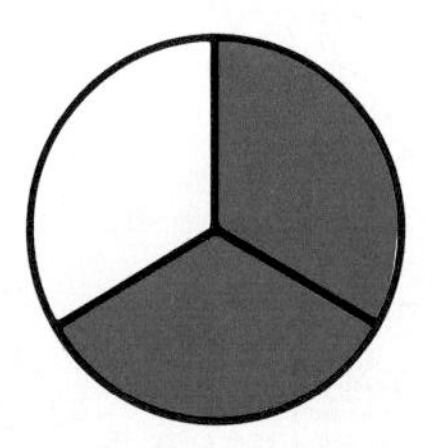

Think of the whole circle as one unit.
It is divided into 3 equal parts.
2 out of 3 parts are shaded.
So, $\frac{2}{3}$ of the circle is shaded.

EXAMPLE 2

Write a fraction for the part of the group that is pennies.

There are 5 coins.
4 of them are pennies.
So, $\frac{4}{5}$ of the coins are pennies.

EXAMPLE 3

What part of an inch is the measure of the paper clip? Write a fraction.

The inch is divided into 4 equal parts.
The paper clip is as long
as 3 of the parts.
So, the paper clip measures $\frac{3}{4}$ in.

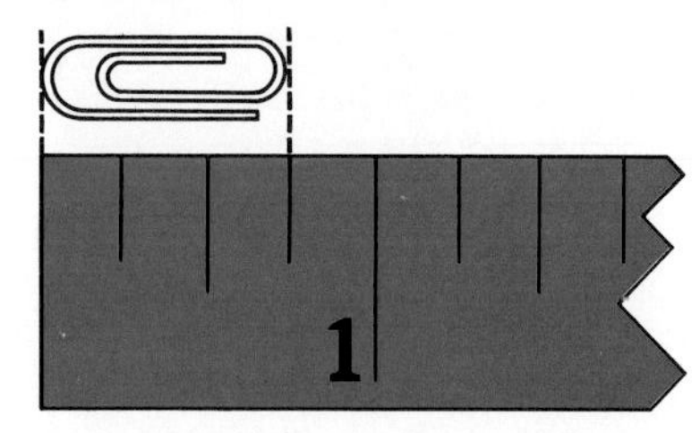

GETTING READY

Write a fraction for the part that is shaded.

Write a fraction for the part of the group that is circled.

1. 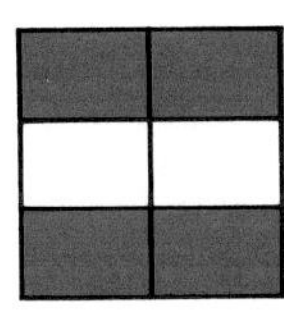

2. 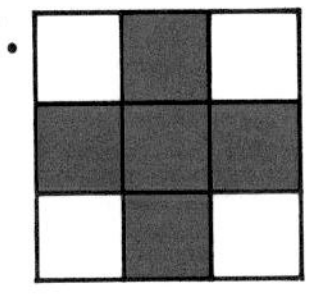

3. 

4. 

5. Write a fraction for the part of the cup that is filled.

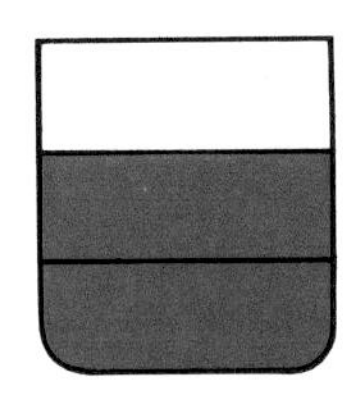

EXERCISES

Write a fraction for the part that is shaded.

1. 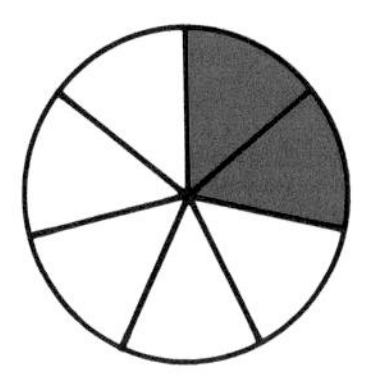

2. 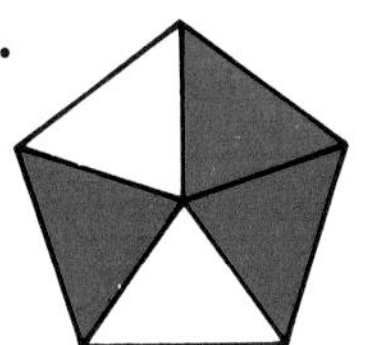

3. 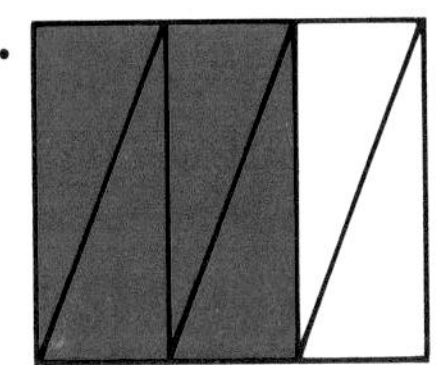

4. Write a fraction for the part of the group that is circled.

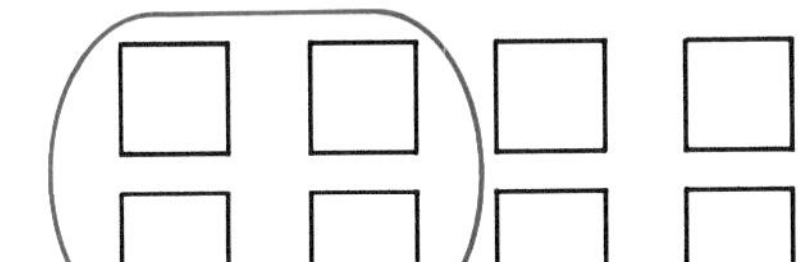

5. Write a fraction for the part of the set of pencils that has erasers.

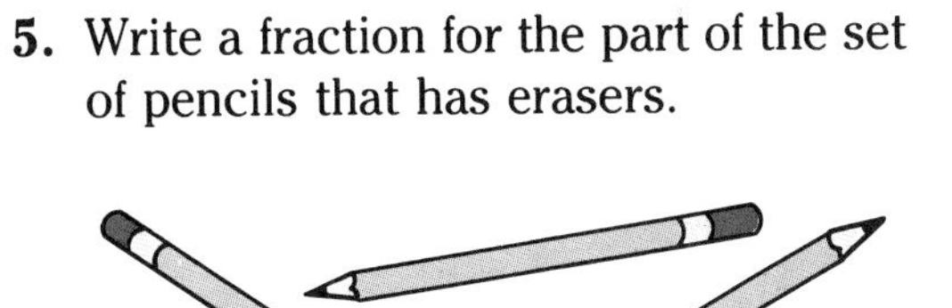

6. Write a fraction for the part of a pound that the cheese weighs.

EQUIVALENT FRACTIONS

To use diagrams to recognize equivalent fractions

EXAMPLE 1

Draw a diagram to show 3 equivalent fractions for $\frac{1}{2}$.

halves $\frac{1}{2}$

fourths $\frac{2}{4}$

sixths $\frac{3}{6}$

Each rectangle shows the same amount shaded.

So, $\frac{1}{2}$, $\frac{2}{4}$, and $\frac{3}{6}$ are equivalent fractions.

> Equivalent fractions are fractions that name the same number.
> **Example:** $\frac{1}{3}$, $\frac{2}{6}$, and $\frac{3}{9}$ are equivalent fractions.

EXAMPLE 2

Write 3 fractions for the part that is shaded.

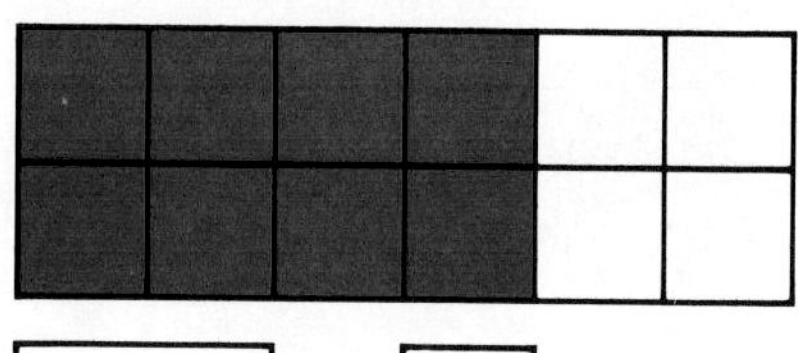

The rectangle is divided into twelfths. $\frac{8}{12}$ is shaded. Also $\frac{2}{3}$ is shaded and $\frac{4}{6}$ is shaded.

So, $\frac{2}{3}$, $\frac{4}{6}$, and $\frac{8}{12}$ name the shaded part.

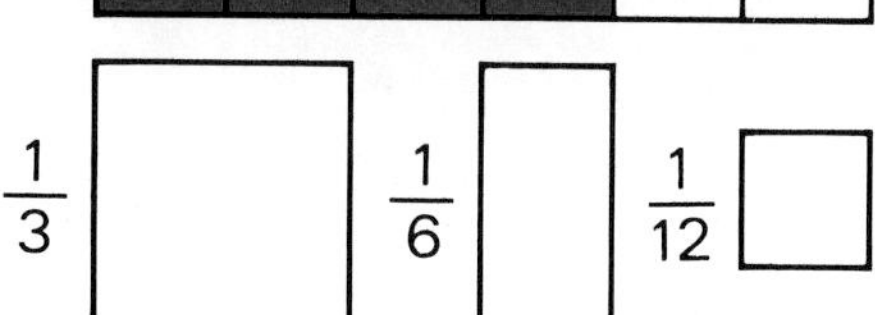

EXAMPLE 3

Draw a number line diagram to show that $\frac{3}{4} = \frac{6}{8}$.

Divide the distance between 0 and 1 into 4 equal parts.
Now, divide that same distance into 8 equal parts.
The arrows show that $\frac{3}{4}$ and $\frac{6}{8}$ name the same part of the line.

So, $\frac{3}{4} = \frac{6}{8}$.

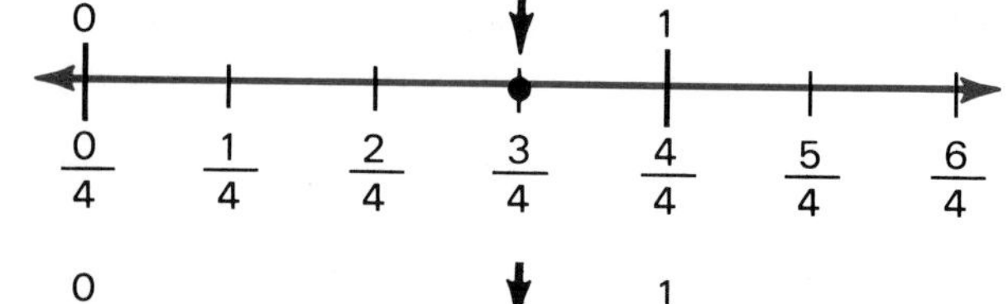

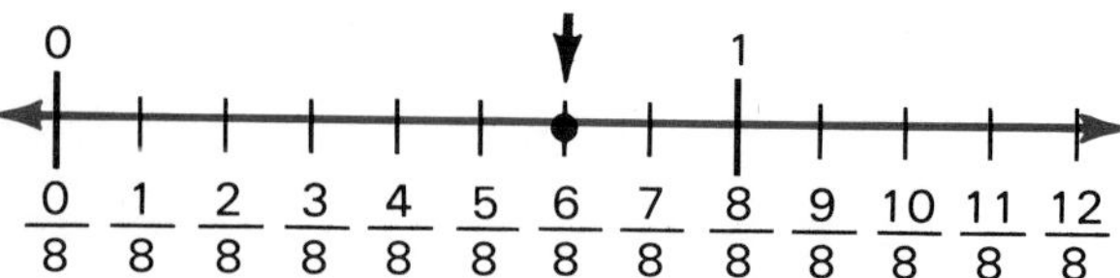

GETTING READY

Draw a diagram to show 2 equivalent fractions.

1. $\frac{1}{3}$ **2.** $\frac{1}{4}$ **3.** $\frac{2}{5}$ **4.** $\frac{5}{6}$

Write 2 fractions for the part that is shaded.

5.

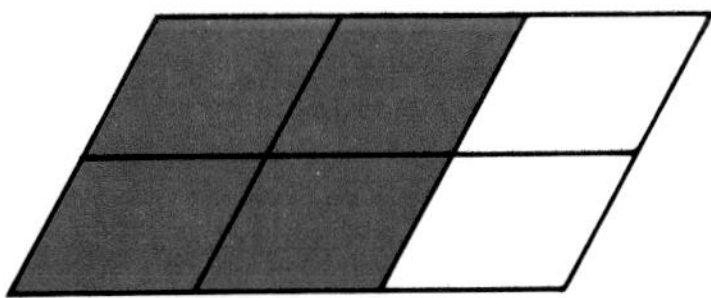

6.

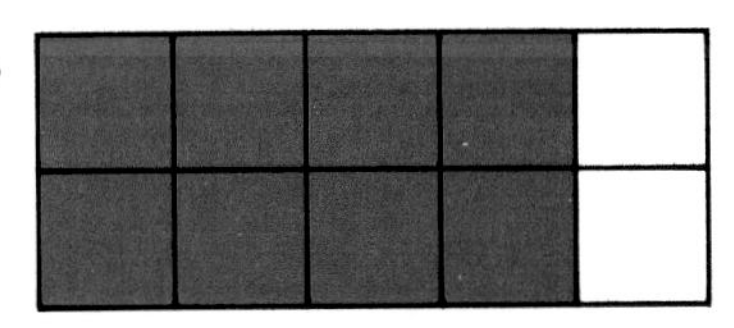

Draw a number line diagram. Tell whether the sentence is true.

7. $\frac{1}{2} = \frac{3}{6}$ **8.** $\frac{3}{5} = \frac{6}{10}$ **9.** $\frac{2}{3} = \frac{5}{6}$ **10.** $\frac{3}{4} = \frac{6}{8}$

EXERCISES

Draw a diagram to show 2 equivalent fractions.

1. $\frac{3}{4}$ **2.** $\frac{4}{5}$ **3.** $\frac{1}{6}$ **4.** $\frac{7}{8}$

Write 2 fractions for the part that is shaded.

5.

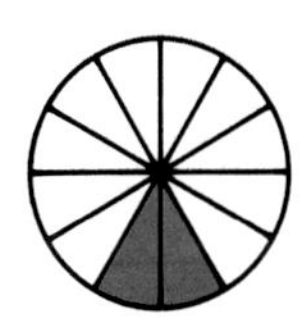

6.

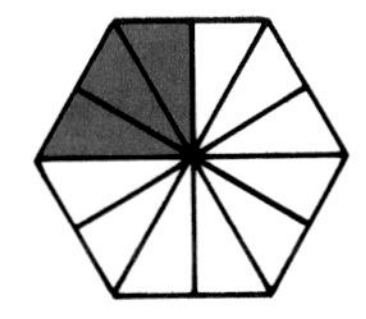

Draw a number line diagram. Tell whether the sentence is true.

7. $\frac{2}{5} = \frac{4}{10}$ **8.** $\frac{2}{8} = \frac{1}{4}$ **9.** $\frac{4}{5} = \frac{2}{3}$ **10.** $\frac{1}{5} = \frac{3}{10}$

11. $\frac{2}{2} = \frac{4}{4}$ **12.** $\frac{5}{5} = \frac{1}{1}$ **13.** $\frac{1}{2} = \frac{4}{10}$ **14.** $\frac{3}{5} = \frac{7}{10}$

15. $\frac{4}{12} = \frac{1}{3}$ **16.** $\frac{2}{3} = \frac{6}{9}$ **17.** $\frac{4}{8} = \frac{1}{2}$ **18.** $\frac{3}{3} = \frac{6}{6}$

19. $\frac{2}{4} = \frac{4}{9}$ **20.** $\frac{6}{12} = \frac{5}{10}$ **21.** $\frac{18}{24} = \frac{6}{8}$ **22.** $\frac{20}{25} = \frac{8}{10}$

Return to Form B on page 185.

FINDING EQUIVALENT FRACTIONS

To find an equivalent fraction with a given denominator

- Multiply the numerator and denominator by the same number.
- Divide the numerator and denominator by the same number.

EXAMPLE 1

Write 4 fractions equivalent to $\frac{2}{5}$.

Multiply the numerator and the denominator by the same number. Any number but 0 can be used. Notice that 2, 3, 4, and 5 have been used.

$$\frac{2 \times 2}{5 \times 2} = \frac{4}{10}$$

$$\frac{2 \times 3}{5 \times 3} = \frac{6}{15}$$

$$\frac{2 \times 4}{5 \times 4} = \frac{8}{20}$$

$$\frac{2 \times 5}{5 \times 5} = \frac{10}{25}$$

So, $\frac{2}{5}$, $\frac{4}{10}$, $\frac{6}{15}$, $\frac{8}{20}$, and $\frac{10}{25}$ are equivalent fractions.

> In a fraction, the top number is the numerator. The bottom number is the denominator.
>
> **Example:**
>
> 1 ⟵ numerator
> 2 ⟵ denominator

EXAMPLE 2

Complete. $\frac{1}{2} = \frac{?}{8}$

Since $2 \times 4 = 8$, 4 is the only multiplier that will give you 8.

$$\frac{1}{2} = \frac{1 \times 4}{2 \times 4} = \frac{4}{8}$$

So, $\frac{1}{2} = \frac{4}{8}$.

EXAMPLE 3

Write 4 fractions equivalent to $\frac{24}{36}$.

Divide the numerator and the denominator by the same number. Notice that 2, 3, 6, and 12 have been used.

$$\frac{24 \div 2}{36 \div 2} = \frac{12}{18} \qquad \frac{24 \div 3}{36 \div 3} = \frac{8}{12}$$

$$\frac{24 \div 6}{36 \div 6} = \frac{4}{6} \qquad \frac{24 \div 12}{36 \div 12} = \frac{2}{3}$$

So, $\frac{24}{36}$, $\frac{12}{18}$, $\frac{8}{12}$, $\frac{4}{6}$, and $\frac{2}{3}$ are equivalent fractions.

EXAMPLE 4

Complete. $\frac{15}{20} = \frac{?}{4}$

Since $20 \div 5 = 4$, 5 is the only divisor that will give you 4.

$$\frac{15}{20} = \frac{15 \div 5}{20 \div 5} = \frac{3}{4}$$

So, $\frac{15}{20} = \frac{3}{4}$.

GETTING READY

Write 4 equivalent fractions.

1. $\frac{1}{3}$ **2.** $\frac{2}{5}$ **3.** $\frac{3}{4}$ **4.** $\frac{1}{10}$

Complete.

5. $\frac{1}{6} = \frac{?}{18}$ **6.** $\frac{3}{4} = \frac{?}{20}$ **7.** $\frac{7}{8} = \frac{?}{40}$ **8.** $\frac{5}{9} = \frac{?}{54}$ **9.** $\frac{7}{10} = \frac{?}{100}$

Write 4 equivalent fractions. Use only division.

10. $\frac{16}{80}$ **11.** $\frac{20}{100}$ **12.** $\frac{40}{120}$ **13.** $\frac{48}{72}$

Complete.

14. $\frac{90}{100} = \frac{?}{10}$ **15.** $\frac{15}{30} = \frac{?}{2}$ **16.** $\frac{15}{18} = \frac{?}{6}$ **17.** $\frac{42}{48} = \frac{?}{8}$ **18.** $\frac{32}{72} = \frac{?}{9}$

EXERCISES

Write 4 equivalent fractions.

1. $\frac{1}{6}$ **2.** $\frac{2}{9}$ **3.** $\frac{7}{10}$ **4.** $\frac{3}{8}$

Complete.

5. $\frac{1}{2} = \frac{?}{18}$ **6.** $\frac{1}{3} = \frac{?}{9}$ **7.** $\frac{3}{4} = \frac{?}{24}$ **8.** $\frac{2}{3} = \frac{?}{24}$ **9.** $\frac{1}{8} = \frac{?}{72}$

10. $\frac{1}{6} = \frac{?}{24}$ **11.** $\frac{3}{5} = \frac{?}{40}$ **12.** $\frac{5}{9} = \frac{?}{27}$ **13.** $\frac{3}{8} = \frac{?}{40}$ **14.** $\frac{7}{10} = \frac{?}{50}$

Write 4 equivalent fractions. Use only division.

15. $\frac{48}{64}$ **16.** $\frac{54}{108}$ **17.** $\frac{60}{100}$ **18.** $\frac{80}{120}$

Complete.

19. $\frac{7}{63} = \frac{?}{9}$ **20.** $\frac{7}{21} = \frac{?}{3}$ **21.** $\frac{14}{35} = \frac{?}{5}$ **22.** $\frac{5}{25} = \frac{?}{5}$ **23.** $\frac{20}{24} = \frac{?}{6}$

24. $\frac{63}{54} = \frac{?}{18}$ **25.** $\frac{50}{80} = \frac{?}{8}$ ***26.** $\frac{48}{56} = \frac{6}{?}$ ***27.** $\frac{36}{48} = \frac{3}{?}$ ***28.** $\frac{90}{100} = \frac{9}{?}$

29. Len baked a cake and gave $\frac{1}{2}$ to Jed. The cake had been cut into 16 pieces. Jed got how many pieces out of 16?

SIMPLIFYING FRACTIONS

To write the simplest form of a fraction

- Divide the numerator and denominator by the same largest factor.

EXAMPLE 1

Write $\frac{4}{8}$ in simplest form.

4 is the largest factor that divides 4 and 8.

$$\frac{4 \div 4}{8 \div 4} = \frac{1}{2}$$

So, $\frac{4}{8}$ is $\frac{1}{2}$ in simplest form.

> A **factor** of a number divides it evenly with 0 as the remainder.
> **Example:** 3 is a factor of 12, since $12 \div 3 = 4$, with 0 remainder.

EXAMPLE 2

Simplify $\frac{24}{36}$.

12 is the largest factor that divides 24 and 36.

$$\frac{24 \div 12}{36 \div 12} = \frac{2}{3}$$

So, $\frac{24}{36}$ is $\frac{2}{3}$ in the simplest form.

EXAMPLE 3

Simplify $\frac{9}{16}$.

1 is the largest factor that divides 9 and 16.

So, $\frac{9}{16}$ is in simplest form.

> When the largest factor that divides both numerator and denominator is 1, the fraction is in simplest form.

GETTING READY

Simplify.

1. $\frac{2}{3}$ **2.** $\frac{5}{15}$ **3.** $\frac{4}{12}$ **4.** $\frac{5}{10}$

5. $\frac{21}{28}$ **6.** $\frac{15}{24}$ **7.** $\frac{12}{15}$ **8.** $\frac{9}{21}$

EXERCISES

Simplify.

1. $\frac{3}{9}$	**2.** $\frac{4}{6}$	**3.** $\frac{5}{20}$	**4.** $\frac{6}{12}$	**5.** $\frac{8}{15}$	**6.** $\frac{10}{20}$
7. $\frac{20}{28}$	**8.** $\frac{12}{27}$	**9.** $\frac{14}{35}$	**10.** $\frac{10}{12}$	**11.** $\frac{8}{12}$	**12.** $\frac{4}{10}$
13. $\frac{9}{15}$	**14.** $\frac{4}{24}$	**15.** $\frac{4}{14}$	**16.** $\frac{15}{20}$	**17.** $\frac{4}{16}$	**18.** $\frac{40}{60}$
19. $\frac{10}{14}$	**20.** $\frac{8}{32}$	**21.** $\frac{60}{80}$	**22.** $\frac{10}{30}$	**23.** $\frac{13}{26}$	**24.** $\frac{6}{36}$
25. $\frac{9}{24}$	**26.** $\frac{15}{65}$	**27.** $\frac{20}{22}$	**28.** $\frac{12}{90}$	**29.** $\frac{54}{63}$	**30.** $\frac{48}{72}$

Solve.

31. In a class of 28 students, 21 play musical instruments. What is the fraction in simplest form for the part of the class that plays musical instruments?

***32.** Steve Gold gave away 3 of his 12 new cassette tapes. Write the fraction in simplest form for the part of the tapes that he kept.

ON YOUR OWN

A message is hidden below. Follow these directions. Find the fraction of the word in each problem. Write the letters in order in the boxes at the bottom of the page. The first two letters have been done for you.

1. the first $\frac{1}{3}$ of *France*

2. the last $\frac{3}{4}$ of *fact*

3. the last $\frac{4}{9}$ of *solutions*

4. the first $\frac{1}{4}$ of *artistic*

5. the last $\frac{1}{4}$ of *love*

6. the first $\frac{3}{5}$ of *frame*

7. the last $\frac{2}{3}$ of *act*

8. the middle $\frac{1}{3}$ of *tub*

9. $\frac{3}{3}$ of *red*

F	R							

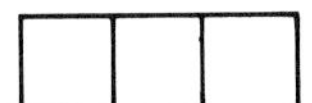

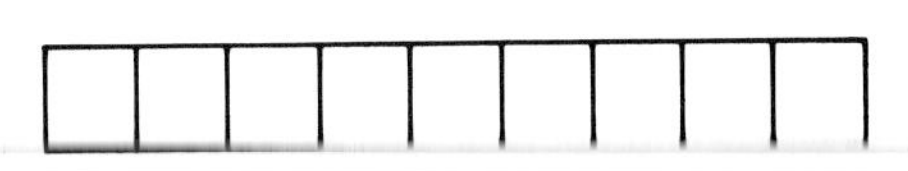

COMPARING FRACTIONS

To compare two fractions

- With like denominators, compare numerators.
- With unlike denominators, find equivalent fractions with like denominators. Then compare.

EXAMPLE 1

Compare $\frac{4}{7}$ and $\frac{6}{7}$.

The denominators are the same. $\quad \frac{4}{7} \quad \frac{6}{7}$

Compare numerators. $\quad \mathbf{4 < 6}$

So, $\frac{4}{7} < \frac{6}{7}$.

> $<$ means **is less than,** and $>$ means **is greater than.** The symbol always points to the smaller number. $6 > 4$ and $4 < 6$.

EXAMPLE 2

Compare $\frac{5}{6}$ and $\frac{3}{4}$.

Change to equivalent fractions with like denominators. $\quad \frac{5 \times 2}{6 \times 2} = \frac{10}{12}, \quad \frac{3 \times 3}{4 \times 3} = \frac{9}{12}$

Compare numerators. $\quad$ Since $\mathbf{10 > 9}$, $\frac{10}{12} > \frac{9}{12}$. So, $\frac{5}{6} > \frac{3}{4}$.

GETTING READY

Compare. Use $<$ or $>$.

1. $\frac{3}{4}, \frac{1}{4}$ **2.** $\frac{5}{8}, \frac{7}{8}$ **3.** $\frac{1}{2}, \frac{1}{3}$ **4.** $\frac{5}{8}, \frac{3}{4}$

EXERCISES

Compare. Use $<$ or $>$.

1. $\frac{2}{3}, \frac{1}{3}$ **2.** $\frac{2}{4}, \frac{3}{4}$ **3.** $\frac{3}{5}, \frac{4}{5}$ **4.** $\frac{3}{6}, \frac{5}{6}$ **5.** $\frac{8}{10}, \frac{7}{10}$

6. $\frac{1}{2}, \frac{3}{5}$ **7.** $\frac{5}{7}, \frac{3}{6}$ **8.** $\frac{8}{9}, \frac{4}{5}$ **9.** $\frac{2}{3}, \frac{7}{9}$ **10.** $\frac{3}{5}, \frac{8}{10}$

11. $\frac{7}{10}, \frac{4}{8}$ **12.** $\frac{5}{9}, \frac{1}{3}$ **13.** $\frac{4}{5}, \frac{7}{10}$ **14.** $\frac{4}{5}, \frac{79}{100}$ **15.** $\frac{7}{10}, \frac{15}{20}$

MAKING MATH COUNT

MILEAGE CHART

	Atlanta	Charlotte	Chicago	Dallas	Memphis	Miami	Norfolk	Omaha
Birmingham	150	391	642	645	246	754	708	898
Charleston	289	210	877	1072	660	590	440	1266
Denver	1398	1580	996	781	1040	2184	1766	537
El Paso	1415	1642	1430	620	1072	1947	1953	1007
Little Rock	509	778	640	314	138	1161	1023	590
Memphis	371	630	530	452	—	1064	881	652
Phoenix	1793	2080	1713	998	1442	2348	2349	1290

EXAMPLE

The Parker family lives in Denver. Soon they will move to Dallas. How many miles is it from Denver to Dallas?

A 1,072 mi **B** 996 mi
C 781 mi **D** 620 mi

Look at the chart. Find Denver in the row on the left. Find the column marked Dallas. Read the number where the row and column meet.
So, **C** is the correct answer.

Choose the correct answer.

1. Mrs. Baynes plans to drive from El Paso to Omaha. How far is it?
A 537 mi **B** 590 mi
C 652 mi **D** 1,007 mi

2. A bus is going from Atlanta to Little Rock. How far is it to Little Rock?
A 371 mi **B** 509 mi
C 1,415 mi **D** 1,434 mi

***3.** If Arnold drives from Birmingham to Memphis and then to Chicago, how many miles is the entire trip?
A 246 mi **B** 776 mi
C 778 mi **D** 1,713 mi

***4.** How much closer is it from Memphis to Miami than from Memphis to Phoenix?
A 378 mi **B** 1,064 mi
C 1,442 mi **D** 2,506 mi

Problem Solving Skill

To solve combination or ordering problems

- Use a tree diagram.

EXAMPLE 1

At the Washington School snack bar, sandwiches are made from 2 kinds of bread: rye or wheat. Students can choose from 3 kinds of filling: ham, cheese, or tuna. How many different kinds of sandwiches can they buy?

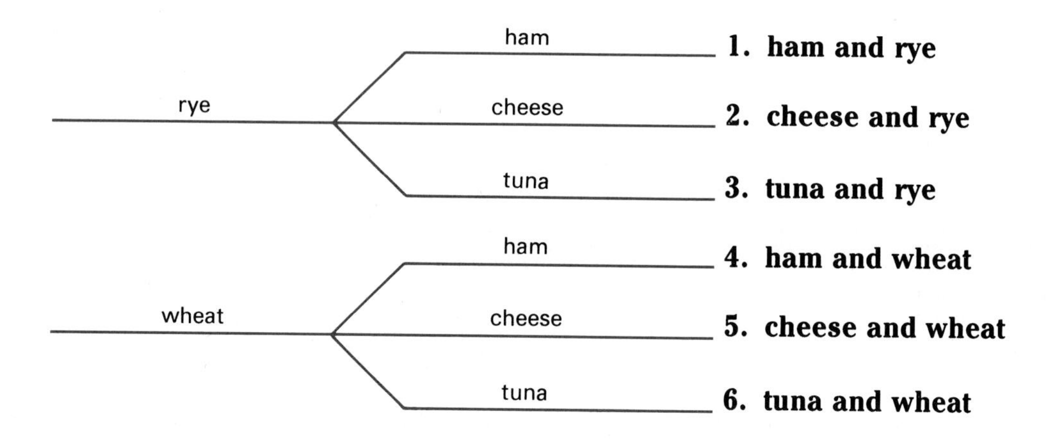

The tree diagram shows the possible sandwiches they can buy. Follow the top branch. This shows a ham sandwich on rye bread. Count the branches on the right. So, there are **6** possible sandwiches.

EXAMPLE 2

In order to save money and have more outfits, Otis buys clothes that can be mixed. This fall he bought 3 pairs of slacks: tan, grey, and blue; 2 jackets: tweed and navy; 2 shirts: white and yellow. Name 2 different outfits of slacks, jackets, and shirts that Otis might wear.

How many possible outfits are there?

Look at the tree diagram at the top of the next page. Count the branches on the right. There are 12 possible outfits. The top branch shows that Otis might wear the **tan slacks, tweed jacket,** and **white shirt;** or the **tan slacks, tweed jacket,** and **yellow shirt.**

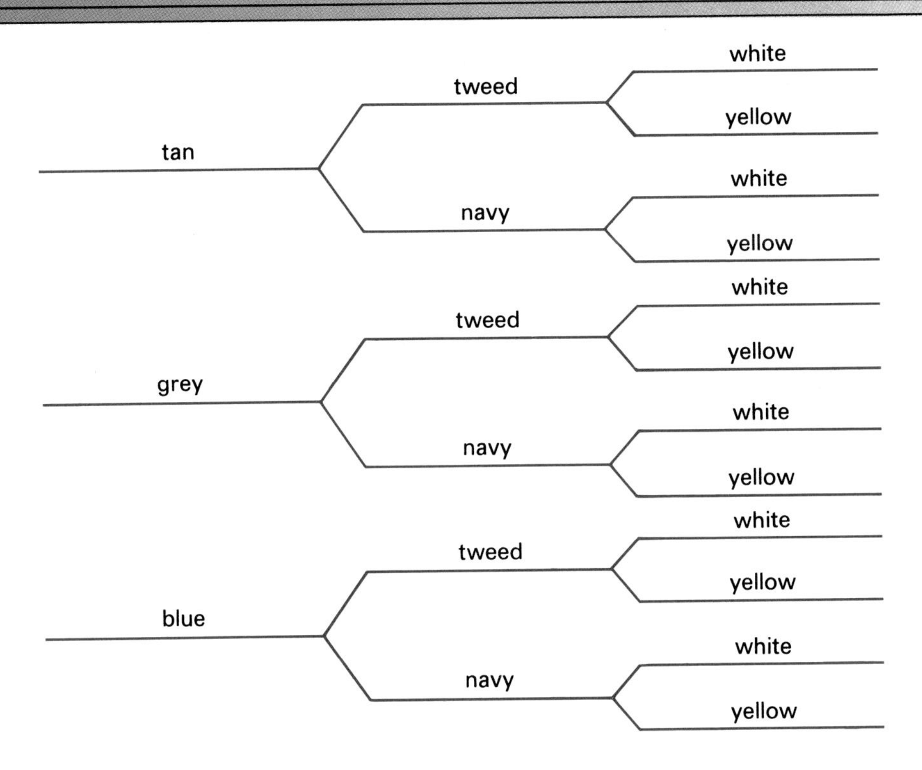

EXERCISES

Draw a tree diagram to solve.

1. A cold drink machine charges 40¢ for a canned drink and accepts only quarters, nickels, and dimes. If Sam has 1 quarter, 1 nickel and 1 dime, in how many ways can he put the coins into the machine to get a drink?

2. Karen has a red blouse and a white blouse that she can wear with three different skirts: navy, white, and red striped. But Karen does not like to wear red and navy together. How many different outfits will she really wear?

3. How many ways can 1, 3, 6, and 7 be arranged to make a four-digit number? What is the largest number that can be made? What is the smallest four-digit number possible?

*4. Four teams, the Jays, the Rockets, the Frogs, and the Lions, play in a neighborhood tournament. If a team loses, it is out of the competition. The Lions win 2 games, the Rockets win 1 game, and the Jays and the Frogs do not win any. The Jays do not play the Lions. What teams played each other and in what order? (Hint: Work the problem backwards.)

Problem Solving Applications

READ • PLAN • SOLVE • CHECK

Students in the electronics classes at Edison Vocational High School explore the basics of electricity and develop skills with power tools. They make simple electrical repairs as part of their training.

AS YOU READ

- amperage—strength of an electric current, as measured in amperes
- parallel circuit—a circuit with more than one path for the current
- voltage—force of an electric current, as measured in volts
- wattage—the rate by which current is used, as measured in watts

Answer these questions about the electronics class.

1. Randolph is checking the fuse box in the laboratory. There are four 125-volt fuses. One is burned out. What part of the total *voltage* is out?

2. Elaine has wired a simple *parallel circuit.* The green wire carries $\frac{3}{5}$ of the current, while the black wire carries $\frac{12}{30}$ of it. Express in fifths the part carried by the black wire.

3. Reggie has cut $\frac{3}{8}$ in. of insulation from a wire. His teacher recommends cutting $\frac{5}{16}$ in. Did he cut too much or too little?

4. A portable radio being repaired uses 4 batteries of equal voltage. If the total voltage is 6 volts, how many volts are supplied by each battery?

5. An alarm clock has a 0.025 *amperage* and a voltage of 120. What is the *wattage*? (Use voltage $\times$ amperage $=$ wattage.)

*6. Pat is calculating the amperage of a 60-watt light bulb. The voltage is 120. What is the amperage?

Career: Electronics Workers

READ • PLAN • SOLVE • CHECK

There are a variety of jobs in electronics. Small businesses as well as large manufacturers employ electronics workers. Training varies from vocational school to 4- or 5-year apprenticeships.

AS YOU READ

- coil—a series of loops of wire
- step-down (-up) transformer—a device that changes voltage down (or up)

Solve.

1. Tracy Conner is testing strings of decorative lights. Each string contains 30 lights made in a parallel circuit. If 4 bulbs do not light, what part of the total has to be replaced?

2. Wallace Humphrey has been an inspector with a technologies company for 4 y. He has been with the company for 18 y. What part of his total years with the company has he been an inspector?

3. To assemble receiving tubes, microscopes are used. If one microscope magnifies 50 times and another magnifies 300 times, by what factor is the second microscope more powerful?

4. Diane Hopper tests transistors. The current gain is equal to the change in current output divided by the current input. If the change is 196 and the input is 4, what is the gain?

5. In a *step-down transformer*, the number of turns in the primary *coil* is 120. The number of turns in the secondary coil is 6. By what factor has the secondary coil been reduced?

6. In a *step-up transformer*, the number of turns in the primary coil is 4. The number of turns in the secondary coil is 90. By what factor has the secondary coil been increased?

CHAPTER TEST 10

Write a fraction for the part that is not shaded.

1.

2. 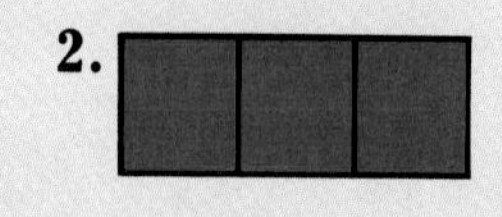

3. 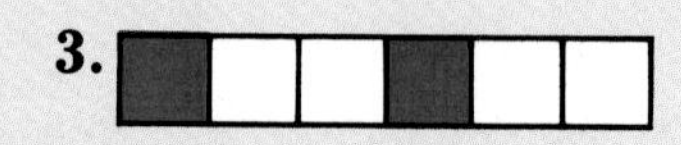

4. 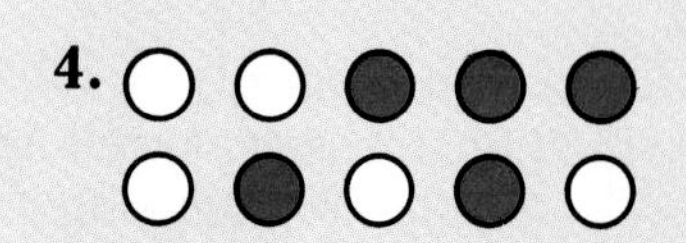

Write a fraction for the part of the group that is circled.

5. 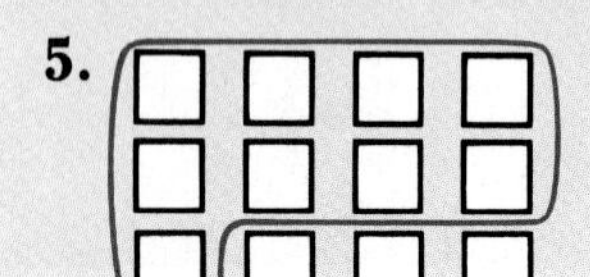

6. 

7. 

Write 2 fractions for the part that is shaded.

8. 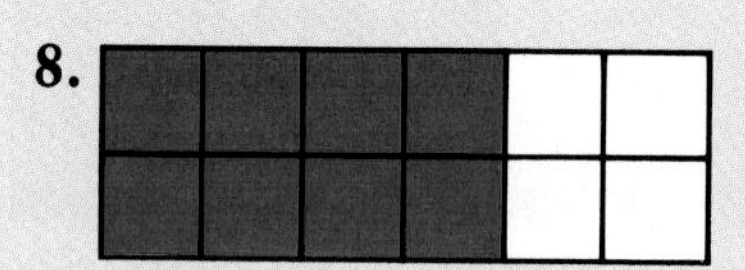

9. 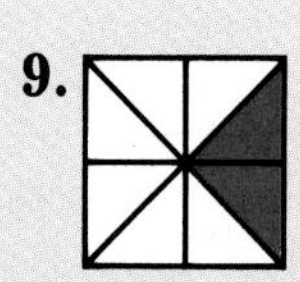

Draw a number line diagram. Tell whether the sentence is true.

10. $\frac{2}{6} = \frac{1}{3}$ **11.** $\frac{1}{2} = \frac{2}{4}$ **12.** $\frac{2}{5} = \frac{6}{12}$ **13.** $\frac{3}{7} = \frac{9}{21}$

Complete.

14. $\frac{2}{3} = \frac{?}{6}$ **15.** $\frac{5}{6} = \frac{?}{18}$ **16.** $\frac{10}{40} = \frac{?}{8}$ **17.** $\frac{75}{125} = \frac{?}{5}$

Simplify.

18. $\frac{6}{10}$ **19.** $\frac{7}{63}$ **20.** $\frac{9}{24}$ **21.** $\frac{12}{16}$

Compare. Use < or >.

22. $\frac{2}{3}, \frac{1}{3}$ **23.** $\frac{3}{10}, \frac{29}{100}$ **24.** $\frac{1}{3}, \frac{2}{5}$

Draw a tree diagram to solve.

25. On Mondays the Middle School cafeteria offers tomato, vegetable, or chicken soup; roast beef, cheese, or egg salad sandwiches; and orange juice or milk. How many combinations of soup, sandwich, and beverage are possible?

Addition and Subtraction of Fractions

MAKING MATH COUNT

October 27

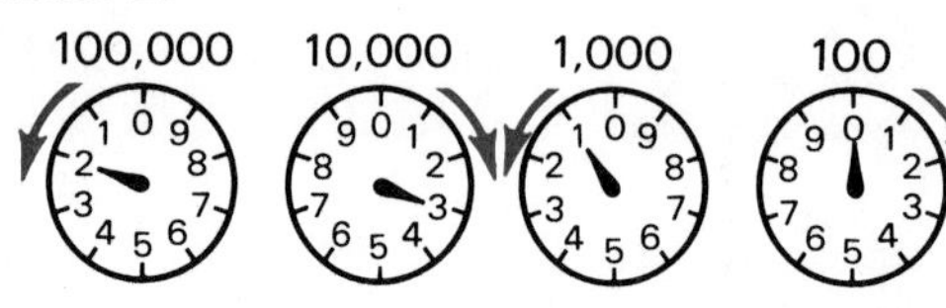

December 29

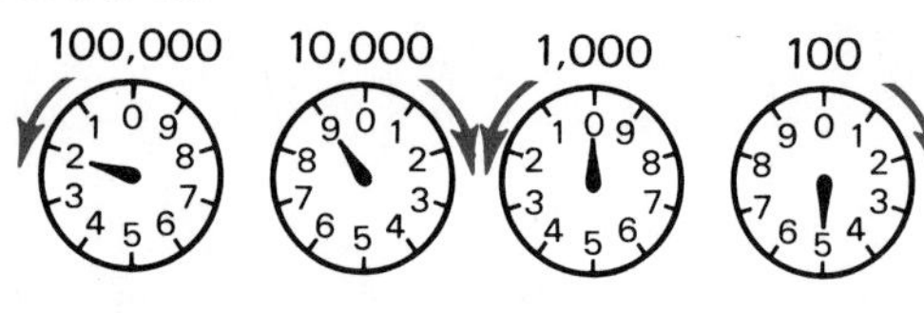

November 24

January 26

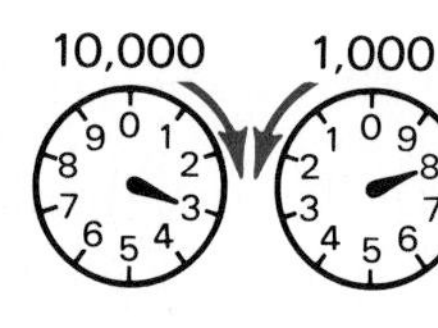

EXAMPLE

How many cubic feet (ft^3) of gas were used from October 27 to November 24?

A 274,000 ft^3 **B** 27,400 ft^3
C 2,740 ft^3 **D** 274 ft^3

Read the dials for October. **2310**
Read the dials for November. **2584**
Note that when the indicator is between 2 numbers, you read the smaller number.
Then subtract. **2584 − 2310 = 274**
Multiply by 100 to get the number of cubic feet. **274 × 100 = 27,400**
So, **B** is the correct answer.

Choose the correct answer.

1. What does the meter read on December 29?
A 205 **B** 915
C 2915 **D** 2905

2. How many cubic feet of gas were used from November 24 to December 29?
A 3,210 ft^3 **B** 3,310 ft^3
C 32,100 ft^3 **D** 33,100 ft^3

3. How many cubic feet of gas were used from December 29 to January 26?
A 49,500 ft^3 **B** 47,500 ft^3
C 4,950 ft^3 **D** 4,750 ft^3

***4.** If Mr. Ford pays \$0.18001 for 100 ft^3 of gas, how much would he pay for gas from December 29 to January 26?
A \$89.11 **B** \$89.10
C \$85.50 **D** \$85.51

DIAGNOSTIC TEST: FORM A (30 MINUTES)

Add. Simplify answers, when possible.

1. $\frac{2}{5} + \frac{1}{5}$ **2.** $\frac{5}{10} + \frac{2}{10}$ **3.** $\frac{1}{9} + \frac{5}{9}$ **4.** $\frac{5}{6} + \frac{1}{6}$ **5.** $\frac{5}{12} + \frac{7}{12}$

6. $\frac{21}{100} + \frac{69}{100}$ **7.** $\frac{37}{50} + \frac{11}{50}$ **8.** $\frac{16}{25} + \frac{4}{25}$ **9.** $\frac{7}{18} + \frac{5}{18}$ **10.** $\frac{11}{16} + \frac{3}{16}$

Subtract. Simplify answers, when possible.

11. $\frac{12}{13} - \frac{4}{13}$ **12.** $\frac{7}{8} - \frac{2}{8}$ **13.** $\frac{11}{12} - \frac{3}{12}$ **14.** $\frac{3}{4} - \frac{1}{4}$ **15.** $\frac{27}{100} - \frac{7}{100}$

16. $\frac{15}{17} - \frac{4}{17}$ **17.** $\frac{24}{25} - \frac{4}{25}$ **18.** $\frac{17}{18} - \frac{5}{18}$ **19.** $\frac{11}{16} - \frac{3}{16}$ **20.** $\frac{39}{50} - \frac{11}{50}$

19 or more right? Go to page 208. *Less than 19 right? Go to page 204.*

DIAGNOSTIC TEST: FORM B (30 MINUTES)

Add. Simplify answers, when possible.

1. $\frac{2}{7} + \frac{4}{7}$ **2.** $\frac{9}{16} + \frac{5}{16}$ **3.** $\frac{7}{10} + \frac{1}{10}$ **4.** $\frac{3}{8} + \frac{5}{8}$ **5.** $\frac{3}{5} + \frac{2}{5}$

6. $\frac{19}{100} + \frac{11}{100}$ **7.** $\frac{47}{50} + \frac{1}{50}$ **8.** $\frac{12}{25} + \frac{8}{25}$ **9.** $\frac{5}{18} + \frac{11}{18}$ **10.** $\frac{5}{16} + \frac{9}{16}$

Subtract. Simplify answers, when possible.

11. $\frac{9}{11} - \frac{3}{11}$ **12.** $\frac{17}{19} - \frac{2}{19}$ **13.** $\frac{5}{9} - \frac{2}{9}$ **14.** $\frac{7}{15} - \frac{1}{15}$ **15.** $\frac{7}{10} - \frac{2}{10}$

16. $\frac{81}{100} - \frac{53}{100}$ **17.** $\frac{21}{25} - \frac{6}{25}$ **18.** $\frac{11}{18} - \frac{5}{18}$ **19.** $\frac{13}{16} - \frac{1}{16}$ **20.** $\frac{47}{50} - \frac{27}{50}$

19 or more right? Go to page 208. *Less than 19 right? See your teacher.*

ADDING LIKE FRACTIONS

To add fractions with like denominators

- Add the numerators.
- Keep the same denominator.
- Simplify answers, when possible.

EXAMPLE 1

Draw a diagram to show $\frac{3}{7} + \frac{1}{7} = \frac{4}{7}$.

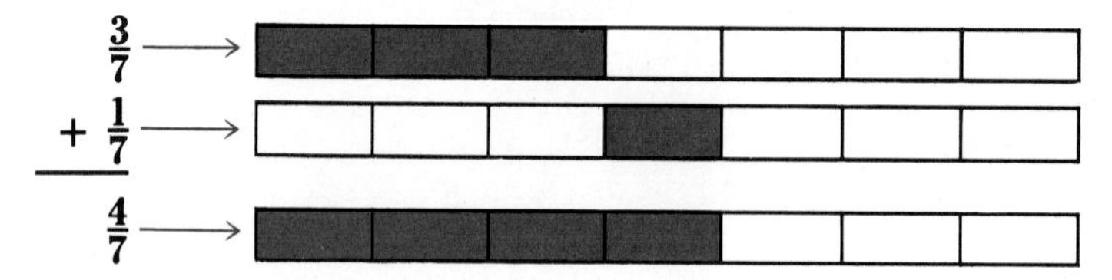

Like fractions are fractions that have the same denominator.

EXAMPLE 2

Add. $\frac{3}{5} + \frac{2}{5}$

Add the numerators. Keep the same denominator, 5.

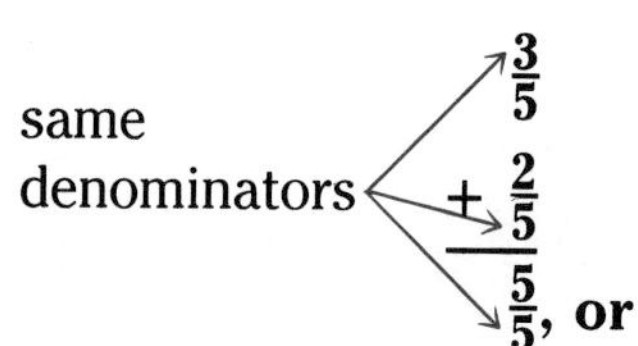

$$\begin{array}{r} \frac{3}{5} \\ + \frac{2}{5} \\ \hline \frac{5}{5}, \text{ or } 1 \end{array}$$

So, $\frac{3}{5} + \frac{2}{5} = \frac{5}{5}$, or **1**.

A fraction with the same numerator and denominator (except 0) is equal to 1.

EXAMPLE 3

Add. $\frac{1}{6} + \frac{2}{6}$

Add the numerators. Keep the same denominator, 6.

$$\begin{array}{r} \frac{1}{6} \\ + \frac{2}{6} \\ \hline \frac{3}{6} \end{array}$$

$\frac{3}{6}$ can be simplified. $\frac{3 \div 3}{6 \div 3} = \frac{1}{2}$

So, $\frac{1}{6} + \frac{2}{6} = \frac{3}{6}$, or $\mathbf{\frac{1}{2}}$.

GETTING READY

1. Draw a diagram to show $\frac{2}{5} + \frac{1}{5} = \frac{3}{5}$.

Add. Simplify answers, when possible.

2. $\frac{1}{3} + \frac{1}{3}$ **3.** $\frac{1}{10} + \frac{4}{10}$ **4.** $\frac{1}{6} + \frac{5}{6}$ **5.** $\frac{3}{16} + \frac{7}{16}$

EXERCISES

Draw a diagram to show that each sentence is true.

1. $\frac{4}{7} + \frac{2}{7} = \frac{6}{7}$ **2.** $\frac{1}{8} + \frac{3}{8} = \frac{1}{2}$ **3.** $\frac{1}{4} + \frac{3}{4} = 1$ **4.** $\frac{2}{9} + \frac{1}{9} = \frac{1}{3}$

Add. Simplify answers, when possible.

5. $\frac{5}{9} + \frac{2}{9}$ **6.** $\frac{1}{25} + \frac{2}{25}$ **7.** $\frac{7}{17} + \frac{8}{17}$ **8.** $\frac{7}{20} + \frac{10}{20}$ **9.** $\frac{1}{10} + \frac{7}{10}$ **10.** $\frac{3}{100} + \frac{7}{100}$

11. $\frac{9}{16} + \frac{5}{16}$ **12.** $\frac{1}{8} + \frac{5}{8}$ **13.** $\frac{3}{10} + \frac{2}{10}$ **14.** $\frac{5}{33} + \frac{7}{33}$ **15.** $\frac{2}{3} + \frac{1}{3}$ **16.** $\frac{8}{15} + \frac{7}{15}$

17. $\frac{3}{10} + \frac{5}{10}$ **18.** $\frac{2}{27} + \frac{4}{27}$ **19.** $\frac{1}{5} + \frac{4}{5}$ **20.** $\frac{7}{12} + \frac{3}{12}$

21. $\frac{4}{21} + \frac{8}{21}$ **22.** $\frac{13}{16} + \frac{3}{16}$ **23.** $\frac{15}{24} + \frac{7}{24}$ **24.** $\frac{2}{9} + \frac{7}{9}$

25. $\frac{3}{8} + \frac{4}{8}$ **26.** $\frac{7}{15} + \frac{2}{15}$ **27.** $\frac{9}{20} + \frac{7}{20}$ **28.** $\frac{5}{12} + \frac{6}{12}$

Solve.

29. A bolt is covered by a washer $\frac{3}{16}$ in. thick and by a nut $\frac{7}{16}$ in. thick. How much of the length of the bolt is covered?

30. Sam has two small electric motors. One is $\frac{1}{4}$ horsepower and one is $\frac{3}{4}$ horsepower. What is the total horsepower available if both motors are used?

ON YOUR OWN

Nine toothpicks form 3 triangles.

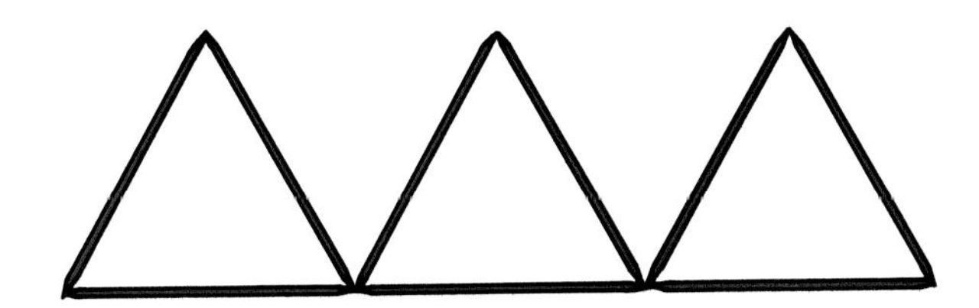

Move only 3 toothpicks to form 5 triangles.

SUBTRACTING LIKE FRACTIONS

To subtract fractions with like denominators

- Subtract the numerators.
- Keep the same denominator.
- Simplify answers, when possible.

EXAMPLE 1

Draw a diagram to show $\frac{4}{5} - \frac{2}{5} = \frac{2}{5}$.

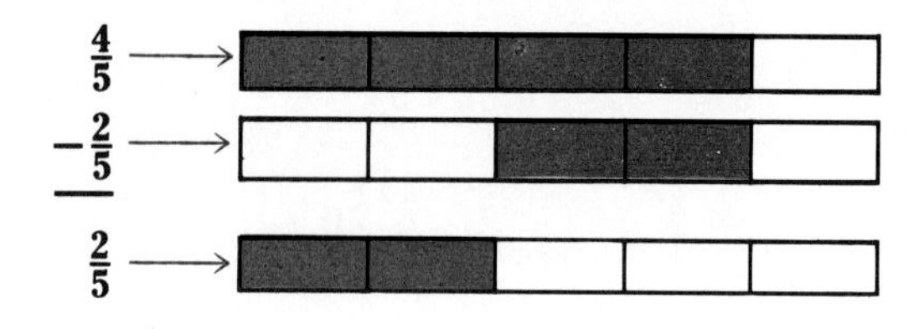

So, $\frac{4}{5} - \frac{2}{5} = \mathbf{\frac{2}{5}}$.

EXAMPLE 2

Subtract. $\frac{8}{9} - \frac{2}{9}$

The denominators are the same.
Subtract the numerators.

$$\begin{array}{r} \frac{8}{9} \\ -\frac{2}{9} \\ \hline \frac{6}{9} \end{array}$$

$\frac{6}{9}$ can be simplified. $\mathbf{\frac{6 \div 3}{9 \div 3} = \frac{2}{3}}$

So, $\frac{8}{9} - \frac{2}{9} = \frac{6}{9}$, or $\mathbf{\frac{2}{3}}$.

EXAMPLE 3

Subtract. $\frac{47}{100} - \frac{22}{100}$

$$\begin{array}{r} \frac{47}{100} \\ -\ \frac{22}{100} \\ \hline \frac{25}{100} \end{array}$$

$\frac{25}{100}$ can be simplified. $\mathbf{\frac{25 \div 25}{100 \div 25} = \frac{1}{4}}$

So, $\frac{47}{100} - \frac{22}{100} = \frac{25}{100}$, or $\frac{1}{4}$.

GETTING READY

1. Draw a diagram to show $\frac{5}{7} - \frac{2}{7} = \frac{3}{7}$.

Subtract. Simplify answers, when possible.

2. $\frac{5}{11} - \frac{2}{11}$
3. $\frac{7}{10} - \frac{7}{10}$
4. $\frac{5}{6} - \frac{1}{6}$
5. $\frac{9}{10} - \frac{3}{10}$

EXERCISES

Draw a diagram to show that each sentence is true.

1. $\frac{2}{3} - \frac{1}{3} = \frac{1}{3}$
2. $\frac{3}{5} - \frac{1}{5} = \frac{2}{5}$
3. $\frac{3}{4} - \frac{1}{4} = \frac{1}{2}$
4. $\frac{8}{9} - \frac{5}{9} = \frac{1}{3}$

Subtract. Simplify answers, when possible.

5. $\frac{9}{11} - \frac{3}{11}$
6. $\frac{8}{9} - \frac{4}{9}$
7. $\frac{7}{12} - \frac{5}{12}$
8. $\frac{11}{24} - \frac{7}{24}$
9. $\frac{1}{2} - \frac{1}{2}$
10. $\frac{57}{100} - \frac{33}{100}$
11. $\frac{7}{10} - \frac{3}{10}$
12. $\frac{13}{16} - \frac{5}{16}$
13. $\frac{17}{22} - \frac{13}{22}$
14. $\frac{13}{15} - \frac{4}{15}$
15. $\frac{5}{6} - \frac{3}{6}$
16. $\frac{5}{9} - \frac{5}{9}$
17. $\frac{9}{14} - \frac{3}{14}$
18. $\frac{20}{21} - \frac{5}{21}$
19. $\frac{15}{18} - \frac{3}{18}$
20. $\frac{13}{20} - \frac{7}{20}$
21. $\frac{19}{30} - \frac{4}{30}$
22. $\frac{21}{25} - \frac{16}{25}$
23. $\frac{19}{30} - \frac{7}{30}$
24. $\frac{7}{8} - \frac{3}{8}$
25. $\frac{59}{100} - \frac{49}{100}$

Solve.

26. Henry is making chili. The recipe calls for $\frac{3}{4}$ lb of ground beef. He has $\frac{1}{4}$ lb. How much more does he need?

27. Juanita lives $\frac{7}{10}$ mi from school, and Mike lives $\frac{5}{10}$ mi from school. How much farther does Juanita live from school than Mike?

*28. Shawn ate $\frac{3}{8}$ of a pizza. How much pizza remained?

Return to Form B on page 203.

IMPROPER FRACTIONS AND MIXED NUMBERS

To write an improper fraction as a mixed number
To write a mixed number as an improper fraction

A fraction can be expressed as a division problem. $\frac{4}{3}$ means $4 \div 3$.

EXAMPLE 1

Write $\frac{23}{5}$ as a mixed number.

$\frac{23}{5}$ is an improper fraction. You can rename an improper fraction as a mixed number by dividing.

$$23 \div 5, \text{ or } \begin{array}{r} 4 \\ 5\overline{)23} \\ \underline{20} \\ 3 \end{array}$$

> An **improper fraction** is a fraction in which the numerator is greater than the denominator.

The quotient 4 r 3 can be written as $4\frac{3}{5}$ ← **remainder** (3), ← **divisor** (5)

So, $\frac{23}{5} = 4\frac{3}{5}$.

EXAMPLE 2

Write $2\frac{4}{9}$ as an improper fraction.

Multiply the denominator by the whole number. $9 \times 2 = 18$

Add the numerator. Keep the same denominator. $18 + 4 = 22$

So, $2\frac{4}{9} = \frac{22}{9}$.

> A **mixed number** is the sum of a whole number and a fraction.

GETTING READY

Write as mixed numbers.

1. $\frac{9}{2}$ **2.** $\frac{13}{7}$ **3.** $\frac{29}{6}$

Write as improper fractions.

4. $1\frac{3}{5}$ **5.** $3\frac{5}{7}$ **6.** $19\frac{7}{8}$

EXERCISES

Write as mixed numbers.

1. $\frac{8}{5}$ **2.** $\frac{13}{8}$ **3.** $\frac{19}{2}$ **4.** $\frac{18}{3}$ **5.** $\frac{19}{9}$ **6.** $\frac{27}{12}$

Write as improper fractions.

7. $1\frac{2}{3}$ **8.** $1\frac{1}{16}$ **9.** $2\frac{3}{7}$ **10.** $6\frac{2}{3}$ **11.** $8\frac{3}{10}$ **12.** $7\frac{5}{9}$

ADDING FRACTIONS AND MIXED NUMBERS

To add fractions and mixed numbers

- Add the fractional parts.
- Add any whole number parts.
- Simplify answers, when possible.

EXAMPLE 1

Add. $\frac{5}{6} + \frac{4}{6}$

Add the like fractions. $\mathbf{\frac{5}{6} + \frac{4}{6} = \frac{9}{6}}$

Simplify $\frac{9}{6}$. $\mathbf{\frac{9}{6} = \frac{9 \div 3}{6 \div 3} = \frac{3}{2}}$ So, $\frac{5}{6} + \frac{4}{6} = \frac{3}{2}$, or $\mathbf{1\frac{1}{2}}$.

EXAMPLE 2

Add. $1\frac{7}{8} + 2\frac{5}{8}$

Add the fractional parts. $\mathbf{1\frac{7}{8}}$

Add the whole number parts. $\mathbf{+\ 2\frac{5}{8}}$

Change $\frac{12}{8}$ to a mixed number. $\mathbf{3\frac{12}{8} = 3 + 1\frac{4}{8}}$

Simplify. $\mathbf{= 4\frac{4}{8}}$**, or** $\mathbf{4\frac{1}{2}}$ So, $1\frac{7}{8} + 2\frac{5}{8} = \mathbf{4\frac{1}{2}}$.

GETTING READY

Add. Simplify answers, when possible.

1. $\frac{3}{5} + \frac{4}{5}$

2. $\frac{4}{10} + 1\frac{1}{10}$

3. $\frac{5}{8} + 2\frac{1}{8}$

4. $3\frac{5}{9} + 4\frac{2}{9}$

5. $6\frac{11}{15} + 2\frac{8}{15}$

EXERCISES

Add. Simplify answers, when possible.

1. $\frac{11}{20} + \frac{13}{20}$

2. $\frac{77}{100} + \frac{32}{100}$

3. $5\frac{1}{4} + 6\frac{1}{4}$

4. $4\frac{1}{2} + 3\frac{1}{2}$

5. $2\frac{9}{10} + \frac{3}{10}$

6. $3\frac{6}{21} + 5\frac{20}{21}$

7. $9\frac{13}{24} + 2\frac{17}{24}$

8. $11\frac{3}{10} + 7\frac{8}{10}$

9. $2\frac{5}{8} + 11\frac{7}{8}$

10. $8\frac{4}{15} + 10\frac{14}{15}$

FINDING THE LEAST COMMON DENOMINATOR

To write equivalent fractions with the least common denominator

- Use multiples to find the LCD.
- Write each fraction using the LCD as the new denominator.

EXAMPLE 1

Write equivalent fractions with the LCD for $\frac{2}{4}$ and $\frac{3}{8}$.

The denominators are 4 and 8. 8, the larger denominator, is a multiple of 4, since $4 \times 2 = 8$.

$\frac{2}{4} = \frac{?}{8}$ $\qquad \frac{2 \times 2}{4 \times 2} = \frac{4}{8}$

So, $\frac{4}{8}$ and $\frac{3}{8}$ are the equivalent fractions.

Least Common Denominator (LCD) is the smallest number that is a common multiple of the denominators.

EXAMPLE 2

Write equivalent fractions with the LCD for $\frac{3}{4}$ and $\frac{2}{5}$.

Look at the multiples of the larger denominator, 5.

Multiples of 5
5, 10, 15, 20

Look at the multiples of 4.

Multiples of 4
4, 8, 12, 16, 20

20 is the first **common** multiple of 4 and 5.

Write equivalent fractions with a denominator of 20.

$\frac{3}{4} = \frac{?}{20}$ $\qquad \frac{3 \times 5}{4 \times 5} = \frac{15}{20}$

$\frac{2}{5} = \frac{?}{20}$ $\qquad \frac{2 \times 4}{5 \times 4} = \frac{8}{20}$

So, $\frac{15}{20}$ and $\frac{8}{20}$ are the equivalent fractions with the LCD.

EXAMPLE 3

Write equivalent fractions with the LCD for $\frac{1}{4}$, $\frac{2}{3}$, and $\frac{5}{9}$.
Since 9 is the largest denominator, list multiples of 9 until you find the smallest multiple that both 3 and 4 divide evenly, 9, 18, 27, 36, . . .
Both 3 and 4 divide 36. So, 36 is the smallest multiple of 3, 4, and 9.

Write equivalent fractions with a denominator of 36.

$\frac{1}{4} = \frac{?}{36}$ $\qquad \frac{1 \times 9}{4 \times 9} = \frac{9}{36}$

$\frac{2}{3} = \frac{?}{36}$ $\qquad \frac{2 \times 12}{3 \times 12} = \frac{24}{36}$

$\frac{5}{9} = \frac{?}{36}$ $\qquad \frac{5 \times 4}{9 \times 4} = \frac{20}{36}$

So, $\frac{9}{36}$, $\frac{24}{36}$, and $\frac{20}{36}$ are the equivalent fractions with the LCD.

GETTING READY

Write equivalent fractions with the LCD.

1. $\frac{1}{3}, \frac{4}{9}$ **2.** $\frac{3}{4}, \frac{1}{10}$ **3.** $\frac{5}{8}, \frac{7}{12}$ **4.** $\frac{1}{3}, \frac{5}{6}$

5. $\frac{3}{4}, \frac{5}{8}$ **6.** $\frac{1}{7}, \frac{3}{5}$ **7.** $\frac{1}{5}, \frac{3}{4}, \frac{1}{10}$ **8.** $\frac{1}{3}, \frac{3}{4}, \frac{4}{15}$

EXERCISES

Write equivalent fractions with the LCD.

1. $\frac{1}{2}, \frac{1}{4}$ **2.** $\frac{1}{3}, \frac{5}{6}$ **3.** $\frac{1}{2}, \frac{2}{5}$ **4.** $\frac{1}{3}, \frac{7}{9}$ **5.** $\frac{1}{4}, \frac{1}{6}$

6. $\frac{2}{5}, \frac{3}{4}$ **7.** $\frac{1}{3}, \frac{3}{5}$ **8.** $\frac{1}{6}, \frac{2}{9}$ **9.** $\frac{1}{4}, \frac{7}{10}$ **10.** $\frac{2}{7}, \frac{2}{3}$

11. $\frac{1}{2}, \frac{5}{8}$ **12.** $\frac{1}{6}, \frac{3}{8}$ **13.** $\frac{3}{5}, \frac{7}{10}$ **14.** $\frac{2}{3}, \frac{1}{8}$ **15.** $\frac{1}{3}, \frac{8}{15}$

16. $\frac{3}{4}, \frac{3}{8}$ **17.** $\frac{4}{5}, \frac{7}{10}$ **18.** $\frac{1}{6}, \frac{3}{4}$ **19.** $\frac{3}{4}, \frac{5}{16}$ **20.** $\frac{3}{10}, \frac{9}{20}$

21. $\frac{1}{3}, \frac{1}{4}, \frac{1}{12}$ **22.** $\frac{1}{4}, \frac{3}{20}, \frac{7}{20}$ **23.** $\frac{7}{8}, \frac{2}{3}, \frac{3}{4}$

24. $\frac{3}{4}, \frac{1}{8}, \frac{9}{10}$ **25.** $\frac{1}{7}, \frac{3}{14}, \frac{5}{21}$ **26.** $\frac{9}{100}, \frac{3}{10}, \frac{17}{1{,}000}$

Solve. Use equivalent fractions with the LCD.

27. On Monday, the rainfall was $\frac{5}{8}$ in. On Tuesday, the rainfall was $\frac{3}{4}$ in. On which day did it rain more?

28. Janet got $\frac{9}{10}$ of the problems correct on her math test. She got $\frac{11}{15}$ of the answers on her history test. On which test did she get a better grade?

ADDING UNLIKE FRACTIONS AND MIXED NUMBERS

To add fractions or mixed numbers

- Change to equivalent fractions with the LCD.
- Add the numerators and add any whole numbers.

EXAMPLE 1

Add. $\frac{1}{2} + \frac{3}{8}$

Change to equivalent fractions with the LCD.

Add the numerators.

$$\begin{array}{rcrcr} \frac{1}{2} & \longrightarrow & \frac{1 \times 4}{2 \times 4} & = & \frac{4}{8} \\ +\frac{3}{8} & \longrightarrow & & & +\frac{3}{8} \\ \hline & & & & \frac{7}{8} \end{array}$$

So, $\frac{1}{2} + \frac{3}{8} = \mathbf{\frac{7}{8}}$.

EXAMPLE 2

Add. $\frac{1}{2} + \frac{2}{3} + \frac{3}{4}$

LCD for 2, 3, and 4 is 12.

$$\begin{array}{rcrcr} \frac{1}{2} & \longrightarrow & \frac{1 \times 6}{2 \times 6} & = & \frac{6}{12} \\ \frac{2}{3} & \longrightarrow & \frac{2 \times 4}{3 \times 4} & = & \frac{8}{12} \\ +\frac{3}{4} & \longrightarrow & \frac{3 \times 3}{4 \times 3} & = & +\frac{9}{12} \\ \hline & & & & \frac{23}{12} \end{array}$$

So, $\frac{1}{2} + \frac{2}{3} + \frac{3}{4} = \mathbf{\frac{23}{22}}$, or $\mathbf{1\frac{11}{12}}$.

EXAMPLE 3

Add. $4\frac{5}{6} + 5\frac{3}{10}$

LCD for 6 and 10 is 30.

Add.

$$\begin{array}{rcr} 4\frac{5}{6} & \longrightarrow & 4\frac{25}{30} \\ +5\frac{3}{10} & \longrightarrow & +5\frac{9}{30} \\ \hline & & 9\frac{34}{30} \end{array} = 9 + 1\frac{4}{30}$$

$$9 + 1\frac{4}{30} = 10\frac{4}{30}.$$

Simplify $\frac{4}{30}$. $\frac{4 \div 2}{30 \div 2} = \frac{2}{15}$

So, $4\frac{5}{6} + 5\frac{3}{10} = \mathbf{10\frac{2}{15}}$.

GETTING READY

Add. Simplify answers, when possible.

1. $\frac{2}{3} + \frac{1}{4}$
2. $\frac{7}{9} + \frac{1}{3}$
3. $\frac{1}{4} + \frac{7}{10}$
4. $\frac{3}{8} + \frac{1}{2} + \frac{1}{4}$
5. $2\frac{2}{3} + 1\frac{1}{5}$
6. $4\frac{7}{12} + 6\frac{1}{3}$

EXERCISES

Add. Simplify answers, when possible.

1. $\frac{8}{15} + \frac{1}{3}$
2. $\frac{1}{8} + \frac{7}{16}$
3. $\frac{7}{6} + \frac{2}{5}$
4. $\frac{4}{5} + \frac{1}{4}$
5. $\frac{2}{5} + \frac{1}{6}$
6. $3\frac{3}{8} + 2\frac{2}{3}$
7. $5\frac{11}{20} + 4\frac{3}{8}$
8. $10\frac{2}{5} + 3\frac{3}{10}$
9. $3\frac{1}{2} + 1\frac{3}{8}$
10. $4\frac{1}{6} + 2 + 1\frac{1}{3}$
11. $2\frac{2}{3} + 7 + 1\frac{1}{4}$
12. $3\frac{2}{5} + 4\frac{1}{10} + \frac{1}{2}$
13. $2\frac{1}{3} + 8\frac{1}{9} + 9\frac{5}{9}$
14. $9\frac{2}{5} + 3\frac{2}{9}$
15. $8\frac{3}{10} + 1\frac{43}{100}$
16. $2\frac{2}{3} + 9\frac{5}{9} + 8\frac{2}{18}$
17. $4\frac{3}{8} + 7\frac{2}{5} + 8\frac{1}{2}$
18. $3\frac{1}{3} + 5\frac{3}{4} + 7\frac{5}{6}$

Solve.

19. Bob is making slacks and a vest. The pattern calls for $2\frac{1}{2}$ yd of material for the slacks and $\frac{3}{4}$ yd for the vest. How much material should he buy?

*20. Beth wants to put a fence around her garden. The sides of her garden measure $2\frac{1}{2}$ yd, $2\frac{1}{8}$ yd, $3\frac{1}{2}$ yd, and $2\frac{1}{16}$yd. How much fencing is needed?

SUBTRACTING UNLIKE FRACTIONS AND MIXED NUMBERS

To subtract fractions and mixed numbers

- Change to equivalent fractions with the LCD.
- Subtract the numerators and subtract any whole numbers.

EXAMPLE 1

Subtract. $\frac{5}{6} - \frac{7}{10}$

Change to equivalent fractions.

Subtract and simplify.

$$\frac{5}{6} \longrightarrow \frac{5 \times 5}{6 \times 5} = \frac{25}{30}$$

$$-\frac{7}{10} \longrightarrow \frac{7 \times 3}{10 \times 3} = -\frac{21}{30}$$

$$\frac{4}{30} = \frac{4 \div 2}{30 \div 2} = \frac{2}{15}$$

So, $\frac{5}{6} - \frac{7}{10} = \mathbf{\frac{2}{15}}$.

EXAMPLE 2

Subtract. $3 - 1\frac{2}{3}$

Rename 3. $3 \longrightarrow 2 + 1 \longrightarrow 2\frac{3}{3}$

Subtract. $-1\frac{2}{3} \longrightarrow -1\frac{2}{3}$

$$1\frac{1}{3}$$

So, $3 - 1\frac{2}{3} = \mathbf{1\frac{1}{3}}$.

EXAMPLE 3

Subtract. $5\frac{1}{4} - 1\frac{1}{3}$

Find the LCD. $5\frac{1}{4} \longrightarrow 5\frac{3}{12}$

Change to equivalent fractions. $-1\frac{1}{3} \longrightarrow -1\frac{4}{12}$

Rename 5. $5\frac{3}{12} \longrightarrow 4 + 1 + \frac{3}{12} \longrightarrow 4 + \frac{12}{12} + \frac{3}{12} \longrightarrow 4\frac{15}{12}$

Subtract. $-1\frac{4}{12} \longrightarrow -1\frac{4}{12}$

$$3\frac{11}{12}$$

So, $5\frac{1}{4} - 1\frac{1}{3} = \mathbf{3\frac{11}{12}}$.

GETTING READY

Subtract. Simplify answers, when possible.

1. $\frac{5}{6} - \frac{1}{4}$ **2.** $2\frac{1}{2} - 1\frac{2}{5}$ **3.** $5 - 3\frac{1}{4}$ **4.** $6\frac{1}{4} - 2\frac{3}{4}$ **5.** $8\frac{2}{3} - 3\frac{1}{6}$ **6.** $11\frac{1}{3} - 7\frac{3}{4}$

EXERCISES

Subtract. Simplify answers, when possible.

1. $\frac{5}{8} - \frac{1}{4}$ **2.** $\frac{7}{12} - \frac{1}{3}$ **3.** $\frac{2}{3} - \frac{3}{5}$ **4.** $\frac{5}{6} - \frac{5}{8}$ **5.** $16\frac{8}{9} - 13\frac{4}{9}$

6. $7\frac{20}{29} - 6\frac{12}{29}$ **7.** $8 - 4\frac{1}{3}$ **8.** $6 - 1\frac{3}{4}$ **9.** $9 - 6\frac{1}{6}$ **10.** $14 - 2\frac{4}{7}$

11. $7 - 2\frac{3}{5}$ **12.** $8 - 1\frac{7}{10}$ **13.** $6\frac{11}{12} - 1\frac{2}{3}$ **14.** $7\frac{3}{4} - 2\frac{1}{2}$ **15.** $6\frac{5}{6} - 2\frac{2}{3}$

16. $10\frac{1}{4} - 10\frac{1}{8}$ **17.** $7\frac{2}{5} - 2\frac{3}{10}$ **18.** $8\frac{5}{6} - 2\frac{1}{4}$ **19.** $14\frac{3}{8} - 8\frac{7}{8}$ **20.** $6\frac{1}{3} - 4\frac{2}{3}$

21. $4\frac{1}{4} - 2\frac{3}{4}$ **22.** $3\frac{1}{8} - 1\frac{7}{8}$ **23.** $8\frac{1}{6} - 3\frac{5}{6}$ **24.** $9\frac{1}{7} - 3\frac{6}{7}$ **25.** $6\frac{1}{4} - 3\frac{1}{2}$

26. $5\frac{1}{4} - 2\frac{7}{8}$ **27.** $8\frac{1}{6} - 2\frac{7}{8}$ **28.** $7\frac{1}{6} - 2\frac{1}{3}$ **29.** $18\frac{1}{5} - 5\frac{3}{10}$ **30.** $46\frac{2}{5} - 19\frac{5}{8}$

Solve.

31. A roll of wire fencing contains 100 ft of wire. If Jacob used a piece $23\frac{3}{4}$ ft long, how much was left on the roll?

***32.** The rain gauge in Josephine's back yard measured $1\frac{1}{2}$ in. yesterday morning. Today the gauge measured $2\frac{3}{4}$ in. What was the amount of rainfall since yesterday?

Problem Solving Skill

To read and interpret a pictograph

- Find the row you need.
- Look at the key for the number of items.
- Multiply by the number of picture symbols.

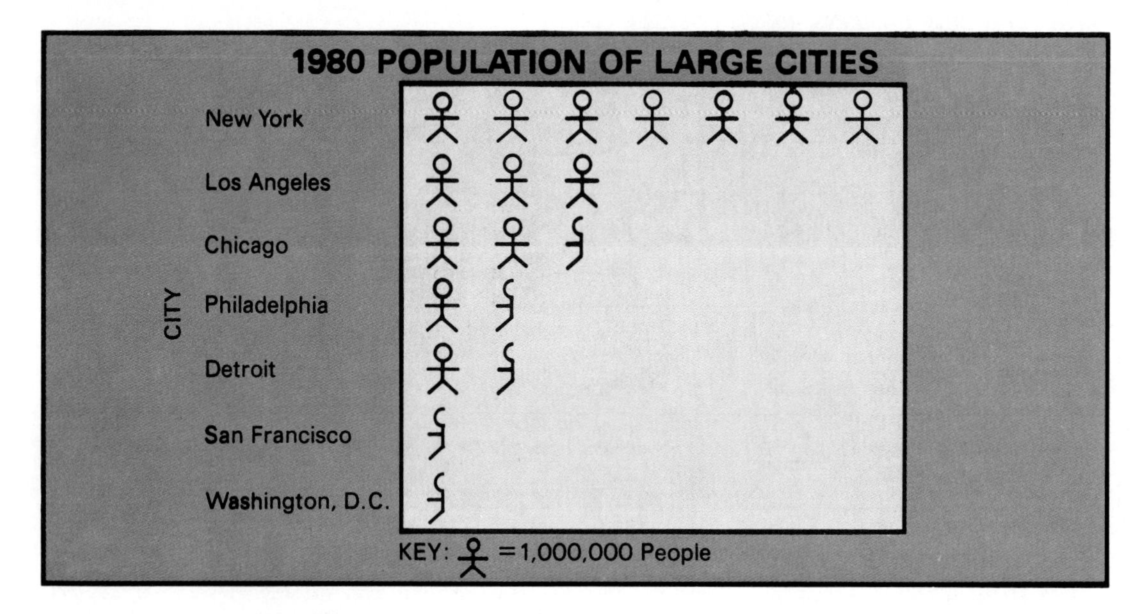

EXAMPLE

What is the population of Los Angeles?
The row for Los Angeles has 3 symbols. The key shows that each represents 1,000,000 people.
Multiply. **1,000,000 × 3 = 3,000,000.**
So, the population of Los Angeles is **3,000,000.**

Answer each question about the pictograph.

1. What is the title of the graph?

2. What does half a symbol stand for?

3. What is the population of New York?

4. Which city has the largest population?

5. What is the difference in population between New York and Chicago?

6. Which city has the second largest population?

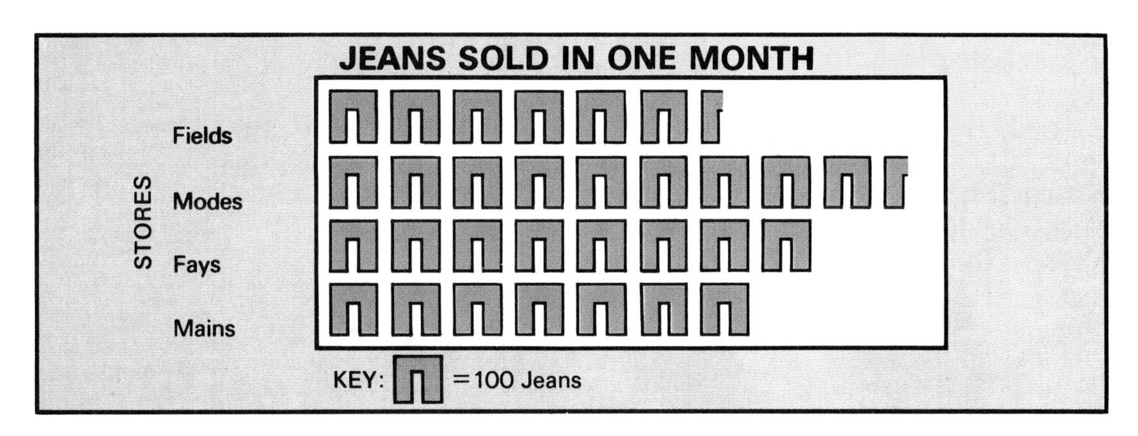

Answer each question about the pictograph.

1. How many jeans did Fays sell in one month?
2. How many jeans did Modes sell in one month?
3. Which store sold the most jeans in one month?
4. Which store sold the fewest jeans in one month?
5. How many more jeans did Fays sell than Mains?
6. How many fewer jeans did Fields sell than Modes?
7. Suppose the key is [symbol] = 50 jeans. Draw the pictograph row that represents the number of jeans sold at Mains.
8. Suppose the key is [symbol] = 200 jeans. Draw the pictograph row that represents the number of jeans sold at Fays.

CALCULATOR

Use your calculator.

1. Pick a four-digit number.
2. Add the next higher number.
3. Add 9 to this sum.
4. Divide by 2.
5. Subtract the original number.
6. The answer is __?__!
7. Try another number.
8. What is the pattern?

Problem Solving Applications

READ • PLAN • SOLVE • CHECK

Students in home economics classes at Thomaston High School study nutrition and cooking. They also learn about textiles and their care and about dressmaking and tailoring.

AS YOU READ

- calorie—a measure of the amount of energy in food
- food value—the amount of vitamins and minerals in food

Answer these questions about the home economics classes.

1. A certain vegetable loses $\frac{3}{20}$ of its *food value* during the first 20 min of cooking and then $\frac{1}{10}$ for each additional 15 min. If the vegetable is cooked 35 min, how much of its food value is lost?

2. Roscoe is making a three-piece suit. The pattern calls for $2\frac{3}{4}$ yd of material for the jacket, $2\frac{7}{8}$ yd for the pants, and $\frac{3}{4}$ yd for the vest. How much material should he buy for the three pieces?

3. The teacher said that girls 15 years old need 2,400 *calories* a day, while women 60 years old need 1,600 calories a day. How many more calories does the 15-year-old need than the 60-year-old?

4. Tim is using a washable wool blend to make slacks. He estimates that he will dry-clean the slacks 6 times a year. If it costs \$2.35 for dry cleaning, how much can he save by washing the slacks?

5. The class is having a party where they will serve natural foods. They bought sesame seeds for \$2.79, dates for \$2.69, soybeans for \$0.99, and sunflower seeds for \$1.99. Find the total cost.

6. Marcella is making four rectangular place mats. Each measures $17\frac{1}{2}$ in. long and $12\frac{1}{2}$ in. wide. How much seam binding does she need to trim the edges? (Draw a diagram.)

Career: Upholsterers and Tailors

READ • PLAN • SOLVE • CHECK

Upholsterers and tailors use measuring, cutting, and sewing skills. Some work in department stores, interior decorating shops, and ready-to-wear stores, while others own their own shops. Accuracy in measuring to ensure a minimum of waste and attention to detail are needed.

AS YOU READ

- alterations—changes made in clothing
- ottoman—a footstool

Solve.

1. Calvin Brown uses the following formula to determine the width of pleated drapes: 2 × the width of the window + 20 in. Use this formula to determine the drapery width for a window 80 in. wide.

2. Ernestine Evans charged the following for *alterations:* 2 hems for \$6.00 each; alterations on the sleeves for \$6.50; and alterations on the waist for \$7.50. What are the total charges?

3. Sally Stone is making custom drapes. If the drapery material costs \$7.88 a yard and the lining \$2.75 a yard, what is the total cost for 15 yd of each?

4. Ronnie Epstein reupholstered two chairs. The cost for material was \$83.50, sales tax was \$4.18, and labor was \$210.00. What were the total charges?

5. Deborah Vogel reupholsters furniture. Find the total amount of material needed for an overstuffed chair requiring $6\frac{3}{4}$ yd and an *ottoman* requiring $1\frac{1}{2}$ yd.

***6.** Jeff Lee is making the following alteration to let out the waist on slacks: $\frac{3}{8}$ in. in each of the 2 side seams and $\frac{5}{8}$ in. in the back seam. What is the new waist measurement if it was 38 in.?

CHAPTER TEST 11

Write as a mixed number or a whole number.

1. $\frac{7}{2}$ **2.** $\frac{18}{6}$ **3.** $\frac{35}{4}$ **4.** $\frac{83}{10}$ **5.** $\frac{19}{3}$

Write as an improper fraction.

6. $6\frac{2}{3}$ **7.** $8\frac{2}{5}$ **8.** $4\frac{3}{10}$ **9.** $10\frac{5}{6}$ **10.** $11\frac{5}{7}$

Add. Simplify answers, when possible.

11. $\frac{7}{10} + \frac{2}{10}$ **12.** $\frac{3}{4} + \frac{1}{12}$ **13.** $4\frac{8}{9} + 3\frac{7}{9}$ **14.** $8\frac{2}{3} + 7\frac{1}{6}$ **15.** $7\frac{3}{10} + 5\frac{7}{100}$

16. $\frac{7}{9} + \frac{2}{3} + \frac{1}{6}$ **17.** $8\frac{1}{4} + 6\frac{1}{3} + 9\frac{5}{12}$ **18.** $1\frac{5}{6} + 2\frac{7}{8} + 2\frac{1}{4}$

Subtract. Simplify answers, when possible.

19. $\frac{6}{7} - \frac{3}{7}$ **20.** $\frac{3}{4} - \frac{1}{3}$ **21.** $5\frac{3}{4} - 2\frac{1}{4}$ **22.** $4 - 2\frac{8}{15}$ **23.** $3\frac{1}{3} - 1\frac{2}{3}$

24. $8\frac{2}{3} - 4\frac{1}{4}$ **25.** $4\frac{2}{7} - 1\frac{1}{3}$ **26.** $8\frac{19}{100} - 5\frac{3}{10}$

Answer each question about the pictograph.

27. What is the title of the pictograph?

28. In which month were the most burgers sold?

29. How many more burgers were sold in July than in September?

30. If [burger symbol] = 200 burgers, draw a pictograph row for the month of September.

Multiplication and Division of Fractions

DIAGNOSTIC TEST: FORM A (45 MINUTES)

Multiply. Simplify answers, when possible.

1. $\frac{3}{4} \times \frac{1}{2}$ **2.** $\frac{4}{5} \times \frac{3}{8}$ **3.** $\frac{5}{8} \times \frac{2}{5}$ **4.** $\frac{4}{9} \times \frac{3}{2}$ **5.** $\frac{3}{4} \times 26$

6. $\frac{9}{8} \times \frac{4}{5}$ **7.** $\frac{4}{5} \times 15$ **8.** $3 \times \frac{20}{3}$ **9.** $\frac{11}{4} \times 8$ **10.** $\frac{9}{10} \times 2$

11. $3\frac{1}{2} \times 4$ **12.** $7 \times 5\frac{2}{3}$ **13.** $6\frac{7}{8} \times 3$ **14.** $2\frac{1}{4} \times 4\frac{2}{5}$ **15.** $8\frac{7}{10} \times 2\frac{3}{4}$

Divide. Simplify answers, when possible.

16. $\frac{3}{4} \div \frac{1}{2}$ **17.** $\frac{9}{10} \div \frac{3}{5}$ **18.** $12 \div \frac{2}{3}$ **19.** $\frac{7}{8} \div 4$ **20.** $\frac{11}{20} \div \frac{7}{10}$

21. $3\frac{4}{5} \div 9$ **22.** $1\frac{1}{2} \div \frac{5}{6}$ **23.** $8\frac{2}{3} \div 2\frac{1}{3}$ **24.** $6\frac{1}{4} \div 1\frac{5}{8}$ **25.** $4\frac{3}{10} \div 2\frac{4}{5}$

24 or more right? Go to page 232. *Less than 24 right? Go to page 223.*

DIAGNOSTIC TEST: FORM B (45 MINUTES)

Multiply. Simplify answers, when possible.

1. $\frac{5}{2} \times \frac{3}{2}$ **2.** $\frac{5}{9} \times \frac{3}{4}$ **3.** $\frac{3}{5} \times \frac{5}{6}$ **4.** $\frac{11}{16} \times 8$ **5.** $14 \times \frac{3}{4}$

6. $\frac{3}{8} \times \frac{1}{6}$ **7.** $\frac{9}{16} \times \frac{4}{5}$ **8.** $\frac{2}{9} \times \frac{3}{4}$ **9.** $\frac{6}{7} \times \frac{14}{15}$ **10.** $\frac{3}{5} \times \frac{7}{18}$

11. $2\frac{7}{10} \times 4$ **12.** $6 \times 5\frac{1}{3}$ **13.** $4\frac{5}{6} \times 3$ **14.** $1\frac{7}{8} \times 3\frac{9}{10}$ **15.** $2\frac{5}{7} \times 4\frac{2}{3}$

Divide. Simplify answers, when possible.

16. $\frac{7}{8} \div \frac{1}{4}$ **17.** $\frac{5}{6} \div \frac{9}{10}$ **18.** $10 \div \frac{5}{7}$ **19.** $\frac{3}{4} \div 2$ **20.** $\frac{17}{20} \div \frac{4}{5}$

21. $3\frac{2}{3} \div 6$ **22.** $5\frac{1}{2} \div \frac{7}{10}$ **23.** $7\frac{1}{3} \div 2\frac{4}{5}$ **24.** $9\frac{1}{10} \div 2\frac{3}{5}$ **25.** $4\frac{5}{6} \div 2\frac{2}{3}$

24 or more right? Go to page 232. *Less than 24 right? See your teacher.*

MULTIPLYING BY A WHOLE NUMBER

To multiply whole numbers and fractions

- Write whole numbers as fractions.
- Multiply numerators and then denominators.
- Simplify answers, when possible.

EXAMPLE

Multiply. $\frac{3}{8} \times 4$

Write 4 as a fraction.

Then multiply numerators and denominators.

$$\frac{3}{8} \times \frac{4}{1} = \frac{3 \times 4}{8 \times 1} = \frac{12}{8}$$

Simplify.

$$\frac{12}{8} = \frac{12 \div 4}{8 \div 4} = \frac{3}{2}$$

So, $\frac{3}{8} \times 4 = \frac{3}{2}$ or $1\frac{1}{2}$.

> A whole number can be written as a fraction with a denominator of 1. **Example:** $6 = \frac{6}{1}$.

GETTING READY

Multiply. Simplify answers, when possible.

1. $2 \times \frac{3}{4}$	**2.** $5 \times \frac{1}{3}$	**3.** $\frac{5}{6} \times 4$	**4.** $\frac{2}{5} \times 5$	**5.** $4 \times \frac{3}{10}$

EXERCISES

Multiply. Simplify answers, when possible.

1. $5 \times \frac{3}{4}$	**2.** $\frac{2}{3} \times 3$	**3.** $\frac{3}{10} \times 4$	**4.** $6 \times \frac{3}{10}$	**5.** $\frac{2}{5} \times 4$
6. $8 \times \frac{2}{3}$	**7.** $\frac{3}{4} \times 6$	**8.** $2 \times \frac{3}{5}$	**9.** $\frac{1}{2} \times 9$	**10.** $\frac{3}{5} \times 8$
11. $\frac{7}{10} \times 4$	**12.** $10 \times \frac{3}{8}$	**13.** $\frac{3}{4} \times 3$	**14.** $\frac{1}{7} \times 8$	**15.** $\frac{1}{8} \times 16$
16. $12 \times \frac{3}{5}$	**17.** $9 \times \frac{2}{3}$	**18.** $\frac{3}{4} \times 8$	**19.** $\frac{2}{7} \times 7$	**20.** $\frac{3}{10} \times 5$

Solve.

21. A micrometer measures very small lengths. The thread section of a micrometer moves $\frac{1}{40}$ in. in one complete turn. How far will it move in 20 turns?

22. Steve can jog to school in $\frac{3}{4}$ h. A year ago it took him twice as long. How long did it take him to jog to school a year ago?

MULTIPLYING FRACTIONS

To multiply fractions

- Multiply numerators.
- Multiply denominators.
- Simplify answers, when possible.

EXAMPLE 1

Multiply. $\frac{1}{2} \times \frac{3}{4}$

Multiply numerators.
Multiply denominators.

$$\frac{1 \times 3}{2 \times 4} = \frac{3}{8}$$

So, $\frac{1}{2} \times \frac{3}{4} = \mathbf{\frac{3}{8}}$.

EXAMPLE 2

Multiply. $\frac{5}{8} \times \frac{2}{3}$

Multiply numerators.
Multiply denominators.
Simplify.

$$\frac{5 \times 2}{8 \times 3} = \frac{10}{24}$$

$$\frac{10 \div 2}{24 \div 2} = \frac{5}{12}$$

So, $\frac{5}{8} \times \frac{2}{3} = \mathbf{\frac{5}{12}}$.

There is a shorter way to multiply fractions when there are common factors.

Divide the numerator and denominator by 2. $\frac{5}{\overset{}{\underset{4}{\cancel{8}}}} \times \frac{\overset{1}{\cancel{2}}}{3}$ ⟵ 2 is a common factor.

Multiply numerators.
Multiply denominators.

$$\frac{5 \times 1}{4 \times 3} = \frac{5}{12}$$

GETTING READY

Multiply.

1. $\frac{1}{2} \times \frac{5}{6}$ **2.** $\frac{1}{5} \times \frac{3}{4}$ **3.** $\frac{3}{10} \times \frac{2}{3}$ **4.** $\frac{1}{2} \times \frac{4}{5}$ **5.** $\frac{2}{3} \times \frac{3}{4}$

6. $\frac{1}{2} \times \frac{5}{8}$ **7.** $\frac{3}{8} \times \frac{4}{9}$ **8.** $\frac{3}{4} \times \frac{5}{6}$ **9.** $\frac{7}{8} \times \frac{4}{7}$ **10.** $\frac{9}{10} \times \frac{4}{7}$

EXERCISES

Multiply. Simplify answers, when possible.

1. $\frac{3}{5} \times \frac{1}{6}$ **2.** $\frac{2}{3} \times \frac{1}{4}$ **3.** $\frac{2}{5} \times \frac{3}{4}$ **4.** $\frac{3}{8} \times \frac{3}{4}$ **5.** $\frac{1}{2} \times \frac{7}{10}$

6. $\frac{3}{8} \times \frac{2}{5}$ **7.** $\frac{1}{3} \times \frac{3}{10}$ **8.** $\frac{2}{3} \times \frac{3}{5}$ **9.** $\frac{5}{8} \times \frac{8}{16}$ **10.** $\frac{1}{4} \times \frac{7}{8}$

11. $\frac{3}{4} \times \frac{5}{6}$ **12.** $\frac{2}{5} \times \frac{4}{7}$ **13.** $\frac{1}{3} \times \frac{5}{6}$ **14.** $\frac{2}{9} \times \frac{1}{5}$ **15.** $\frac{1}{6} \times \frac{2}{9}$

16. $\frac{7}{10} \times \frac{3}{7}$ **17.** $\frac{3}{16} \times \frac{4}{9}$ **18.** $\frac{8}{15} \times \frac{5}{6}$ **19.** $\frac{7}{9} \times \frac{18}{35}$ **20.** $\frac{7}{10} \times \frac{7}{10}$

21. $\frac{5}{6} \times \frac{5}{6}$ **22.** $\frac{3}{8} \times \frac{7}{8}$ **23.** $\frac{2}{3} \times \frac{5}{14}$ **24.** $\frac{2}{9} \times \frac{3}{5}$ **25.** $\frac{2}{3} \times \frac{1}{9}$

Solve.

***26.** Timothy spent $\frac{3}{4}$ h working in his vegetable garden. He spent $\frac{1}{2}$ of the time fertilizing the plants. How much time did Timothy spend fertilizing the plants?

***27.** A recipe for a sauce requires $\frac{2}{3}$ lb of apples. Joella wants to make $\frac{1}{2}$ as much sauce. How many pounds of apples should she use?

CALCULATOR

Try these with your calculator. Remember $\frac{3}{4}$ means $3 \div 4$.

1. $\frac{3}{15} \times 20{,}460$ **2.** $\frac{381}{591} \times 157{,}403$ **3.** $\frac{22}{25} \times 9{,}950$

4. $\frac{476}{488} \times 11{,}590$ **5.** $\frac{89}{125} \times 62{,}000$ **6.** $\frac{195}{5{,}307} \times 281{,}271$

MULTIPLYING FRACTIONS AND MIXED NUMBERS

To multiply fractions and mixed numbers

- Change mixed numbers to improper fractions.
- Then multiply. Simplify answers, when possible.

EXAMPLE 1

Multiply. $\frac{3}{5} \times 4\frac{5}{6}$

Change $4\frac{5}{6}$ to an improper fraction.

Think: $6 \times 4 = 24$, $24 + 5 = 29$ $\longrightarrow$ $\mathbf{4\frac{5}{6} = \frac{29}{6}}$

Divide by 3. Then multiply.

$$\frac{\overset{1}{\cancel{3}}}{5} \times \frac{29}{\underset{2}{\cancel{6}}} = \frac{1 \times 29}{5 \times 2} = \frac{29}{10}, \text{ or } 2\frac{9}{10}$$

So, $\frac{3}{5} \times 4\frac{5}{6} = \mathbf{\frac{29}{10}}$, **or** $\mathbf{2\frac{9}{10}}$.

EXAMPLE 2

Multiply. $2\frac{2}{3} \times 3\frac{1}{8}$

Change $2\frac{2}{3}$ and $3\frac{1}{8}$ to improper fractions.

Think: $3 \times 2 = 6$, $6 + 2 = 8$ $\longrightarrow$ $\mathbf{2\frac{2}{3} = \frac{8}{3}}$

Think: $8 \times 3 = 24$, $24 + 1 = 25$ $\longrightarrow$ $\mathbf{3\frac{1}{8} = \frac{25}{8}}$

Divide by 8. Then multiply.

$$\frac{\overset{1}{\cancel{8}}}{3} \times \frac{25}{\underset{1}{\cancel{8}}} = \frac{1 \times 25}{3 \times 1} = \frac{25}{3}, \text{ or } 8\frac{1}{3}$$

So, $2\frac{2}{3} \times 3\frac{1}{8} = \mathbf{\frac{25}{3}}$, **or** $\mathbf{8\frac{1}{3}}$.

GETTING READY

Multiply. Simplify answers, when possible.

1. $\frac{2}{3} \times 3\frac{3}{4}$ **2.** $2\frac{1}{7} \times \frac{3}{5}$ **3.** $6 \times 2\frac{3}{8}$ **4.** $3\frac{1}{4} \times 8$

EXERCISES

Multiply. Simplify answers, when possible.

1. $\frac{3}{5} \times 2\frac{2}{3}$	**2.** $\frac{5}{7} \times 1\frac{3}{4}$	**3.** $\frac{3}{4} \times 2\frac{5}{8}$	**4.** $2\frac{3}{8} \times \frac{1}{4}$
5. $1\frac{2}{7} \times \frac{3}{5}$	**6.** $3\frac{1}{2} \times \frac{2}{5}$	**7.** $\frac{5}{6} \times 2\frac{2}{7}$	**8.** $5\frac{1}{3} \times \frac{4}{5}$
9. $2\frac{3}{4} \times 6$	**10.** $4\frac{1}{5} \times 15$	**11.** $4 \times 2\frac{1}{3}$	**12.** $9 \times 2\frac{1}{3}$
13. $3\frac{1}{5} \times \frac{3}{4}$	**14.** $\frac{2}{3} \times 3\frac{3}{5}$	**15.** $\frac{1}{2} \times 2\frac{2}{5}$	**16.** $1\frac{1}{12} \times 1\frac{1}{4}$
17. $2\frac{1}{3} \times 1\frac{1}{5}$	**18.** $2\frac{1}{4} \times 1\frac{4}{5}$	**19.** $3\frac{1}{2} \times 7\frac{1}{3}$	**20.** $6\frac{1}{8} \times 2\frac{2}{7}$
21. $2\frac{1}{4} \times 4\frac{1}{2}$	**22.** $1\frac{5}{7} \times 2\frac{1}{6}$	**23.** $3\frac{1}{9} \times \frac{6}{7}$	**24.** $2\frac{1}{6} \times 5\frac{1}{3}$
25. $2\frac{5}{8} \times 1\frac{5}{7}$	**26.** $8\frac{1}{3} \times 2\frac{1}{10}$	**27.** $1\frac{5}{16} \times 1\frac{1}{3}$	**28.** $2\frac{1}{8} \times 1\frac{1}{4}$
29. $2\frac{3}{16} \times 3\frac{3}{7}$	**30.** $3\frac{1}{4} \times 2\frac{7}{8}$	**31.** $2\frac{4}{5} \times 3\frac{1}{2}$	**32.** $4\frac{1}{3} \times 2\frac{1}{4}$

Solve.

33. Dennis is making a recipe that calls for $1\frac{3}{4}$ c of flour. He wants to make a portion that is $2\frac{1}{2}$ times the recipe. How much flour should he use?

34. Abraham is $5\frac{1}{2}$ ft tall. His sister Agatha is $\frac{2}{3}$ as tall as Abraham. How tall is Agatha?

***35.** Michael's tractor gets $15\frac{3}{4}$ mi to a gallon of gas. The gas tank holds $9\frac{1}{3}$ gal. How many miles can Michael's tractor travel on one tankful of gas?

RECIPROCALS

To find the reciprocal of a fraction, mixed number, or whole number

- Change the mixed number to an improper fraction.
- Change the whole number to a fraction.
- Interchange the numerator and the denominator.

$\frac{3}{5}$ and $\frac{5}{3}$ are reciprocals. $\frac{3}{5} \times \frac{5}{3} = \frac{15}{15}$, or 1

8 and $\frac{1}{8}$ are reciprocals. $\frac{8}{1} \times \frac{1}{8} = \frac{8}{8}$, or 1

The product of a number and its reciprocal is 1.

EXAMPLE 1

Find the reciprocal of $\frac{7}{9}$.

Think: $\frac{7}{9} \times \frac{?}{?} = 1$ $\quad \mathbf{\frac{7}{9} \times \frac{9}{7} = \frac{63}{63} = 1}$

So, the reciprocal of $\frac{7}{9}$ is $\mathbf{\frac{9}{7}}$.

EXAMPLE 2

Find the reciprocal of $2\frac{2}{3}$.

Change to a fraction. $\quad \mathbf{2\frac{2}{3} = \frac{8}{3}} \quad \mathbf{\frac{8}{3} \times \frac{?}{?} = 1} \quad \mathbf{\frac{8}{3} \times \frac{3}{8} = \frac{24}{24} = 1}$

So, the reciprocal of $2\frac{2}{3}$ is $\mathbf{\frac{3}{8}}$.

GETTING READY

Find the reciprocal.

1. 7 **2.** $\frac{2}{9}$ **3.** $\frac{6}{5}$ **4.** $5\frac{2}{3}$ **5.** $3\frac{1}{3}$ **6.** $1\frac{7}{8}$

EXERCISES

Find the reciprocal.

1. $\frac{3}{16}$ **2.** $\frac{1}{6}$ **3.** $\frac{3}{7}$ **4.** $\frac{4}{15}$ **5.** $\frac{2}{11}$ **6.** $\frac{5}{6}$

7. $\frac{6}{23}$ **8.** 5 **9.** 12 **10.** $\frac{10}{7}$ **11.** $\frac{18}{5}$ **12.** $1\frac{1}{2}$

13. $1\frac{5}{8}$ **14.** $3\frac{5}{8}$ **15.** 33 **16.** 5 **17.** $2\frac{1}{4}$ **18.** $3\frac{3}{5}$

DIVIDING BY A WHOLE NUMBER

To divide fractions and mixed numbers by whole numbers

- Multiply by the reciprocal of the divisor.
- Check by multiplying the quotient by the divisor.

EXAMPLE

Divide. $3\frac{1}{5} \div 8$

Change $3\frac{1}{5}$ to a fraction. $\mathbf{3\frac{1}{5} \div 8 = \frac{16}{5} \div \frac{8}{1}}$

Multiply by the reciprocal of the divisor. $\mathbf{\frac{\overset{2}{\cancel{16}}}{5} \times \frac{1}{\underset{1}{\cancel{8}}} = \frac{2}{5}}$ Check: $\frac{2}{5} \times 8 = \frac{16}{5}$, or $3\frac{1}{5}$

So, $3\frac{1}{5} \div 8 = \mathbf{\frac{2}{5}}$.

GETTING READY

Divide and check. Simplify answers, when possible.

1. $\frac{5}{8} \div 6$ **2.** $\frac{7}{10} \div 5$ **3.** $\frac{3}{5} \div 8$ **4.** $1\frac{7}{16} \div 7$ **5.** $3\frac{4}{5} \div 8$

EXERCISES

Divide. Simplify answers, when possible.

1. $\frac{7}{8} \div 10$ **2.** $\frac{4}{5} \div 2$ **3.** $\frac{2}{3} \div 4$ **4.** $1\frac{1}{2} \div 3$ **5.** $\frac{8}{9} \div 8$

6. $5\frac{1}{3} \div 5$ **7.** $\frac{11}{20} \div 5$ **8.** $1\frac{7}{8} \div 12$ **9.** $3\frac{2}{3} \div 7$ **10.** $9\frac{1}{2} \div 8$

11. $\frac{9}{10} \div 10$ **12.** $2\frac{7}{10} \div 8$ **13.** $5\frac{3}{5} \div 3$ **14.** $\frac{7}{8} \div 12$ **15.** $4\frac{3}{5} \div 3$

16. $\frac{5}{8} \div 7$ **17.** $4\frac{3}{4} \div 9$ **18.** $6\frac{7}{10} \div 5$ **19.** $3\frac{1}{2} \div 15$ **20.** $9\frac{1}{10} \div 2$

Solve.

21. Jed had $5\frac{1}{2}$ yd of cloth. He cut it into 6 equal pieces. How long was each piece?

22. Terry separated $1\frac{1}{2}$ lb of hamburger meat into 4 equal patties. How much did each patty weigh?

DIVIDING FRACTIONS AND MIXED NUMBERS

To divide fractions and mixed numbers

- Change mixed numbers to fractions.
- Multiply by the reciprocal of the divisor.
- Check by multiplying the quotient by the divisor.

EXAMPLE 1

Divide. $\frac{9}{10} \div \frac{2}{5}$

Multiply by the reciprocal of the divisor. $\frac{9}{10} \div \frac{2}{5} = \frac{9}{10} \times \frac{5}{2}$

Divide by 5. Then multiply. $\frac{9}{\cancel{10}_{2}} \times \frac{\cancel{5}^{1}}{2} = \frac{9}{4}$, or $2\frac{1}{4}$

To check, multiply the quotient by the divisor. $\frac{9}{\cancel{4}_{2}} \times \frac{\cancel{2}^{1}}{5} = \frac{9}{10}$

So, $\frac{9}{10} \div \frac{2}{5} = \mathbf{\frac{9}{4}}$, or $\mathbf{2\frac{1}{4}}$.

EXAMPLE 2

Divide. $8 \div 3\frac{3}{4}$

Change 8 and $3\frac{3}{4}$ to fractions. $8 \div 3\frac{3}{4} = \frac{8}{1} \div \frac{15}{4}$

Multiply by the reciprocal of the divisor. $\frac{8}{1} \times \frac{4}{15} = \frac{32}{15}$, or $2\frac{2}{15}$

Check: $2\frac{2}{15} \times 3\frac{3}{4} = \frac{\cancel{32}^{8}}{\cancel{15}_{1}} \times \frac{\cancel{15}^{1}}{\cancel{4}_{1}} = \frac{8}{1}$, or 8

So, $8 \div 3\frac{3}{4} = \mathbf{\frac{32}{15}}$, or $\mathbf{2\frac{2}{15}}$.

EXAMPLE 3

Divide. $3\frac{2}{3} \div 5\frac{1}{2}$

Change $3\frac{2}{3}$ and $5\frac{1}{2}$ to fractions. $3\frac{2}{3} \div 5\frac{1}{2} = \frac{11}{3} \div \frac{11}{2}$

Multiply by the reciprocal of the divisor. $\frac{\cancel{11}^{1}}{3} \times \frac{2}{\cancel{11}_{1}} = \frac{2}{3}$

Check: $\frac{2}{3} \times 5\frac{1}{2} = \frac{\cancel{2}^{1}}{3} \times \frac{11}{\cancel{2}_{1}} = \frac{11}{3}$ or $3\frac{2}{3}$

So, $3\frac{2}{3} \div 5\frac{1}{2} = \mathbf{\frac{2}{3}}$.

GETTING READY

Divide and check. Simplify answers, when possible.

1. $\frac{3}{8} \div \frac{2}{3}$
2. $\frac{7}{16} \div \frac{3}{4}$
3. $6 \div \frac{5}{8}$
4. $5 \div \frac{9}{10}$
5. $5\frac{3}{5} \div 8$
6. $\frac{1}{3} \div 4\frac{2}{3}$
7. $3\frac{1}{2} \div \frac{3}{4}$
8. $4\frac{1}{6} \div 3\frac{7}{8}$

EXERCISES

Divide. Simplify answers, when possible.

1. $\frac{1}{3} \div \frac{3}{5}$
2. $\frac{2}{9} \div \frac{2}{4}$
3. $\frac{5}{12} \div \frac{4}{7}$
4. $\frac{2}{15} \div \frac{2}{3}$
5. $\frac{5}{8} \div 10$
6. $1 \div \frac{2}{3}$
7. $7 \div 14$
8. $1\frac{5}{8} \div 7$
9. $1 \div 1\frac{3}{4}$
10. $\frac{7}{8} \div \frac{5}{12}$
11. $\frac{8}{9} \div \frac{5}{6}$
12. $\frac{4}{9} \div \frac{2}{3}$
13. $\frac{13}{15} \div 7\frac{4}{5}$
14. $\frac{5}{8} \div 1\frac{7}{16}$
15. $4\frac{1}{2} \div \frac{3}{4}$
16. $\frac{1}{2} \div 2\frac{2}{3}$
17. $\frac{5}{6} \div 3\frac{1}{3}$
18. $3 \div \frac{1}{3}$
19. $\frac{5}{8} \div 6$
20. $3\frac{5}{8} \div 2$
21. $\frac{7}{15} \div 2\frac{1}{3}$
22. $1\frac{1}{2} \div 2\frac{1}{4}$
23. $3\frac{1}{2} \div 1\frac{3}{4}$
24. $2\frac{2}{5} \div 3\frac{1}{3}$
25. $6 \div 4\frac{5}{6}$
26. $4\frac{2}{3} \div 3\frac{1}{9}$
27. $6\frac{2}{3} \div 1\frac{1}{3}$
28. $5\frac{3}{5} \div 2\frac{1}{3}$

Solve.

29. Sharon is packing corn into $1\frac{1}{2}$ kg packages. How many packages can she get from 60 kg of corn?

***30.** Wendell ran $1\frac{1}{2}$ mi in 6 min. How far can he run in 1 min?

Return to Form B on page 222.

FINDING A PART OF A NUMBER

To find a fractional part of a number

- Multiply.

EXAMPLE 1

Find $\frac{1}{3}$ of 12.

Think: $\frac{1}{3}$ of a dozen is 4.

$\frac{1}{3}$ of 12 means $\frac{1}{3} \times 12$.

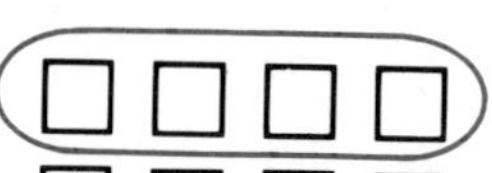

Multiply. $\quad \mathbf{\frac{1}{\cancel{3}_{1}} \times \frac{\cancel{12}^{4}}{1} = \frac{4}{1}, \text{ or } 4}$

So, $\frac{1}{3}$ of 12 is **4.**

EXAMPLE 2

Find $\frac{3}{5}$ of \$35.

Think: $\frac{3}{5}$ of \$35 = $\frac{3}{5} \times 35$.

Multiply. $\quad \mathbf{\frac{3}{\cancel{5}_{1}} \times \frac{\cancel{35}^{7}}{1} = 21}$

So, $\frac{3}{5}$ of \$35 is **\$21.**

EXAMPLE 3

Find $\frac{3}{4}$ of $2\frac{5}{6}$.

Change $2\frac{5}{6}$ to a fraction. $\quad \mathbf{2\frac{5}{6} = \frac{17}{6}}$

Think: $\frac{3}{4}$ of $\frac{17}{6} = \frac{3}{4} \times \frac{17}{6}$

Multiply. $\quad \mathbf{\frac{\cancel{3}^{1}}{4} \times \frac{17}{\cancel{6}_{2}} = \frac{17}{8}, \text{ or } 2\frac{1}{8}}$

So, $\frac{3}{4}$ of $2\frac{5}{6}$ is $\mathbf{\frac{17}{8}}$, or $\mathbf{2\frac{1}{8}}$.

GETTING READY

Compute.

1. $\frac{1}{2}$ of 38 **2.** $\frac{2}{3}$ of 81 **3.** $\frac{3}{8}$ of $1\frac{6}{7}$ **4.** $\frac{3}{5}$ of $2\frac{1}{2}$

EXERCISES

Compute.

1. $\frac{1}{4}$ of 72
2. $\frac{1}{5}$ of 45
3. $\frac{1}{9}$ of 144
4. $\frac{1}{8}$ of 88
5. $\frac{1}{7}$ of 770
6. $\frac{5}{12}$ of 132
7. $\frac{3}{4}$ of 70
8. $\frac{4}{9}$ of 90
9. $\frac{5}{6}$ of $5\frac{1}{2}$
10. $\frac{4}{5}$ of $3\frac{2}{3}$
11. $\frac{9}{10}$ of $2\frac{2}{3}$
12. $\frac{2}{3}$ of $1\frac{1}{9}$
13. $\frac{2}{9}$ of $2\frac{3}{5}$
14. $\frac{4}{7}$ of $1\frac{7}{12}$
15. $\frac{3}{10}$ of $2\frac{5}{8}$
16. $\frac{5}{9}$ of $3\frac{7}{10}$
17. $\frac{3}{4}$ of 24
18. $\frac{3}{4}$ of $1\frac{8}{9}$
19. $\frac{7}{8}$ of $4\frac{3}{4}$
20. $\frac{4}{9}$ of 27
21. $\frac{8}{9}$ of \$45
22. $\frac{2}{3}$ of \$1.20
23. $\frac{1}{2}$ of \$12
24. $\frac{3}{4}$ of \$8
25. $\frac{1}{5}$ of 40
26. $\frac{2}{5}$ of 70
27. $\frac{1}{8}$ of $3\frac{2}{5}$
28. $\frac{5}{6}$ of $2\frac{3}{4}$

*29. $4\frac{2}{7}$ of 98

*30. $3\frac{1}{6}$ of 54

*31. $9\frac{2}{3}$ of 81

*32. $8\frac{1}{2}$ of 42

Solve.

33. One-fourth of the students in Mr. Green's class belong to the school band. There are 32 students in Mr. Green's class. How many of his students belong to the school band?

34. Annette bought $3\frac{3}{4}$ yd of material for a suit. She used half the material to make a skirt. How much was left for the jacket?

35. Paul jogs $4\frac{1}{2}$ mi each morning. Today he jogged only $\frac{3}{4}$ of his usual distance. How far did Paul jog today?

DECIMALS AND FRACTIONS

To change decimals to fractions

- Use place value.
- Simplify, where possible.

To change fractions to decimals

- Divide the numerator by the denominator.
- Round to the nearest thousandth, where necessary.

EXAMPLE 1

Write 0.45 as a fraction.

Think: 0.45 means 45 hundredths   45 hundredths = $\frac{45}{100}$

Simplify. $\frac{45}{100} = \frac{9}{20}$

So, 0.45 = $\frac{9}{20}$.

EXAMPLE 2

Write 1.7 as a fraction.
Think: 1.7 means 1 and 7 tenths   1 and 7 tenths = $1\frac{7}{10}$

So, 1.7 = $1\frac{7}{10}$ or $\frac{17}{10}$.

EXAMPLE 3

Write $\frac{3}{4}$ as a decimal.

Divide the numerator by the denominator.

$$\begin{array}{r} 0.75 \\ 4\overline{)3.00} \\ \underline{2\,8} \\ 20 \\ \underline{20} \end{array}$$

So, $\frac{3}{4}$ = **0.75.**

EXAMPLE 4

Write a decimal to the nearest thousandth for $2\frac{4}{9}$.

Change $2\frac{4}{9}$ to a fraction. $2\frac{4}{9} = \frac{22}{9}$

Divide to 4 places and round.

$$\begin{array}{r} 2.4444 \\ 9\overline{)22.0000} \end{array}$$

So, $2\frac{4}{9}$ = **2.444.**

GETTING READY

Write as a fraction in simplest form.

1. 0.4 **2.** 0.25 **3.** 1.3 **4.** 1.04 **5.** 2.65

Write as a decimal. Round to the nearest thousandth, when necessary.

6. $\frac{1}{2}$ **7.** $\frac{3}{5}$ **8.** $\frac{5}{6}$ **9.** $3\frac{9}{10}$ **10.** $2\frac{2}{3}$

EXERCISES

Write as a fraction in simplest form.

1. 0.9 **2.** 0.85 **3.** 0.64 **4.** 0.08 **5.** 0.7

6. 0.416 **7.** 0.625 **8.** 0.8 **9.** 0.750 **10.** 0.48

11. 0.17 **12.** 0.1 **13.** 1.5 **14.** 9.2 **15.** 4.06

16. 1.02 **17.** 3.12 **18.** 3.25 **19.** 2.125 **20.** 5.25

Write as a decimal. Round to the nearest thousandth, where necessary.

21. $\frac{1}{4}$ **22.** $\frac{1}{8}$ **23.** $\frac{3}{8}$ **24.** $\frac{2}{5}$ **25.** $\frac{5}{8}$

26. $\frac{4}{5}$ **27.** $\frac{1}{10}$ **28.** $\frac{9}{12}$ **29.** $\frac{2}{16}$ **30.** $\frac{5}{10}$

31. $\frac{6}{100}$ **32.** $\frac{2}{9}$ **33.** $\frac{9}{50}$ **34.** $\frac{7}{25}$ **35.** $\frac{81}{200}$

36. $2\frac{1}{3}$ **37.** $3\frac{3}{4}$ **38.** $6\frac{3}{5}$ **39.** $2\frac{2}{7}$ **40.** $4\frac{5}{9}$

Write a fraction for the decimal. Solve.

41. Anna Long sold 0.875 of the potholders she made. She made 40 potholders. How many of them did she sell?

Write a decimal for the fraction. Solve.

42. In the Johnson City election, half of the eligible voters actually voted. There are 8,996 eligible voters. How many of the voters actually voted?

Problem Solving Applications

READ • PLAN • SOLVE • CHECK

The General's Inn at Washington High School is run by students in the food services class. Once a week they serve luncheon by reservation only to the staff, central office personnel, and parents. The students plan the menus, buy the groceries, cook, wait on tables, and act as cashiers.

AS YOU READ

- buffet—a meal set out on a table where people serve themselves

Answer these questions about The General's Inn.

1. The General's Inn begins serving at 11:30 am and closes at 2:00 pm. How long is it open?

2. Today the total receipts were \$334.75. If each lunch cost \$3.25, how many people were served?

3. The students work in shifts clearing the tables. If each shift lasts $\frac{1}{2}$ h, how many shifts are there during luncheon? (See Ex. 1.)

*4. Fred's recipe for 4 servings of Potatoes Florentine calls for $1\frac{1}{4}$ c of milk. In order to serve 96 people, how much milk is needed?

5. Charlotte's recipe for Chicken Kiev takes $\frac{1}{4}$ h to deep-fry. If only one basket can be fried at a time, how long will it take to fry 9 baskets?

*6. This week José and Sara are buying the meat for The General's Inn *buffet.* They bought a $10\frac{1}{2}$-lb whole ham and a $9\frac{3}{4}$-lb beef roast. If the beef cost \$23.40, how much did it cost per pound?

Career: Cooks and Chefs

READ • PLAN • SOLVE • CHECK

Cooks and chefs work in such places as restaurants, hotels, schools, hospitals, factories, and airports. Large kitchen staffs include kitchen helpers, specialty cooks, and head cooks who supervise the work. Post–high school training for careers in food service can last from a few months to several years.

AS YOU READ

- baba—a rich cake soaked in a syrup

Solve.

1. Salina Fredrick is short-order cook at the L and S café. She uses $14\frac{1}{2}$ boxes of frozen hamburger patties each week. If each box holds 60 patties, how many hamburgers are used each week?

2. Antonia Perez is the chef at La Maison, and Joe, her husband, is the chef at The Barn. Last week Antonia worked 48 h and Joe worked $9\frac{1}{2}$ h less. How many hours did Joe work?

3. On Saturday $\frac{1}{6}$ of the 384 potatoes baked at Helene's Inn were left over. On Sunday these were used to make potato salad. How many potatoes were used for the salad?

4. The recipe that the pastry chef at the Brant Hotel uses for *baba* calls for $2\frac{2}{3}$ c of flour. He wishes to make 8 babas. How many cups of flour will he need?

5. A gallon of frozen fruit punch lasts $3\frac{1}{2}$ days at the Atlas Industries Cafeteria. Jon Herbert, head cook, orders the supplies. How many gallons of punch should he order for a month? (Use 1 month = 28 days.)

6. Pierre André is taking a 6-wk course in ice sculpture for his catering business. He has completed 4 wk. What part of the total has he completed?

Problem Solving Skill

To read and interpret a line graph

- Find the dot you need.
- Imagine a line from the dot to the scale.
- Then read the value from the scale.

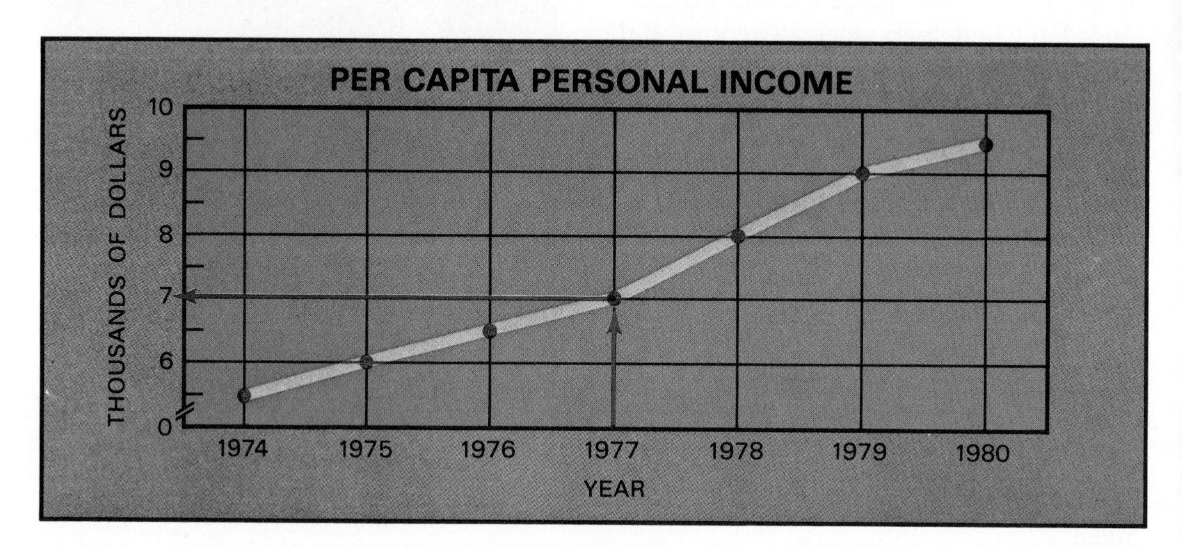

EXAMPLE

What was the per capita personal income in 1977?

The horizontal axis shows years. Find 1977 on the horizontal axis. Move straight up to the dot above 1977. Imagine a line from this dot to the vertical scale. Notice that each number on the vertical scale represents $1,000. Now read the number on the vertical scale.
Multiply. **7 × $1,000 = $7,000.**
So, in 1977 the per capita personal income was **$7,000.**

Use the graph to answer the questions.

1. In the years 1974–1980, did per capita personal income ever fall?

2. During which years did per capita personal income rise the most?

3. During 1974–1980, by how much did per capita personal income rise?

4. The 1980 per capita personal income is how many times as much as the 1974 per capita personal income?

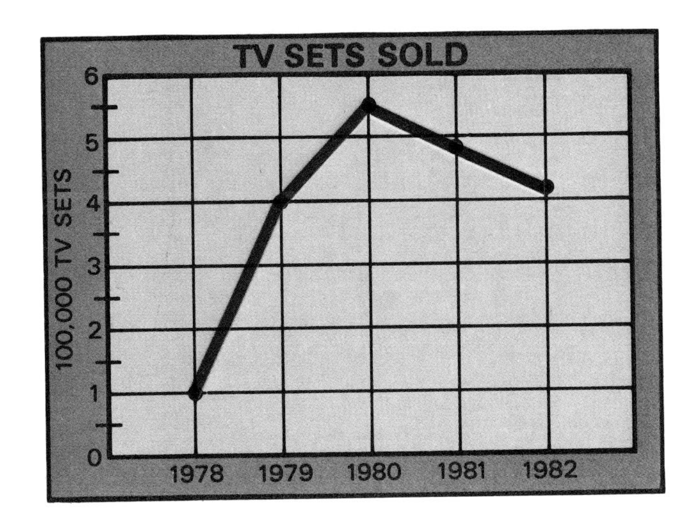

Use the graph above to answer the questions.

5. What is shown on the vertical scale?

6. How many TV sets were sold in 1979?

7. During which year were the sales 475,000?

8. During which year were the most sets sold?

9. During which year were the sales 25,000 more than for 1979?

10. Between which years was there a decrease in sales?

Use the graph at the right to answer the questions.

11. What is shown on the vertical scale?

12. Estimate the number of robots predicted for 1987.

13. What is the estimated increase in the number of robots between 1985 and 1990?

14. By what year is the number of robots estimated to be 25,000?

CHAPTER TEST 12

Multiply. Simplify answers, when possible.

1. $\frac{3}{4} \times \frac{1}{10}$ **2.** $\frac{1}{2} \times 20$ **3.** $7 \times \frac{3}{10}$ **4.** $4\frac{2}{3} \times \frac{3}{4}$

5. $3 \times 2\frac{2}{3}$ **6.** $\frac{5}{11} \times 3\frac{2}{3}$ **7.** $2\frac{1}{4} \times 5\frac{1}{2}$ **8.** $1\frac{5}{7} \times 2\frac{5}{6}$

Find the reciprocal.

9. $\frac{2}{3}$ **10.** $\frac{1}{9}$ **11.** $6\frac{1}{2}$ **12.** 8

Divide. Simplify answers, when possible.

13. $\frac{7}{8} \div \frac{3}{4}$ **14.** $\frac{5}{9} \div \frac{2}{3}$ **15.** $2\frac{1}{2} \div 3$ **16.** $\frac{1}{2} \div 2\frac{2}{3}$

17. $2 \div \frac{1}{4}$ **18.** $7 \div 6\frac{5}{7}$ **19.** $5\frac{5}{6} \div \frac{7}{12}$ **20.** $1\frac{1}{8} \div 1\frac{1}{5}$

Compute.

21. $\frac{3}{4}$ of $\frac{4}{9}$ **22.** $\frac{2}{3}$ of $1.98 **23.** $\frac{4}{5}$ of $30

Write as a fraction in simplest form.

24. 0.18 **25.** 0.375 **26.** 3.2

Write as a decimal to the nearest thousandth, where necessary.

27. $\frac{1}{8}$ **28.** $\frac{2}{3}$ **29.** $1\frac{3}{4}$

Answer each question about the graph.

30. On what date(s) was the temperature twice what it was on November 1st?

31. On what date(s) was the temperature half what it was on November 7th?

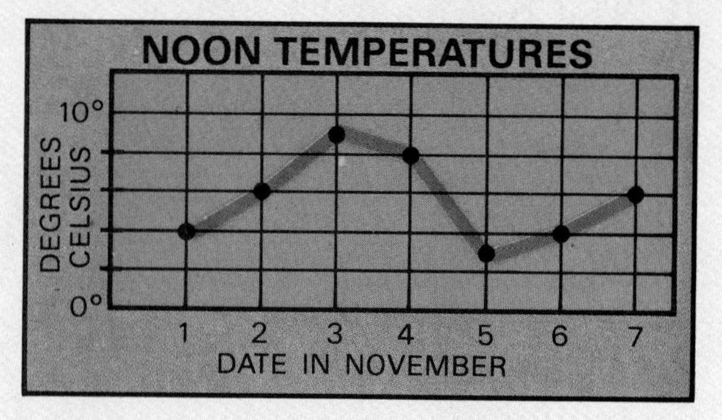

Problem Solving with Formulas

EVALUATING EXPRESSIONS

To evaluate an expression containing a variable

- Replace the variable with the given number.
- Complete the arithmetic.

In algebra, letters are often used as placeholders for numbers. These letters are called **variables.** The expressions $n + 7$, $35 - q$, $\frac{r}{3}$, and $5y$ use the variables n, q, r, and y as placeholders.

EXAMPLE 1

Evaluate the expression $4 + y$ if y is 17.

Replace y with 17. $4 + y$

Then add. $\mathbf{4 + 17 = 21}$

So, the value of the expression is **21.**

> To **evaluate** an expression means to replace each variable with the given number and compute.

EXAMPLE 2

Evaluate the expression $7a$ if a is 6.

Replace a with 6. $7a$

Then multiply. $\mathbf{7 \times 6 = 42.}$

So, the value of the expression is **42.**

> Expressions like $7a$, $3y$, and $2r$ describe multiplication. So, $7a$ means 7 times a.

EXAMPLE 3

Evaluate the expression $\frac{y}{3}$ if y is 18.

Replace y with 18. $\frac{y}{3} \longrightarrow \frac{18}{3}$

Then divide. $\mathbf{18 \div 3 = 6}$

So, the value of the expression is **6.**

> Expressions like $\frac{y}{3}$, $\frac{a}{7}$, and $\frac{18}{r}$ describe division. So, $\frac{y}{3}$ means y divided by 3.

GETTING READY

Evaluate each expression using the given values.

1. $m + 3; m = 8$

2. $y - 7; y = 16$

3. $18 - p; p = 9$

4. $\frac{1}{5} + z; z = \frac{1}{10}$

5. $7r; r = 9$

6. $16s; s = \$2.20$

7. $\frac{n}{8}; n = 56$

8. $\frac{2.7}{m}; m = 0.3$

9. $\frac{1}{3}f; f = \frac{3}{7}$

EXERCISES

Evaluate each expression using the given values.

1. $a + 4; a = 9$

2. $19 + u; u = 14$

3. $15 - s; s = 7\frac{1}{4}$

4. $z - \frac{2}{3}; z = 2$

5. $1.7 - s; s = 0.8$

6. $1.3 + q; q = 8.2$

7. $\$132 + t; t = \$103$

8. $n - \$7.30; n = \$14.50$

9. $3m; m = 8$

10. $\frac{r}{7}; r = 98$

11. $\frac{20}{m}; m = 2.5$

12. $12h; h = \$13.25$

13. $15q; q = 8.1$

14. $21s; s = 0.7$

15. $\frac{p}{0.2}; p = 6.4$

16. $\frac{1}{5}t; t = 43$

17. $20w; w = \frac{2}{5}$

18. $\frac{n}{6}; n = \frac{3}{2}$

19. $\frac{8}{k}; k = \frac{1}{4}$

***20.** $6q^2; q = 3.4$

***21.** $\frac{7}{3n}; n = \frac{1}{6}$

CALCULATOR

Use a calculator to evaluate these expressions.

1. $s + 13.715; s = 6.813$

2. $1.843 - m; m - 0.095$

3. $1.34t; t = 6.45$

4. $\frac{s}{1.2}; s = 4.44$

5. $1.8r + 3.24; r = 1.1$

6. $\frac{r}{0.8} + 2.4; r = 2.88$

EVALUATING FORMULAS

To evaluate a formula

- Replace the variables with the given values.
- Do the arithmetic.

A **formula** is an equation showing how quantities are related. Some formulas contain more than one variable. Some examples of formulas are: $d = 2r$, $P = 2l + 2w$, and $A = lw$.

EXAMPLE 1

Evaluate $g = a + t$ if a is 645 mph and t is 35 mph.

This formula is used to compute the ground speed of an airplane (g) when the air speed of the plane (a) and the speed of the tail wind (t) are given.

Replace a with 645 and t with 35.

$$g = a + t$$

Add to find g.

$$= \mathbf{645 + 35}$$

$$g = \mathbf{680}$$

So, the ground speed of the airplane is **680 mph.**

EXAMPLE 2

Evaluate $t = \frac{1}{4}c + 40$ if c is 120 chirps per minute.

This formula is used to find the temperature in degrees Fahrenheit (t) when the number of chirps per minute (c) of a cricket is given.

Remember $\frac{1}{4}\,c$ means $\frac{1}{4} \times c$.

$$t = \frac{1}{4} \times c + 40$$

Replace c with 120.
Multiply $\frac{1}{4} \times 120$ first.
Then add to find t.

$$= \mathbf{\frac{1}{4} \times 120 + 40}$$

$$= \mathbf{30 + 40}$$

$$t = \mathbf{70}$$

So, the temperature is **70°F.**

EXAMPLE 3

Evaluate $I = prt$ if p is \$100, r is 0.05, and t is 3.

This formula is used to find the interest paid on a principal of p dollars deposited for t years at an annual interest rate r.

Replace p with 100, r with 0.05, and t with 3.

$$I = p \times r \times t$$

$$= \mathbf{100 \times 0.05 \times 3}$$

Multiply to find I. $\quad I = \mathbf{15}$

So, the interest paid will be **\$15.**

GETTING READY

Evaluate each formula using the given values.

1. $s = k + n;\ k = 34$ and $n = 12$

2. $g = a - h;\ a = 500$ and $h = 15$

3. $y = \frac{1}{3}t - 2;\ t = 7$

4. $d = rt;\ r = 55$ and $t = 3\frac{1}{2}$

5. $p = \frac{c}{g};\ c = \$12$ and $g = 10$

6. $k = \frac{tw}{1{,}000};\ t = 2.2$ and $w = 500$

EXERCISES

Evaluate each formula using the given values.

1. $e = Pt;\ P = 860$ and $t = 6$

2. $G = P + Q;\ P = \$175$ and $Q = \$24$

3. $N = G - D;\ G = \$246$ and $D = \$59$

4. $c = kr;\ k = 6$ and $r = 0.5$

5. $s = \frac{c}{p};\ c = 37.2$ and $p = 10$

6. $h = \frac{w}{746};\ w = 2{,}238$

7. $F = N + \frac{1}{8}M;\ N = 4\frac{1}{2}$ and $M = 20$

8. $y = rs;\ r = \frac{2}{3}$ and $s = \frac{3}{5}$

9. $p = 100 + \frac{a}{2};\ a = 42$

10. $t = \frac{1}{3}n - 3\frac{1}{2};\ n = 42$

***11.** $M = 0.9h + \frac{1.2r}{t};\ h = 16,\ r = 10,$ and $t = 4$

***12.** $L = \frac{20}{m} + \frac{100}{n - m};\ m = 10$ and $n = 20$

WRITING RELATED EQUATIONS

To write a related equation

- Think of related sentences.

Suppose a bookshelf has space for 10 books. There are 3 books already on the shelf. How many books can you put on the shelf?

Think: What number plus 3 is 10? Since $7 + 3 = 10$, the number is 7. So, you can put 7 more books on the shelf. You can also find the number by subtraction: $10 - 3 = 7$.

$7 + 3 = 10$ and $10 - 3 = 7$ are related sentences.

Related equations are like related sentences. You could have used the following related equations to show the problem.

$n + 3 = 10$ and $10 - 3 = n$, where n is the number of books needed.

EXAMPLE 1

Write related equations for $y + 9 = 16$ and $s - 2 = 8$.

Sentence	$7 + 9 = 16$	Equation	$y + 9 = 16$
Related sentence	$7 = 16 - 9$	Related equation	$y = 16 - 9$
Sentence	$10 - 2 = 8$	Equation	$s - 2 = 8$
Related sentence	$10 = 8 + 2$	Related equation	$s = 8 + 2$

EXAMPLE 2

Write related equations for $3t = 21$ and $\frac{w}{6} = 4$.

Sentence	$3 \times 7 = 21$	Equation	$3t = 21$
Related sentence	$7 = 21 \div 3$	Related equation	$t = 21 \div 3$
Sentence	$\frac{24}{6} = 4$	Equation	$\frac{w}{6} = 4$
Related sentence	$24 = 4 \times 6$	Related equation	$w = 4 \times 6$

EXAMPLE 3

Write a related equation for each equation.

Equation	$m + \$12.20 = \$27.50$	$\frac{1}{3}r = 10$	$\frac{s}{6.5} = 7.7$
Related equation	$m = \$27.50 - \$12.20$	$r = 10 \div \frac{1}{3}$	$s = 7.7 \times 6.5$

EXAMPLE 4

Write a related equation for each equation.

Equation	$210 = 90 + t$	$1.2 = \frac{h}{0.7}$	$3\frac{1}{2} = 4b$
Related equation	$t = 210 - 90$	$h = 1.2 \times 0.7$	$b = 3\frac{1}{2} \div 4$

GETTING READY

For each equation, write a related equation with the variable on the left.

1. $z - 19 = 21$　**2.** $w + 1.7 = 3.9$　**3.** $4m = 50$

4. $\frac{r}{1.1} = 6.5$　**5.** $\$40.20 = \$19.95 + d$　**6.** $\frac{2}{3}f = 30$

7. $65 = \frac{n}{5}$　**8.** $\frac{7}{2} = 2s$　**9.** $3.62 = a - 1.89$

EXERCISES

For each equation, write a related equation with the variable on the left.

1. $s + 9 = 21$　**2.** $h - 1.25 = 6.75$　**3.** $u + 5.4 = 8.3$

4. $w - 57 = 108$　**5.** $8r = 72$　**6.** $1.2f = 6$

7. $\frac{m}{3.3} = 1.1$　**8.** $\frac{k}{15} = 6$　**9.** $\$18.80 = p - \$20.12$

10. $\frac{4}{5}h = 20$　**11.** $3.8 = \frac{z}{10}$　**12.** $590 = 90 + c$

13. $6\frac{1}{3} = 5q$　**14.** $\$140.00 = \frac{t}{90}$　**15.** $\frac{7}{8} = p + \frac{3}{4}$

***16.** $19 = 37 - a$　***17.** $\frac{3}{y} = 18$　***18.** $\frac{2t}{7} = 16$

SOLVING EQUATIONS: ADDITION AND SUBTRACTION

To solve equations like $n + 12 = 32$ and $43 = y - 17$

- Write a related equation.
- Complete the arithmetic to find the value of the variable.

When you **solve** an equation, you find a value for the variable that makes the equation true. $w = 36$ is the solution of the equation $w - 5 = 31$ because $36 - 5 = 31$ is true.

EXAMPLE 1

Solve and check. $n + 12 = 32$

SOLVE Write a related equation.

Equation $n + 12 = 32$
Related equation $n = 32 - 12$
Subtract to find n.

$\mathbf{32 - 12 = 20}$
So, $n =$ **20.**

CHECK To check, replace n with 20 in the original equation.

$n + 12 = 32$
↓
$\mathbf{20 + 12 = 32}$ (true) So, **20** is the solution.

EXAMPLE 2

Solve and check. $43 = y - 17$

SOLVE Write a related equation.

Equation $43 = y - 17$
Related equation $y = 43 + 17$
Add to find y.

$\mathbf{43 + 17 = 60}$
So, $y =$ **60.**

CHECK To check, replace y with 60 in the original equation.

$$43 = y - 17$$

$$\mathbf{43 = 60 - 17} \quad \text{(true)}$$

So, **60** is the solution.

GETTING READY

Solve and check.

1. $n - 11 = 27$ **2.** $y + 15 = 43$ **3.** $18 = p - 39$

4. $9.7 = 8.4 + s$ **5.** $m + 2\frac{1}{2} = 5\frac{1}{4}$ **6.** $\$37.60 + d = \$99.99$

EXERCISES

Solve and check.

1. $n + 6 = 10$ **2.** $m - 7 = 11$ **3.** $31 + r = 66$

4. $73 = t - 42$ **5.** $4.9 = q + 1.5$ **6.** $y - 0.8 = 2.3$

7. $z + \$2.40 = \$7.65$ **8.** $\$12.25 = h - \$13.50$ **9.** $r - 3\frac{1}{2} = 12\frac{1}{3}$

10. $\frac{3}{5} + f = 1\frac{1}{5}$ **11.** $\frac{19}{3} = 2 + w$ **12.** $\frac{1}{6} = b - \frac{1}{8}$

13. $n + \$135.11 = \$237.86$ **14.** $0.3 + p = 1$ **15.** $\frac{9}{4} = c - \frac{3}{2}$

***16.** $19 + y = 6$ ***17.** $450 = 721 + n$ ***18.** $3.7 = 3.7 + k$

CALCULATOR

Use a calculator to solve these equations.

1. $m + 6.4137 = 9.8142$ **2.** $57{,}314 = r - 46{,}298$

3. $h + \$3{,}210.35 = \$7{,}385.60$ **4.** $0.8370 = 0.9435 - s$

SOLVING EQUATIONS: MULTIPLICATION AND DIVISION

To solve equations like $4a = 24$, $\frac{k}{5} = 7$, and $12 = \frac{2}{3}y$

- Write a related equation.
- Complete the arithmetic to find the value of the variable.

EXAMPLE 1

Solve and check. $4a = 24$

SOLVE Write a related equation.

Equation $4a = 24$

Related equation $a = 24 \div 4$

Divide to find a.

$$\mathbf{24 \div 4 = 6}$$

So, $a =$ **6.**

CHECK To check, replace a with 6 in the original equation.

$4a = 24$

↓

$\mathbf{4 \times 6 = 24}$ (true) So, **6** is the solution.

EXAMPLE 2

Solve and check. $\frac{k}{5} = 7$

SOLVE Write a related equation.

Equation $\frac{k}{5} = 7$

Related equation $k = 7 \times 5$

Multiply to find k.

$$\mathbf{7 \times 5 = 35}$$

So, $k =$ **35.**

CHECK To check, replace k with 35 in the original equation.

$\frac{k}{5} = 7 \longrightarrow \mathbf{\frac{35}{5} = 7}$ (true) So, **35** is the solution.

EXAMPLE 3

Solve and check. $12 = \frac{2}{3}y$

SOLVE Write a related equation.

Equation $12 = \frac{2}{3}y$

Related equation $y = 12 \div \frac{2}{3}$

Complete the arithmetic.

$\mathbf{12 \div \frac{2}{3}}$ **is the same as** $\mathbf{12 \times \frac{3}{2}}$

$$\frac{\overset{6}{\cancel{12}}}{1} \times \frac{3}{\underset{1}{\cancel{2}}} = 18$$

So, $y =$ **18.**

CHECK $12 = \frac{2}{3}y$

$$12 = \frac{2}{3} \times 18 \quad \text{(true)}$$

So, **18** is the solution.

GETTING READY

Solve and check.

1. $8z = 48$ **2.** $\frac{n}{7} = 25$ **3.** $\frac{r}{16} = \$5.21$ **4.** $\frac{3}{7}s = 66$

EXERCISES

Solve and check.

1. $2y = 44$ **2.** $12k = 60$ **3.** $\frac{k}{8} = 7$

4. $\frac{t}{9} = 3$ **5.** $130 = \frac{n}{26}$ **6.** $125 = 25n$

7. $3.9 = 1.3y$ **8.** $3.5 = \frac{s}{1.2}$ **9.** $\frac{k}{10} = \$14.30$

10. $20d = \$48.00$ **11.** $\frac{1}{4}z = 120$ **12.** $108 = \frac{2}{9}w$

13. $181 = \frac{t}{42}$ **14.** $0.34 = \frac{f}{0.71}$ **15.** $\frac{5}{11}m = \frac{25}{33}$

***16.** $\$7.50 = \$1.25p$ ***17.** $\frac{47}{n} = 94$ ***18.** $\frac{y}{\frac{3}{4}} = 60$

WRITING EXPRESSIONS

To translate a phrase into a mathematical expression

- Choose a letter for the quantity.
- Look for key words that suggest the operation to use.
- Write the expression.

EXAMPLE 1

Write an expression for "twice as many albums as Fred has."

Choose a letter for the quantity. Fred's albums ⟶ a

Look for key words. twice as many ⟶ multiply by 2

Write the expression. twice as many albums as Fred has

2 × (twice as many) **a** (albums)

So, the expression is $2 \times a$, or **$2a$**.

Some key words or phrases that suggest multiplication are: **twice as many, times, double, triple.**

EXAMPLE 2

Write an expression for "80 mi less than Eunice drove."

Choose a letter for the quantity. miles Eunice drove ⟶ e

Look for key words. less than ⟶ subtract

Write the expression. Think: To get 80 miles less than e you subtract 80 from e.
So, write **$e - 80$.**

Some key words or phrases that suggest subtraction are: **less than, minus, decreased by, reduced by, difference.**

EXAMPLE 3

Write an expression for "the interest charges divided by 12."

Choose a letter. interest charges ⟶ I

Look for key words. divided by ⟶ divide I by 12

Write the expression. **$I \div 12$, or $\frac{I}{12}$**

So, the expression is $I \div 12$, or $\frac{I}{12}$.

Some key words or phrases that suggest division are: **divided by, half of, half as much.**

EXAMPLE 4

Write an expression for "10 degrees more than the starting temperature."

Choose a letter. starting temperature $\longrightarrow T$
Look for key words. more than $\longrightarrow$ add
Write the expression. $\mathbf{T + 10}$, or $\mathbf{10 + T}$

Some key words or phrases that suggest addition are: **more than, plus, increased by, added, sum.**

GETTING READY

Write an expression for each phrase. **Variables may differ.**

1. Three times the weight of the truck
2. The starting salary increased by $500
3. Half of the population of Tucson
4. The price reduced by a $70 discount

EXERCISES

Write an expression for each phrase.

1. Double the original number of applicants
2. The cost of the car plus a tax of $350
3. A third of the animals in the reservation
4. Six times as many openings as last year
5. The weekly salary with an added bonus of $50
6. The number of school days minus 20 days
7. The height of the ceiling decreased by 2 feet
8. Half as much land as the campground

*9. Half again as many spectators as yesterday

*10. A third fewer inches of snow than predicted

ON YOUR OWN

In some algebra problems, a number is described in words and you must figure out the number from the description. Here are two examples for you to try.

1. Seven more than half a number is 10. What is the number?
2. Six less than three times a number is 30. Find the number.

WRITING FORMULAS

To translate a sentence into a formula

- Choose letters for the quantities.
- Look for key words that suggest the operations.
- Write a formula.

EXAMPLE 1

The cost of mailing a batch of magazines is \$0.20 for each magazine plus a handling charge of \$12.00 for the batch. Write a formula for the mailing cost.

Choose letters for the quantities.	mailing cost ⟶ C number of magazines ⟶ n
Look for key words.	\$0.20 **for each** magazine ⟶ **multiply** \$0.20 by the number of magazines **plus** a handling charge ⟶ **add** \$12.00
Write a formula.	$C = \$0.20n + \$12.00$

EXAMPLE 2

A person's typical body weight in pounds is 5.5 times the difference between his or her height in inches and 40. Write a formula for typical body weight.

Choose letters for the quantities.	typical weight ⟶ w height in inches ⟶ h
Look for key words.	**difference** between height and 40 ⟶ **subtract** 40 from the height 5.5 **times** ⟶ **multiply** by 5.5
Write a formula.	$w = 5.5(h - 40)$

GETTING READY

Write a formula.

1. The cost of installing a carpet is figured at \$15, plus \$1.50 for each square yard of carpet. Write a formula for the installation cost.

2. The average of two numbers is half their sum. Write a formula for the average of two numbers.

EXERCISES

Write a formula.

1. The total amount in a savings account is the sum of the principal and the interest. Write a formula for the total amount.

2. The average word length in a paragraph is the number of letters divided by the number of words. Write a formula for the average word length.

3. The annual precipitation is the sum of the number of inches of rain and 0.1 times the number of inches of snow. Write a formula for the annual precipitation.

4. A sales representative's annual income is her salary plus a commission on her sales plus a bonus. Write a formula for her annual income.

5. A pitcher's earned run average is 9 times the number of earned runs given up divided by the number of innings pitched. Write a formula for the earned run average.

6. The weight on one wheel of a truck is the empty weight plus the weight of the load, all divided by 4. Write a formula for the weight on one wheel.

***7.** Use the formula in Example 1 to find the cost of mailing 500 magazines.

***8.** Use the formula in Example 2 to find the typical weight of a person 65 in. tall.

ON YOUR OWN

The same relationship can often be shown by several equivalent formulas. For example, the distance formula $d = rt$ can also be written $r = \frac{d}{t}$, or $\frac{d}{r} = t$. In each set of formulas below, two are equivalent and one is different. Find the one that is different.

1. $m + 13L = R$ $\quad$ $R + 13L = m$ $\quad$ $R - 13L = m$

2. $\frac{b}{a} = z$ $\quad$ $bz = a$ $\quad$ $\frac{a}{z} = b$

3. $y = 6 - 2s$ $\quad$ $2y + 4s = 12$ $\quad$ $4s - 12 = 2y$

4. $3h + 0.5 = s$ $\quad$ $s \div 3 = h + 0.5$ $\quad$ $(s - 0.5) \div 3 = h$

USING FORMULAS

To use a formula to solve a problem

- Replace the variables with given values.
- Solve the equation or complete the arithmetic.
- If no formula is given, first write one.

EXAMPLE 1

Find Jennifer's hourly wage if she earns $150 for a 40-hour week. Use the formula $HR = W$, where H is hours worked in a week, R is the hourly wage, and W is the weekly pay.

Replace H with 40 and W with 150.	$HR = W$ $\mathbf{40R = 150}$
To solve the equation, write a related equation.	$R = \mathbf{150 \div 40}$
Divide to find R.	$R = \mathbf{3.75}$

So, Jennifer's hourly wage is **$3.75.**

EXAMPLE 2

The formula $f = 1.8c + 32$ converts a Celsius temperature (c) into a Fahrenheit temperature (f). Use the formula to find the temperature in °F when it is 25°C.

Replace c with 25.	$f = \mathbf{1.8 \times 25 + 32}$
Complete the arithmetic.	$= \mathbf{45 + 32}$
	$f = \mathbf{77}$

So, the temperature is **77°F.**

°F is read "degrees Fahrenheit."
°C is read "degrees Celsius."

EXAMPLE 3

Swimmers use this quick rule to change yard-pool times to the equivalent meter-pool times: The meter swim will take about 1.1 times as long as the yard swim. Write a formula for this rule.
If it takes Benji 28 s (seconds) to swim 50 yd, how long will it take him to swim 50 m?

Choose letters for the quantities. — meter-pool time $\longrightarrow M$, yard-pool time $\longrightarrow Y$

Look for key words. — 1.1 times as long $\longrightarrow$ multiply by 1.1

Write the formula. — $\boldsymbol{M = 1.1Y}$

Replace Y with 28. — $= \mathbf{1.1 \times 28}$

Multiply to find M. — $M = \mathbf{30.8}$

So, it will take Benji about **30.8 s** to swim 50 m.

GETTING READY

Solve.

1. Use the formula in Example 1 to find what Jennifer's hourly wage would be if she earned \$210 for a 35-hour week.

2. Use the formula in Example 2 to find the temperature in °F when it is 5°C.

3. A bowler's final score is the handicap added to the scratch score. Write a formula for the final score. Find the final score for a 20-handicap bowler with a scratch score of 185.

EXERCISES

Solve.

1. Use the formula for weekly pay given in Example 1 to find what Jennifer's hourly wage would be if she earned \$360 for a 45-hour week.

2. The formula $d = gm$ gives the distance d in miles a car can go on g gallons of gasoline if it gets m miles per gallon. Find the miles per gallon of a car that travels 126 mi on 7 gal.

3. Use the formula in Example 2 to find the temperature in °F when it is 0°C.

4. Jody can swim 100 yd in 59 s. How long will it take her to swim 100 m? Use the formula in Example 3.

5. A delivery service estimates total job hours (T) by the formula $T = \frac{L}{30} + 0.25S$, where L is the length of the route and S is the number of stops. How long will a 60-mile route with 12 stops take?

6. A person's normal systolic blood pressure, in millimeters, can be expressed as 100 plus half his or her age. Write a formula and find the normal blood pressure for someone 16 years old.

***7.** Mildred can swim 25 m in 22 s. How long should it take her to swim 25 yd? Use the formula in Example 3.

***8.** Use the formula in Example 2 to find the Celsius temperature when it is 95°F.

Problem Solving Applications

READ • PLAN • SOLVE • CHECK

Washington High School offers students vocational training in retail sales. Supervisors help students gain experience in retailing. The school store serves as a training laboratory where students practice skills in retail sales.

AS YOU READ

- distributive education—a vocational program between schools and employers for classroom and on-the-job training

Answer these questions about retail sales training.

1. In the *distributive education* program, Ms. Brockridge has 15 students training at level I, 18 at level II, and 23 at level III. How many students are in the training program altogether?

2. May's training supervisor helped her find a summer job as a salesperson for an electronics supply company. If she works 39 h at $6.25 an hour, how much pay will she receive?

3. Greta worked 20 h last week and 23 h this week at her part-time job after school. Her pay for the 2 wk is $122.55. How much is she paid per hour?

4. A student bought a folder for $0.89, 3 erasers for $0.35 each, and a pen for $0.99 at the school store. He gave the cashier a $5 bill. Find his change.

5. One morning Richard used 50 pennies, 40 nickels, 50 dimes, 40 quarters, 10 ones, and 2 fives to fill the cash box in the school store. How much was that altogether?

6. When the school store closed, Richard counted $184.50 in the cash box. How much money did the store take in that day if $37.50 was in the cash box to start?

Career: Retail Sales Workers

READ • PLAN • SOLVE • CHECK

Salespersons, stock clerks, buyers, department supervisors, and store managers are just a few types of jobs in retail sales. A high school education and good social skills are the basic requirements for many of the jobs in this field.

AS YOU READ

- bonus—something given as an extra
- net—remaining after all charges are deducted

Solve.

1. Terry Rawley receives delivery orders for Wiley Grocers. Last month the orders of potatoes were 16 bu, 14 bu, 10 bu, 17 bu, and 9 bu. Also, 10 crates of lettuce and 2 crates of celery were delivered. How many total bushels of potatoes were delivered?

2. The buyer for Weyland's Hardware needs to restock several items. The order to the wholesaler includes 5 skillets at \$25 each, 3 variable-speed drills at \$19 each, and 2 coffeemakers at \$17 each. Find the total cost of the order.

3. Mara Hurst is a traveling sales representative. She is allowed \$200 per 5-day work week for expenses. If she spends about \$25 a night for lodging, how much does she have left for meals?

4. Last year Mr. Robbins made \$49,000 in total sales for his company. He earns a *bonus* of \$100 on each \$1,000 of his sales over \$30,000. How much did he earn in bonuses last year?

5. A salesperson is figuring the number of British Thermal Units (BTU) needed to cool a room. The formula used is BTU = room area × exposure factor × climate factor. If the room has an area of 238 ft^2, an exposure factor of 30, and a climate factor of 1.05, find the BTU needed.

6. A retail store manager is preparing the store budget for next year. *Net* sales are expected to be \$1,325,000. The total salaries for sales personnel should be 0.056 of net sales. How much should be budgeted for their salaries?

CHAPTER TEST 13

Evaluate each expression using the given values.

1. $a - 17; a = 33$
2. $\frac{y}{12}; y = 72$
3. $1.8r; r = 6.7$
4. $\frac{2}{3} + s; s = \frac{2}{3}$

Evaluate each formula using the given values.

5. $s = c + p; c = \$3.15$ and $p = \$0.80$
6. $h = gw; g = 1.4$ and $w = 20.5$
7. $A = B - 2C; B = 6\frac{1}{2}$ and $C = 1\frac{1}{4}$
8. $r = 5p + \frac{3}{5}; p = \frac{18}{25}$

For each equation, write a related equation with the variable on the left.

9. $3k = 93$
10. $\$35.20 = P + \$18.50$
11. $\frac{w}{7.4} = 12.2$

Solve.

12. $n + 18 = 31$
13. $153 = y - 68$
14. $7.2 = 1.8m$
15. $\frac{t}{5} = 15$
16. $\$13.60 = n + \$10.90$
17. $\frac{3}{4}h = 21$
18. $f - 8\frac{1}{8} = 7\frac{7}{8}$
19. $0.2 = \frac{r}{15}$
20. $8.94 = 16 - z$

Write an expression.

21. The wholesale cost increased by $35.

Write a formula.

22. The sale price is the regular price minus the discount. Write a formula for the sale price.

Solve.

23. Find the ground speed (g) of an airplane when the air speed (a) is 680 mph and the headwind speed (h) is 43 mph. Use the formula $g = a - h$.
24. The formula $W = 70M$ gives the number of words (W) a typist can type in M minutes. How many minutes will it take him to type 840 words?

Ratio and Proportion

RATIO

To write a ratio as a fraction

EXAMPLE 1

There are 22 students and 31 desks in Anita's English class. Write a ratio of the number of students to the number of desks.

> A **ratio** is a comparison of two numbers by division. The ratio $\frac{4}{5}$ is read 4 to 5.

Compare students to desks. **22 to 31**

Write the ratio as a fraction. $\mathbf{\frac{22}{31}}$

So, $\frac{22}{31}$ is the ratio.

EXAMPLE 2

There are 12 dogs and 14 cats in Mrs. Lockard's pet shop. Write these ratios: dogs to total animals; cats to total animals.

Think: 12 dogs + 14 cats = 26 total animals.

$$\frac{\textbf{dogs}}{\textbf{animals}} \begin{matrix} \longrightarrow \\ \longrightarrow \end{matrix} \frac{12}{26} \qquad \frac{\textbf{cats}}{\textbf{animals}} \begin{matrix} \longrightarrow \\ \longrightarrow \end{matrix} \frac{14}{26}$$

So, the ratios are $\frac{12}{26}$ and $\frac{14}{26}$.

EXAMPLE 3

Write each as a ratio.

45 km in 3 h	kilometers to hours	$= \frac{45}{3}$
\$6 for 2 lb of ground round steak	price to pounds	$= \frac{6}{2}$
80 mi on 4 gal of gas	miles to gallons	$= \frac{80}{4}$

GETTING READY

Write each as a ratio.

1. 12 games won to 8 games lost; games won to games lost

2. 2 candles burning out of 16 candles; candles burning to total candles

3. 500 km in 9 h; kilometers to hours

4. Dave had a single and a double for 5 "at bats." What was the ratio of hits to "at bats"?

EXERCISES

Write each as a ratio.

1. 300 pages in 8 h; pages to hours
2. 120 km on 14 L of gasoline; kilometers to liters
3. 1,000 mL ginger ale and 2,000 mL lemonade; ginger ale to lemonade
4. 60 cm long, 35 cm wide; length to width
5. 8 completed passes out of 19 attempts; completed passes to attempts
6. won 9 games and lost 13; games won to games lost
7. 6 cents tax on each dollar spent; tax to dollars
8. 10 triangles and 9 circles; triangles to circles
9. 30 free throws out of 41 attempts; free throws to attempts
10. $36 spent in 5 days; dollars spent to days
11. 19 revolutions in 2 s; revolutions to seconds
12. 82 yd gained in 29 carries; yards gained to carries

Use the chart to write each ratio.

	Singles	Doubles	Triples	Home Runs	Total Hits
Scouts	8	3	1	3	15
Hawks	4	2	2	1	9

13. triples by Hawks to triples by Scouts
14. triples by Scouts to triples by Hawks
15. singles by Scouts to doubles by Scouts
16. total home runs to total hits

Luci's Cycle Shop has 21 bicycles, 12 mopeds, 9 three-wheelers, and 17 full-sized motorcycles. Write each as a ratio.

17. bicycles to total cycles
18. three-wheelers to full-sized motorcycles
19. cycles with two wheels to cycles with three wheels
20. mopeds to full-sized motorcycles

SIMPLIFYING RATIOS

To write a ratio in simplest form

- Write the ratio as a fraction.
- Simplify the fraction.

EXAMPLE 1

Simplify the ratio 15 to 9.
Write a fraction. $\frac{15}{9}$

Simplify the fraction. Divide the numerator and denominator by 3. $\frac{15 \div 3}{9 \div 3} = \frac{5}{3}$

So, 15 to 9 simplified is $\frac{5}{3}$, or **5 to 3.**

EXAMPLE 2

In Saturday's game, the Astros had 6 hits and the Royals had 8 hits. Write a ratio, in simplest form, of the number of hits for the Royals to the total number of hits.

Total hits: $6 + 8 = 14$ Think: $\frac{\textbf{number of Royals' hits}}{\textbf{total number of hits}} \begin{matrix} \longrightarrow \\ \longrightarrow \end{matrix} \frac{8}{14}$

Simplify the fraction. Divide the numerator and denominator by 2. $\frac{8 \div 2}{14 \div 2} = \frac{4}{7}$

So, the ratio is $\frac{4}{7}$, or **4 to 7.**

GETTING READY

Simplify.

1. 8 to 20 **2.** $\frac{40}{18}$ **3.** 27 to 36 **4.** $\frac{85}{100}$

EXERCISES

Simplify.

1. 4 to 16 **2.** 60 to 48 **3.** 12 to 18 **4.** 65 to 100

5. $\frac{250}{900}$ **6.** $\frac{39}{26}$ **7.** $\frac{33}{77}$ **8.** $\frac{10}{25}$

9. Out of every \$100 that Mr. Warshaw earns, \$40 is paid in taxes. Write a ratio, in simplest form, of the amount of taxes to the amount of money earned.

10. If a goalie has about 25 saves in every 5 games that he plays, what is the ratio, in simplest form, of the number of saves to the number of games played?

MAKING MATH COUNT

Northbound Bus 36

Location	Time
24th and Francis	8:10 am
24th and Felix	8:21
24th and Corby	8:33
24th and 8th Ave	8:46
24th and Foley	9:02
Krug Park	9:18

Eastbound Bus 43

Location	Time
3rd and Foley	7:55 am
12th and Foley	8:08
24th and Foley	8:23
35th and Foley	8:37
49th and Foley	8:52
East Hills Mall	9:16

EXAMPLE

Ben arrives at 24th and Foley at 8:15 am. According to the schedule, how long will he have to wait for the bus to East Hills Mall?

A 18 min **B** 6 min
C 8 min **D** 53 min

Bus 43 leaves 24th and Foley for East Hills Mall at 8:23 am.

8:23 − 8:15 = 0:08, or 8 min

So, **C** is the correct answer.

Choose the correct answer.

1. What time does Bus 36 stop at 24th and Foley?
A 8:23 am **B** 9:02 am
C 8:33 am **D** 8:23 pm

2. What time does Bus 43 arrive at 49th and Foley?
A 8:37 am **B** 8:52 pm
C 8:52 am **D** 8:37 pm

3. Alexandra arrives at 24th and Corby at 8:27 am. How long does she have to wait for Bus 36?
A 6 min **B** 7 min
C 8 min **D** 5 min

4. Teri got on Bus 43 at 12th and Foley and got off at East Hills Mall. How long was she on the bus?
A 58 min **B** 62 min
C 68 min **D** 92 min

5. Fernando wants to take Bus 36 to Krug Park. He lives near 24th and Felix. What time can he catch the bus?
A 8:10 am **B** 7:55 am
C 8:08 am **D** 8:21 am

***6.** Jennifer lives near 3rd and Foley. She wants to go to Krug Park. At which stop should she transfer?
A 24th and Felix **B** Krug Park
C 3rd and Foley **D** 24th and Foley

WRITING RATIOS AS DECIMALS

To write a ratio as a decimal

- Write the ratio in fraction form.
- Divide the numerator by the denominator.

EXAMPLE 1

Write the ratio 7 to 8 as a decimal.

Write the ratio in fraction form. $\frac{7}{8}$

Divide the numerator by the denominator.

```
   0.875
8)7.000
  6 4
    60
    56
     40
```

So, the ratio 7 to 8 is equal to **0.875.**

EXAMPLE 2

In 1982 George had 175 hits in 449 times at bat. Write his batting average as a decimal to the nearest thousandth.

Write a ratio for hits to times at bat. $\frac{175}{449}$

Divide.

```
       0.3897
449)175.0000
     134 7
      40 30
      35 92
       4 380
       4 041
         3390
         3143
          247
```

Round 0.3897 to the nearest thousandth.

So, George's batting average was **0.390.**

GETTING READY

Write the ratio as a decimal to the nearest thousandth.

1. 5 to 8 **2.** $\frac{8}{11}$ **3.** 4 to 9 **4.** $\frac{3}{16}$

5. Pete had 140 hits in 431 times at bat. Find his batting average as a decimal to the nearest thousandth.

EXERCISES

Solve by writing each ratio as a decimal to the nearest thousandth.

Who had the better batting average?

1.

Player	At Bat	Hits
Joan	165	47
Edith	176	51

2.

Player	At Bat	Hits
Ken	210	65
Felix	406	116

3.

Player	At Bat	Hits
Jill	125	30
Sue	55	18

4.

Player	At Bat	Hits
Reggie	334	80
Carl	345	82

Who allowed the fewer hits per inning?

5.

Pitcher	Innings	Hits
Tom	166	120
Don	127	102

6.

Pitcher	Innings	Hits
Lois	82	75
Carol	103	88

Who gained more yardage per carry?

7.

Player	Carries	Yards
Otis	263	1,407
OJ	329	1,817

8.

Player	Carries	Yards
Earl	301	1,450
Walter	339	1,852

Who averaged more points per game?

9.

Player	Games	Points
Sally	24	550
Lou	22	532

10.

Player	Games	Points
Pete	73	2,273
Ralph	80	2,365

ON YOUR OWN

To find a pitcher's earned run average, use the formula:

$$\text{ERA} = \frac{\text{number of earned runs}}{\text{number of innings pitched}} \times 9$$

1. Roberto gave up 11 earned runs in 42 innings. What is his ERA?

2. In 75 innings, Sue gave up 28 runs, 5 of them unearned. Find her ERA.

RATES AND RATIOS

To determine which of two or more rates is best

- Write each rate as a ratio and compare.

EXAMPLE 1

Jane and Toby are pitchers. Jane has won 8 games and lost 6 games. Toby has won 6 games and lost 3 games. Who has the better record?

Write each rate as a ratio.

	Jane	Toby
$\frac{\text{games won}}{\text{games pitched}}$	$\frac{8}{14}$, or $\frac{4}{7}$	$\frac{6}{9}$, or $\frac{2}{3}$
Find the LCD and compare the two ratios.	$\frac{4}{7} \rightarrow \frac{4 \times 3}{7 \times 3} = \frac{12}{21}$	$\frac{2}{3} \rightarrow \frac{2 \times 7}{3 \times 7} = \frac{14}{21}$

Since $\frac{14}{21} > \frac{12}{21}$, **Toby** has the better record.

EXAMPLE 2

Maria can type 212 words in 5 min. Carl can type 243 words in 6 min. Whose typing rate is better?

> A rate may be written as a ratio.

Write each typing rate as a ratio.

	Maria's Rate	Carl's Rate
$\frac{\text{number of words}}{\text{number of minutes}}$	$\frac{212}{5}$	$\frac{243}{6}$
Find the number of words per minute. Write as a decimal.	$5\overline{)212.0}$ = 42.4; 20; 12; 10; 2 0; 2 0	$6\overline{)243.0}$ = 40.5; 24; 3 0; 3 0

Since 42.4 > 40.5, **Maria's** typing rate is better.

EXAMPLE 3

Susan can buy 8 apples for 78¢ or 10 apples for 95¢. Which is the better buy?

$\frac{\text{cost}}{\text{number of apples}}$	$\frac{78¢}{8}$	$\frac{95¢}{10}$
Find the cost of one apple.	$8\overline{)78.00}$ = 9.75 = 9.8¢	$10\overline{)95.0}$ = 9.5 = 9.5¢

Since 9.5¢ < 9.8¢, **10 apples for 95¢** is the better buy.

GETTING READY

Solve by comparing ratios.

1. Who averaged more yardage per carry?

	Yards	Carries
Lopez	120	20
Walters	130	13

2. Which team has the better record?

	Won	Games
A	18	27
B	24	32

3. Which is the better buy?

3 pairs of socks for $6.60
4 pairs of socks for $8.75

4. Which is the better buy?

32 oz orange juice for $1.28
6 oz orange juice for $0.27

EXERCISES

Solve by comparing ratios.

1. Which team had the better record?

	Won	Played
Cubs	8	32
Eagles	6	18

2. Which car gets better mileage?

Car A: 150 km on 25 L
Car B: 180 km on 30 L

3. Which is the better rate?

135 km in 3 h
204 km in 4 h

4. Which is the better buy?

$1.50 for 12 oz
$1.20 for 8 oz

5. Which is the better buy?

10 pencils for $1.10
6 pencils for $0.60

6. Who had the better pass completion record?

Harrell: 14 completed out of 35
Jackson: 12 completed out of 32

7. Who worked faster?

Sally: 1,253 parts in 7 h
Joan: 1,500 parts in 8 h

8. Who drove faster in the race?

André: 330 mi in 3 h
Bill: 410 mi in 4 h

9. Which driver had the fastest driving speed? The slowest?

Teri: 200 km in 5 h
Jacob: 225 km in 6 h
Phil: 270 km in 8 h

10. Which driver spent the least per km?

Betty: $35.50 for 250 km
Judi: $39.20 for 280 km
Kathy: $45.00 for 300 km

PROBABILITY

To find the probability of a simple event

- Use the ratio: Probability (P) $= \frac{\text{favorable outcomes}}{\text{possible outcomes}}$.

EXAMPLE 1

When tossing a dime, what is the probability of getting tails in one toss?

There are 2 possible outcomes: heads or tails. The favorable outcome (tails) can occur in 1 way.

$$\frac{\text{favorable outcomes}}{\text{possible outcomes}} = \frac{1}{2}$$

So, P (tails) $= \frac{1}{2}$.

P (tails) is read probability of tails.

EXAMPLE 2

Find the probability of the pointer's stopping on an odd number.

There are 6 possible outcomes: 1, 2, 3, 4, 5, 6.
The favorable outcomes (odd numbers) are 1, 3, and 5.

$$\frac{\text{favorable outcomes}}{\text{possible outcomes}} \longrightarrow \frac{3}{6} = \frac{3 \div 3}{6 \div 3} = \frac{1}{2}$$

So, P (odd number) $= \frac{1}{2}$.

EXAMPLE 3

Eloise has 4 nickels, 6 dimes, and 2 quarters in her wallet. She chooses one coin at random. Find the probability of her choosing a nickel.

There are $4 + 6 + 2 = 12$ coins. There are 4 nickels.

$$\frac{\text{favorable outcomes}}{\text{possible outcomes}} \longrightarrow \frac{4}{12} = \frac{4 \div 4}{12 \div 4} = \frac{1}{3}$$

So, P (nickel) $= \frac{1}{3}$.

GETTING READY

1. Find the probability of spinning a 4.

2. What is the probability of spinning a 5?

3. If the faces of a number cube are labeled 1, 2, 3, 4, 5, 6, find the probability of tossing a 6.

4. You have 2 black socks, 2 blue socks, and 2 white socks in a drawer. One sock is chosen at random. What is the probability of picking a white sock?

EXERCISES

1. A number dial is labeled 1 through 10. Find the probability of spinning an even number.

2. A box contains 2 red, 5 black, and 3 green balls. One ball is drawn at random. What is the probability of drawing a black ball?

3. A number cube with faces labeled 1 through 6 is tossed. What is the probability that an odd number will show facing up?

4. Michael has 3 nickels, 5 dimes, 1 quarter, and 6 pennies in his pocket. One coin is drawn at random. Find each probability: P (nickel), P (dime), P (penny).

5. There are 900 marbles in a box. 500 are blue and 1 is black. One marble is drawn. Find P (blue marble). Find P (black marble).

6. Suppose you write the letters of the word TENNESSEE on cards and place them in a box. What is the probability of drawing an E on one draw?

7. Two nickels are tossed. Here are the possible outcomes:

First nickel	Second nickel
Heads	Tails
Tails	Heads
Heads	Heads
Tails	Tails

What is the probability of two heads?

8. Two pennies are tossed 100 times. Predict how many times both coins will come up tails.

SOLVING PROPORTIONS

To determine whether a statement shows a proportion
To find a missing number in a proportion

EXAMPLE 1

Does this statement show a proportion?

$$\frac{4}{6} = \frac{12}{18}$$

Compare the cross products.

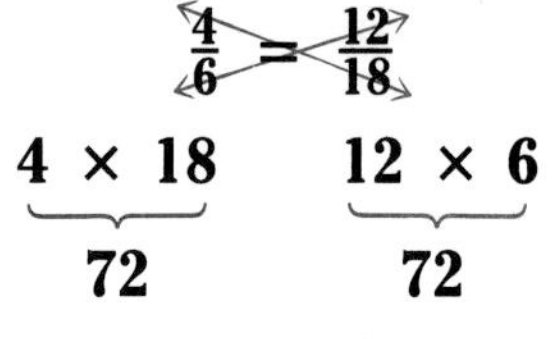

$$\underbrace{4 \times 18}_{72} \qquad \underbrace{12 \times 6}_{72}$$

Since the cross products are equal, $\frac{4}{6} = \frac{12}{18}$ **is a proportion.**

> An equation showing that two ratios are equal is a **proportion.**

> The cross products of a proportion are equal.

EXAMPLE 2

Find the missing number in the proportion.

$$\frac{5}{65} = \frac{2}{n}$$

Write the cross products.

$$\frac{5}{65} \times \frac{2}{n}$$

$$\underbrace{5 \times n}_{5 \times n} \qquad \underbrace{2 \times 65}_{130}$$

Equation: $5n = 130$
Write a related equation. $n = 130 \div 5$
Divide to find n. $n = 26$
So, the missing number in the proportion is **26.**

EXAMPLE 3

Muna hit 3 home runs in 28 times at bat. At this rate, how many home runs would she hit in 84 times at bat?

Let h = number of home runs in 84 times at bat.

Think: $\frac{\text{home runs}}{\text{times at bat}} \longrightarrow \frac{3}{28} = \frac{h}{84}$

Write the cross products. $3 \times 84 \qquad 28 \times h$
Equation: $252 = 28h$
Write a related equation. $h = 252 \div 28$
Divide to find h. $h = 9$
So, Muna would hit **9 home runs.**

GETTING READY

Does the statement show a proportion?

1. $\frac{8}{3} = \frac{24}{9}$

2. $\frac{3}{4} = \frac{9}{15}$

3. $\frac{12}{54} = \frac{4}{18}$

Find the missing number in the proportion.

4. $\frac{9}{6} = \frac{12}{n}$

5. $\frac{n}{10} = \frac{7.5}{25}$

6. $\frac{57}{n} = \frac{19}{3}$

EXERCISES

Does the statement show a proportion?

1. $\frac{6}{8} = \frac{9}{12}$

2. $\frac{15}{10} = \frac{5}{3}$

3. $\frac{2}{90} = \frac{3}{135}$

Find the missing number in the proportion.

4. $\frac{6}{8} = \frac{3}{n}$

5. $\frac{n}{28} = \frac{2}{7}$

6. $\frac{15}{54} = \frac{4}{n}$

7. $\frac{3}{8} = \frac{n}{20}$

8. $\frac{12}{n} = \frac{3}{96¢}$

9. $\frac{5.4}{27} = \frac{n}{100}$

***10.** Kenny completes 5 out of every 9 passes he throws. At this rate, how many passes would he complete in a game in which he threw 27 passes?

***11.** If 14 kg of cement are used to make 70 kg of concrete, how many kilograms of concrete can be made with 30 kg of cement?

CALCULATOR

Sometimes it is easier to use your calculator to find a missing number in a proportion.

$\frac{24}{1.28} = \frac{n}{3.84}$

Multiply 24 by 3.84.
Then divide by 1.28.
Your calculator shows 72.
So, $n = 72$.

Try these on your calculator.

1. $\frac{0.5}{3.14} = \frac{n}{7.85}$

2. $\frac{80}{3,600} = \frac{108}{n}$

3. $\frac{1.75}{1.25} = \frac{n}{14.2}$

Problem Solving Skill

To solve word problems using proportions

EXAMPLE 1

The Andersons use 2 storage cubes for 8 video-game cartridges. They plan to buy 20 more game cartridges. How many more storage cubes will they need?

Write a proportion.
Let n = number of storage cubes needed.

$$\frac{\text{storage cubes}}{\text{game cartridges}} \longrightarrow \frac{2}{8} = \frac{n}{20}$$

Cross multiply. $\mathbf{2 \times 20 = 8 \times n}$

$\mathbf{40 = 8 \times n}$

Write a related equation. $\mathbf{n = 40 \div 8}$

$\mathbf{n = 5}$

So, **5** storage cubes are needed.

EXAMPLE 2

Fruit juice sells at 6 cans for $1.92. How many cans can be bought with $5. 00?

Write a proportion.
Let n = number of cans that can be bought for $5.00.

$$\frac{\text{number of cans}}{\text{cost}} \longrightarrow \frac{6}{\$1.92} = \frac{n}{\$5.00}$$

$\mathbf{6 \times \$5.00 = \$1.92 \times n}$

$\mathbf{\$30.00 = \$1.92 \times n}$

Write a related equation. $\mathbf{n = \$30.00 \div \$1.92}$

Divide. $\mathbf{1.92\overline{)3\,0.0\,0\,0\,0\,0}}$ quotient $\mathbf{1\,5.6\,2\,5}$ $\qquad \mathbf{n = 15.625}$

So, **15** cans of juice can be bought, with some money left over.

Solve.

1. If 20 L of gasoline cost \$7.25, how much would 75 L cost?

2. Erica paid \$57 in sales tax on a used car that sold for \$950. How much sales tax did she pay on \$1?

3. If 5 peanut butter cookies weigh 30 g, about how many cookies are in a package weighing 255 g?

4. Knee-high socks are selling at 3 pairs for \$2.95. How many pairs can be bought with \$5.00?

5. A 180-lb astronaut weighs about 32 lb on the moon. About how much does a 450-lb machine weigh on the moon?

6. If 3 acres yield 90 bu of soybeans, how many bushels will 10 acres yield?

7. A pitcher gave up 18 runs in 54 innings. At this rate, how many runs would he give up in 9 innings?

8. A car can travel 73.5 mi on 3 gal of gas. How far can it travel on a full tank of 16 gal?

9. A recipe for two people calls for $\frac{1}{2}$ c of cheese. How much cheese is needed to serve six people?

10. An airplane can travel 2,850 mi in 6 h. How long will it take to travel 1,900 mi at the same speed?

11. In the first 3 games of the season, Tony gained a total of 420 yd. At the same rate, during which game can he expect to reach 1,000 yd?

12. Jocile earned \$31.50 interest on savings of \$450. At the same rate, how much interest will she earn on \$650?

13. A room addition with a floor area of 30 m^2 costs \$8,900. At that rate, how much would an addition with a floor area of 45 m^2 cost?

14. An oil tank in the form of a cylinder holds 1,500 L when filled to a depth of 3 m. How many liters are in a tank when the gauge reads 1.2 m?

15. Tapes are selling at 3 for \$10.95. How many can you buy with \$20?

16. A basket of apples weighs 20 kg. If 3 apples weigh about 0.5 kg, about how many apples are in the basket?

Problem Solving Applications

READ • PLAN • SOLVE • CHECK

The students enrolled in the World of Construction courses at Carver Vocational High School learn about the variety of jobs in construction work. The courses cover the planning, designing, building, and managing of typical construction projects. Safety on the job and in the classroom is emphasized.

AS YOU READ

- blueprint—a print in white on a bright blue background used for copying construction plans
- patio—a paved area that is attached to a house

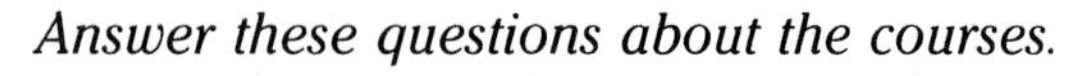

Answer these questions about the courses.

1. Concrete is made by mixing cement, sand, and gravel. The amount of gravel needed is twice the amount of sand needed. Write the ratio of the amount of gravel to the amount of sand.

2. Dana Crandle and Kim Woo are laying a *patio* for their class project. If the ratio of amount of cement to gravel is 1 to 4, how much gravel should they use with 112 yd^3 of cement?

3. Matthew Weeks is making a scale drawing. He decides to use the scale 2 in. = 15 ft. If the length of the room is to be 37.5 ft, what should he make its length on the scale drawing?

4. Don knows that a 500-ft coil of wire weighs 75 lb. He has used 125 ft from the coil. What is the weight of the remaining wire?

5. Katrina Heath is drawing a *blueprint* of a building that is $200' \times 135'$. She decides to use the scale 1 in. = 8 ft. On the blueprint, what is the length and what is the width of the building?

*6. The specific gravity of steel is the ratio of the weight of 1 ft^3 of steel to the weight of 1 ft^3 of water. Water weighs 62.5 lb/ft^3 and steel weighs 490 lb/ft^3. Express the specific gravity of steel as a decimal.

Career: Construction Workers

READ • PLAN • SOLVE • CHECK

Construction workers are employed at a variety of sites, such as houses, highways, dams, and bridges. A high school or vocational school diploma and an apprenticeship are recommended for gaining the variety of skills needed in the construction business.

AS YOU READ

- debris—the remains of something destroyed

Solve.

1. Roy Grant is a construction apprentice. He has completed 3 of the 4 y of training required. What part of the total has he completed?

2. Tony Harris is an independent cement mason. If 3,300 kg of sand makes $3m^3$ of concrete, how much concrete can he make from 7,700 kg of sand?

3. Chad Hawkins owns the Hawkins Company. He rents dump trucks of two sizes. The larger size dump truck holds 10 yd^3 of dirt. If there are 750 yd^3 of dirt to be moved, how many loads would be needed using the larger truck?

4. Heather Green is a construction supervisor with Albright Contractors. On the blueprint she is reading, 1 in. represents 12 ft. What is the actual length of a room that is 2.5 in. on the blueprint?

5. Last week, 3 days were spent demolishing and removing *debris* after the destruction of a house. At that rate, how much time should be allowed for demolition and removal of debris in the destruction of a store 4.5 times as large as the house?

6. Edmund's brother Weymouth drives a front loader for Sweet Construction Company. It takes him 3 buckets to fill the small dump truck, which holds 6 yd^3. How many buckets are needed to fill the large dump truck, which holds 10 yd^3?

CHAPTER TEST 14

Write each ratio as a fraction.

1. 18 completed passes out of 29 attempts; completed passes to attempts

2. A basketball team won 23 games and lost 7; games won to games lost

Simplify.

3. $\frac{12}{36}$

4. 100 to 75

5. $\frac{99}{22}$

Solve by writing each ratio as a decimal to the nearest thousandth.

6. Which is the better completion record?
12 completed passes out of 18 attempts
14 completed passes out of 22 attempts

7. Which is the better batting average?
68 hits out of 200 "at bats"
100 hits out of 350 "at bats"

Solve by comparing ratios.

8. Which is the better buy?
6 pencils for \$0.48
4 pencils for \$0.35

9. Which is the better rate?
462 km for 6 h
616 km for 7 h

10. Find the probability of spinning a 2.

11. A number cube with faces labeled 1 through 6 is tossed. What is the probability of tossing a 5?

Does the statement show a proportion?

12. $\frac{3}{6} = \frac{12}{24}$

13. $\frac{5}{9} = \frac{25}{36}$

Find n in the proportion.

14. $\frac{2}{7} = \frac{2}{n}$

15. $\frac{n}{18} = \frac{3}{6}$

Solve.

16. Earl bowled 14 strikes in his first 10 games. At the same rate, how many strikes will he make in 30 games?

17. Mary works 4 h to earn \$15. At the same rate, how much will she earn for 6 h work?

18. If 3 cups of pancake mix make 24 pancakes, how many cups are needed for 16 pancakes?

19. A package of 24 paper cups costs \$1.25. How much would 36 paper cups cost?

Geometry

ANGLES

To classify and measure angles
To draw angles

- Classify by comparing to a right angle.
- Measure and draw by using a protractor.

Angles can be named by 3 letters. The middle letter names the vertex.

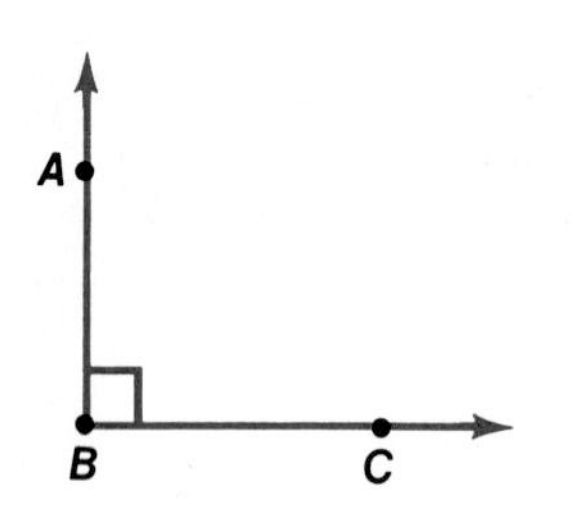

right angle
∠ *ABC* measures exactly 90°.

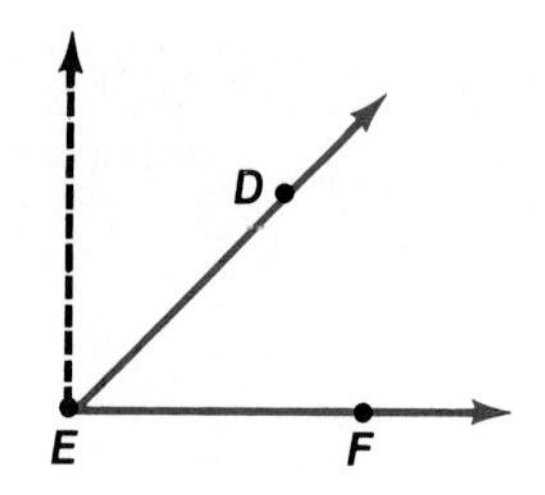

acute angle
∠ *DEF* measures less than 90°.

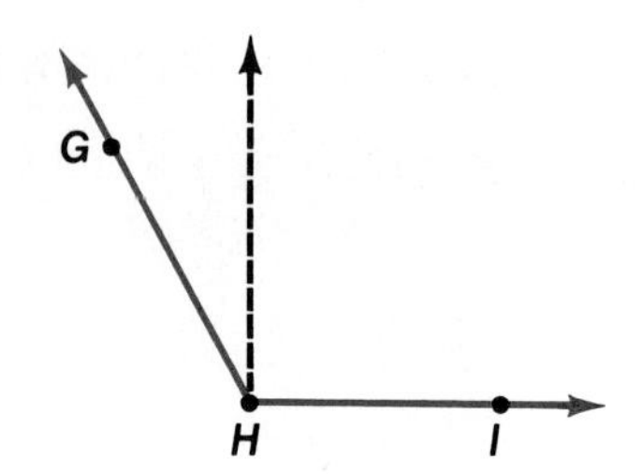

obtuse angle
∠ *GHI* measures more than 90°.

EXAMPLE 1

Classify, then measure ∠ *WRH*.

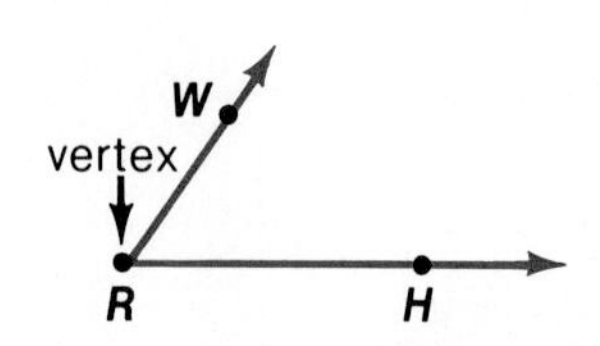

Classify the angle.
∠ *WRH* is **acute.**

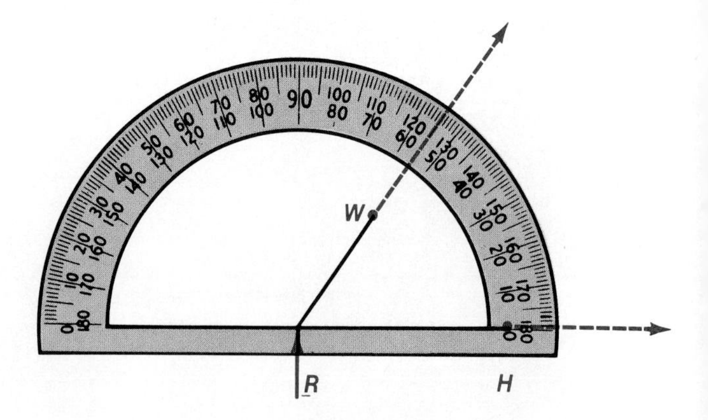

Place the center of the protractor at the vertex *R*. Extend the sides of the angle to cross the scale, if necessary. Since ∠ *WRH* is acute, read the measure of the angle on the lower scale of the protractor.

So, ∠ *WRH* measures **55°.**

EXAMPLE 2

Use a protractor to draw an angle that measures 130°.

Step 1 Draw line AB.

Step 2 Place the protractor center on A with line AB at 0°.

Step 3 Mark C at 130° on the lower scale of the protractor.

Step 4 Draw a line from C to A.

So, $\angle BAC$ measures **130°.**

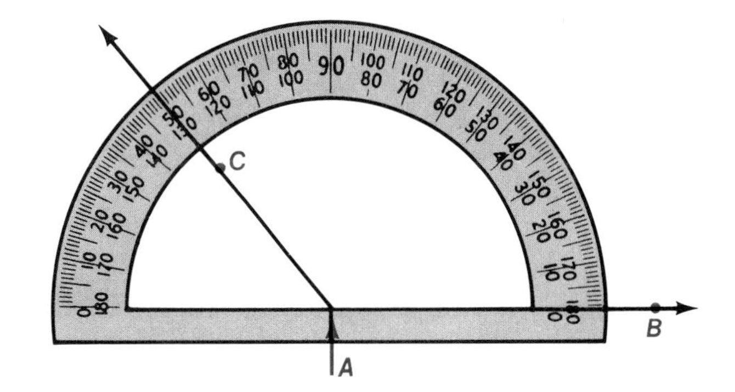

GETTING READY

Classify, then measure these angles.

1.

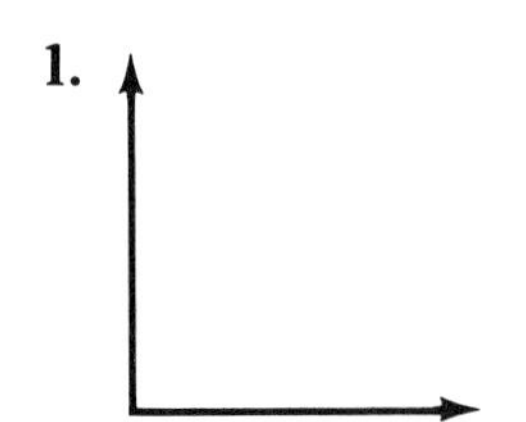

2.

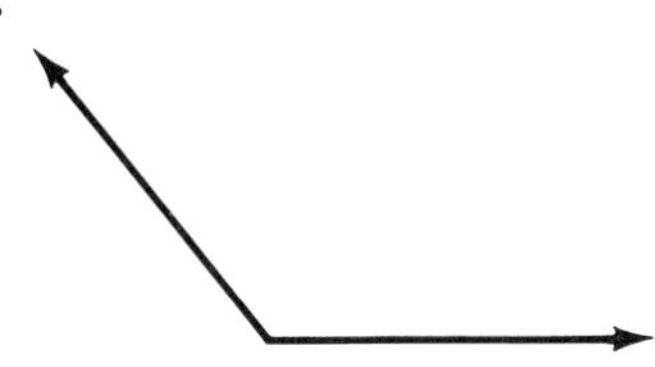

3.

4.

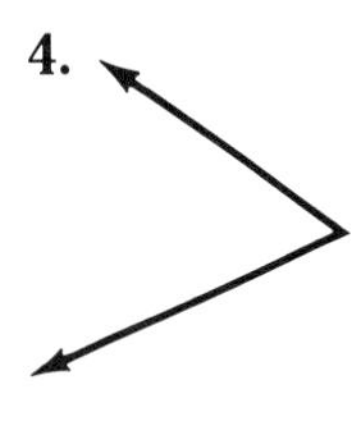

Use a protractor to draw an angle having each given measure.

5. 90° **6.** 40° **7.** 135° **8.** 73° **9.** 15° **10.** 170°

EXERCISES

Classify, then measure these angles.

1. **2.**

3.

4.

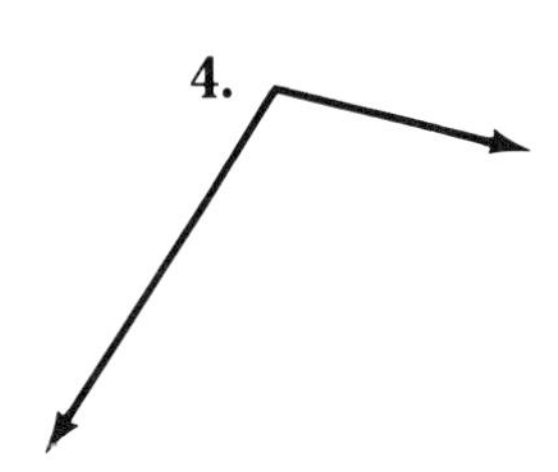

Use a protractor to draw an angle having each given measure.

5. 50° **6.** 140° **7.** 110° **8.** 20° **9.** 86° **10.** 70°

POLYGONS

To identify and name polygons

A **polygon** is a closed figure whose sides are line segments. Polygons are named by the number of sides they have.

Number of sides	Name of polygon
3	triangle
4	quadrilateral
5	pentagon
6	hexagon
8	octagon

EXAMPLE

Name the polygon.
The polygon has 4 sides.
It is a **quadrilateral.**

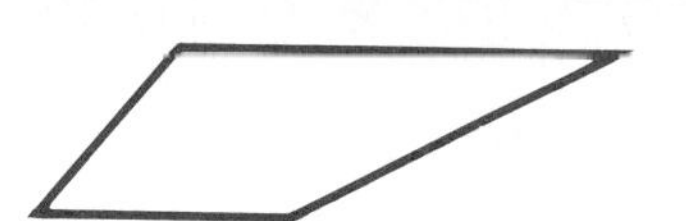

GETTING READY

Name each polygon.

1.

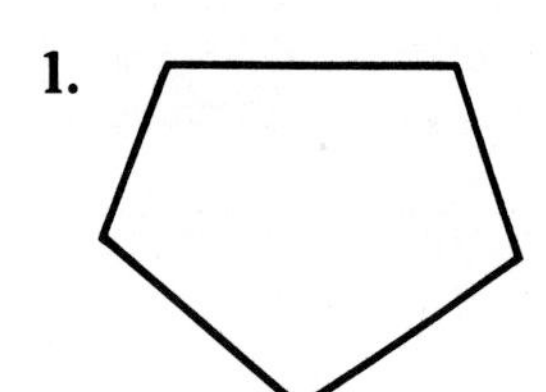

2.

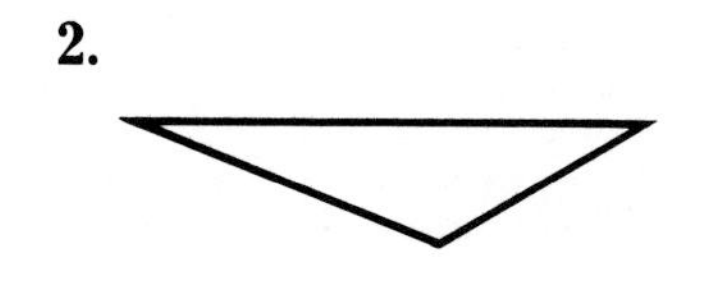

3.

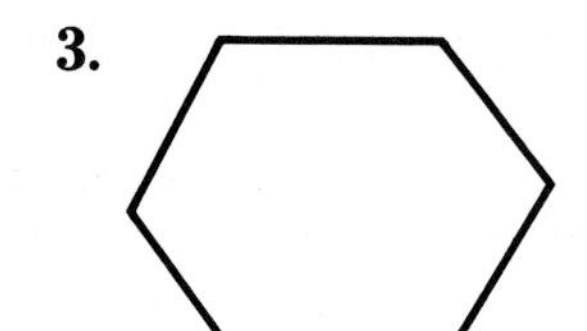

EXERCISES

Name each polygon.

1.

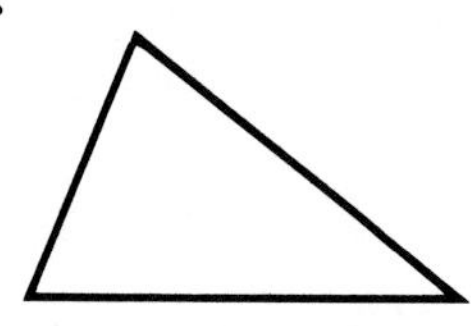

2.

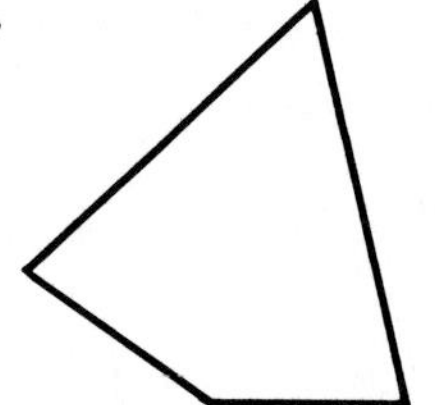

3.

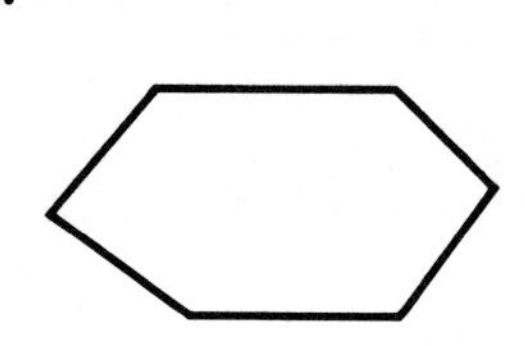

4.

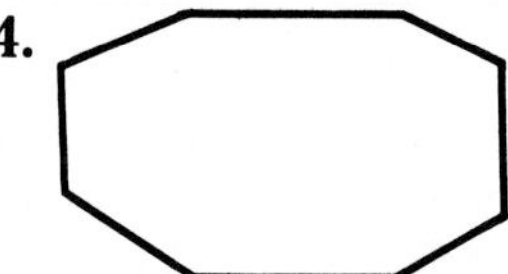

5.

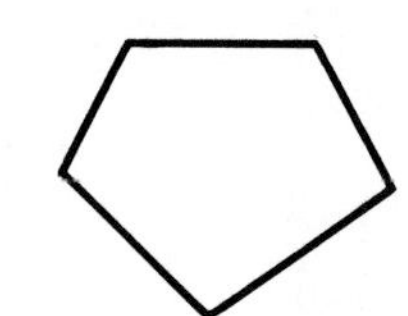

6.

Problem Solving Skill

To solve problems involving scale drawings

- Use the scale to write a proportion.
- Solve the proportion.

EXAMPLE 1

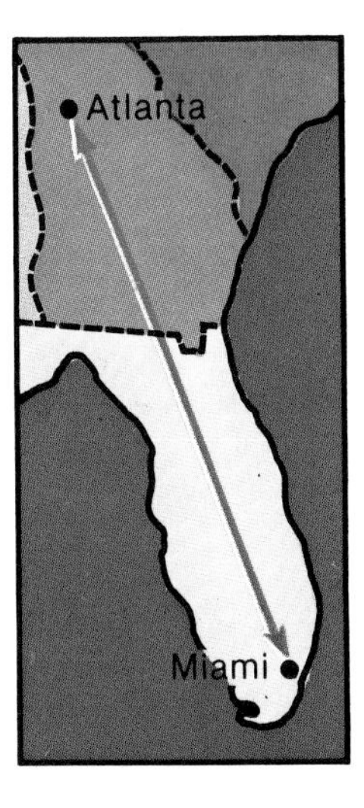

On a map, the scale is 1 in. = 325 mi. If the distance from Atlanta to Miami measures 2 in. on the map, find the actual distance.

Write a proportion. Let n = the actual distance.

Distance on map ⟶ / Actual distance ⟶ $\frac{1}{325} = \frac{2}{n}$

Solve the proportion.

$$\frac{1}{325} = \frac{2}{n}$$
$$n \times 1 = 2 \times 325$$
$$n = 650$$

So, the distance from Atlanta to Miami is **650 mi.**

EXAMPLE 2

A science lab in a scale drawing measures 2.8 cm × 1.9 cm. The scale is 1 cm = 6 m. Find the actual dimensions of the science lab.

Write a proportion to find the actual length and width.

Length on drawing ⟶ / Actual length ⟶ $\frac{1}{6} = \frac{2.8}{l}$ Width in drawing ⟶ / Actual width ⟶ $\frac{1}{6} = \frac{1.9}{w}$

$l \times 1 = 6 \times 2.8$ $l = 16.8$ m $w \times 1 = 6 \times 1.9$ $w = 11.4$ m

So, the actual dimensions of the science lab are **16.8 m × 11.4 m.**

EXERCISES

1. On a map, the distance from Boston to Dallas is 6.3 cm. Find the actual distance if the scale is 1 cm = 500 km.

2. In a scale drawing, a car is 9 cm long. What is the actual length of the car if the scale is 1 cm = 60 cm?

3. On a map, the scale is 1 in. = 350 mi. The distance from Chicago to Cincinnati is 0.8 in. Find the actual distance.

4. In a floor plan, the bedroom is $2\frac{1}{4}$ in. long. The scale is 1 in. = 6 ft. Find the actual length of the bedroom.

TRIANGLES

To classify triangles by the measures of their angles
To classify triangles by the measures of their sides

A polygon with three sides is called a **triangle.** Triangles can be classified according to the measures of their angles.

EXAMPLE 1

Classify these triangles by the measures of their angles.

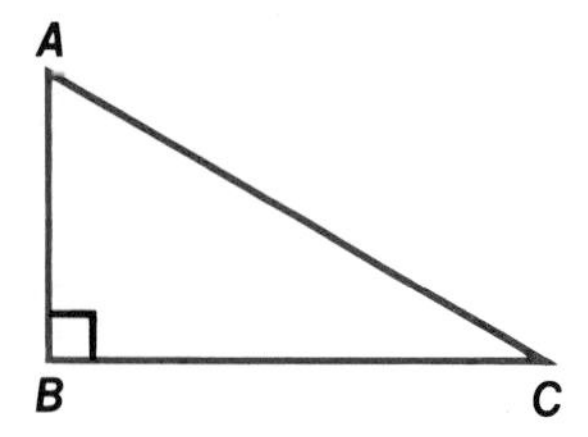

$\triangle$ *ABC* is a **right** triangle. It has a right angle.

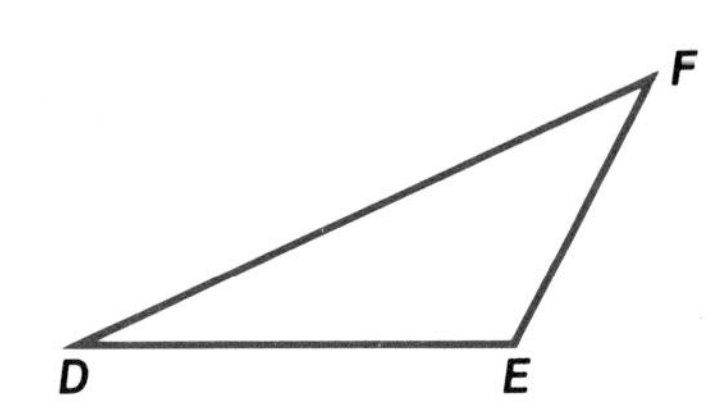

$\triangle$ *DEF* is an **obtuse** triangle. It has one obtuse angle.

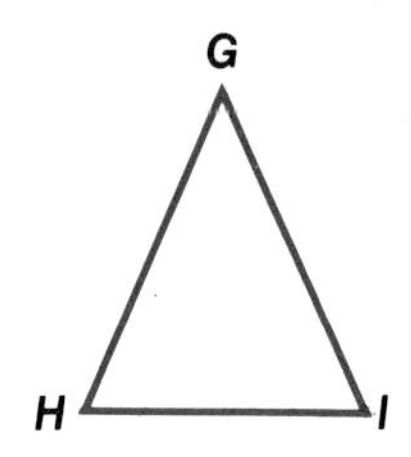

$\triangle$ *GHI* is an **acute** triangle. Each angle is an acute angle.

Triangles can also be classified according to the number of equal sides.

EXAMPLE 2

Classify these triangles by the number of equal sides.

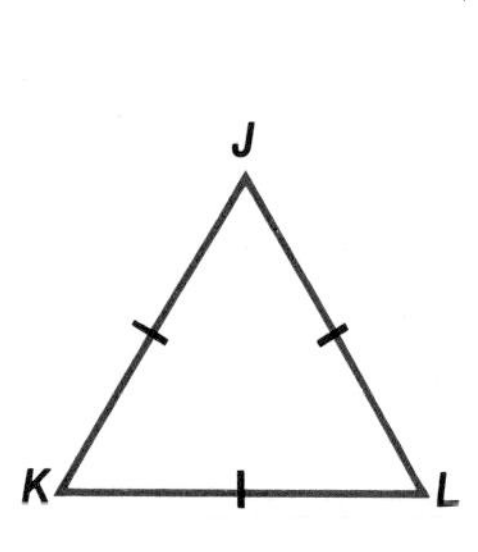

$\triangle$ *KLJ* is an **equilateral** triangle. It has 3 equal sides.

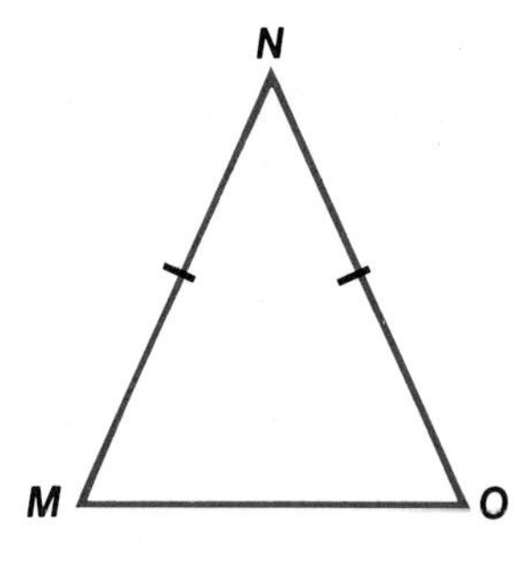

$\triangle$ *MNO* is an **isosceles** triangle. It has 2 equal sides.

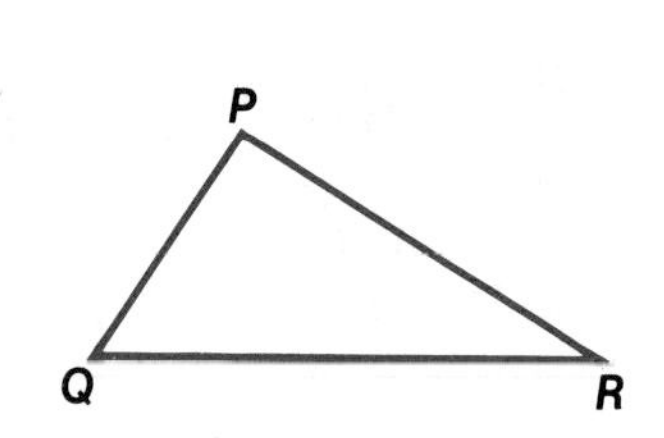

$\triangle$ *PQR* is a **scalene** triangle. It has no equal sides.

GETTING READY

Classify the triangles by the measures of their angles and the number of equal sides.

1. 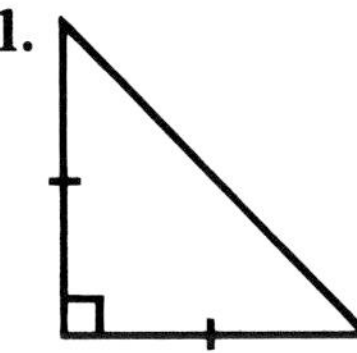

2. 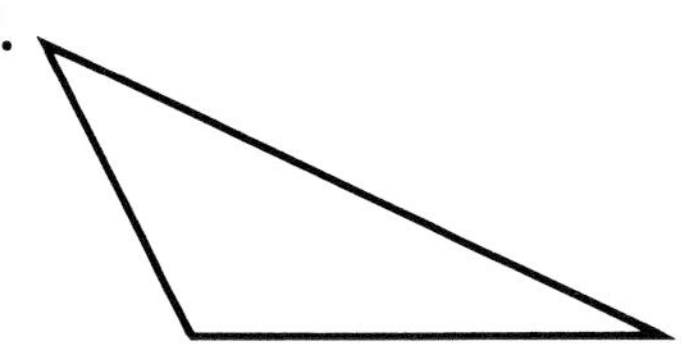

3. 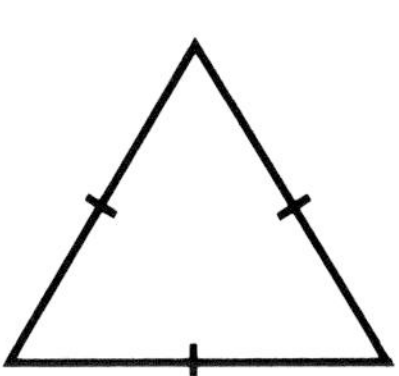

EXERCISES

Classify the triangles by the measures of their angles.

1. 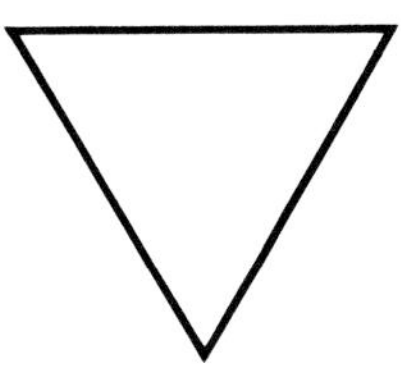

2. 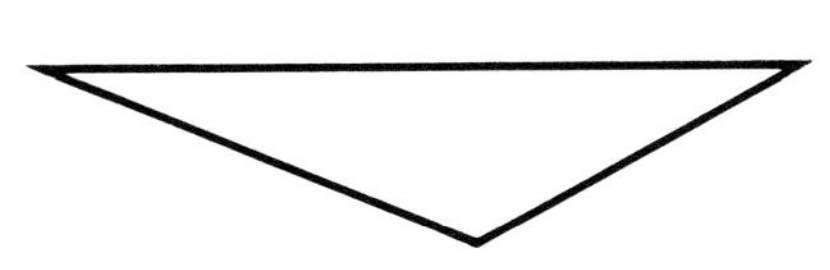

3. 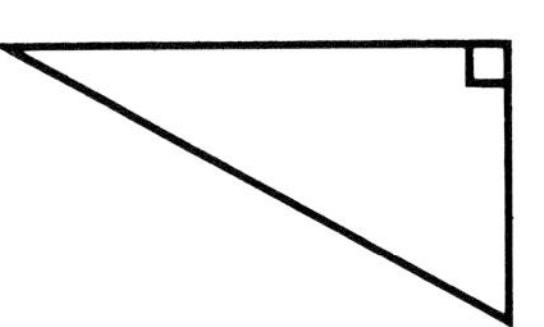

Classify the triangles by the number of equal sides.

4. 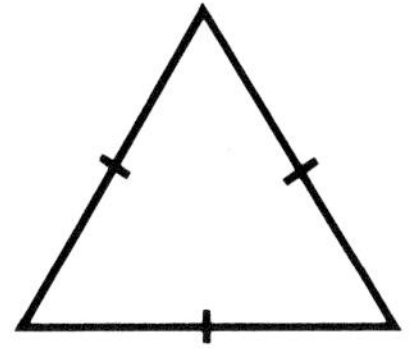

5. 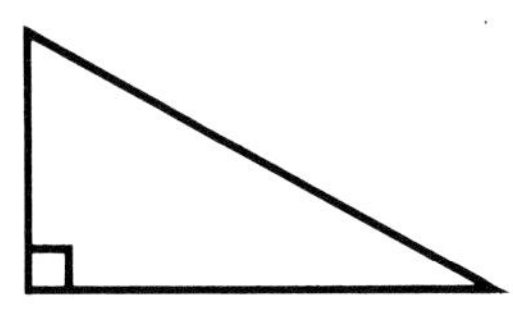

6. 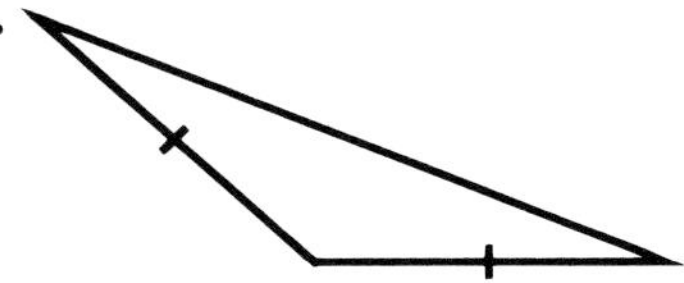

True or false?

***7.** An equilateral triangle is also an isosceles triangle.

***8.** A right triangle can also be an obtuse triangle.

ON YOUR OWN

Use the number pattern in the Example to complete the statements. Remember, $5^2 = 5 \times 5$. You can use a calculator to square numbers.

Example: $(5 - 2) \times (5 + 2) = 5^2 - 2^2 = 25 - 4 = 21$

Complete.

1. $(7 - 1) \times (7 + 1) =$ __?__ $- 1^2 =$ __?__

2. $(13 - 5) \times (13 + 5) =$ __?__ $-$ __?__ $=$ __?__

3. $(189 - 14) \times (189 + 14) =$ __?__ $-$ __?__ $=$ __?__

SPECIAL QUADRILATERALS

To identify and name special quadrilaterals

Polygons with four sides and four angles are called **quadrilaterals.** Here are some special quadrilaterals.

parallelogram

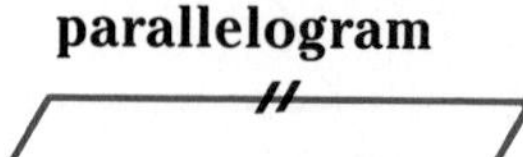

Opposite sides are equal and parallel.

rhombus

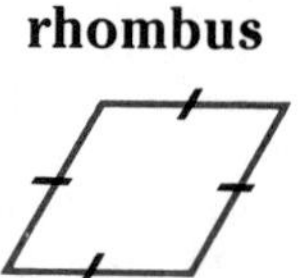

A parallelogram with 4 equal sides

rectangle

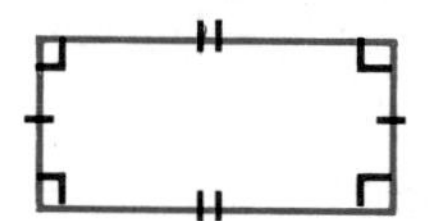

A parallelogram with 4 right angles

square

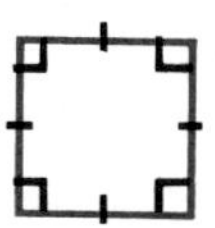

A rectangle with 4 equal sides

EXAMPLE

Name the quadrilateral.

The figure is a parallelogram with 4 equal sides. The angles are not right angles. So, the quadrilateral is a **rhombus.**

GETTING READY

Name each quadrilateral.

1. 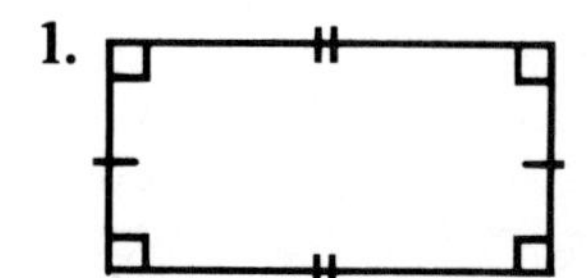

2. 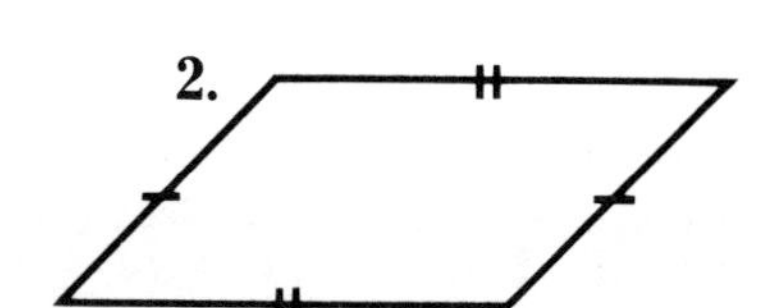

3. 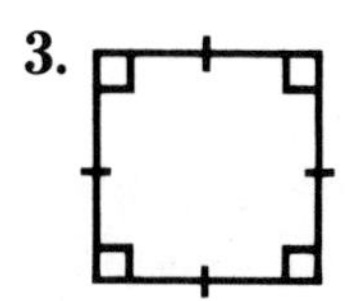

EXERCISES

Name each quadrilateral.

1. 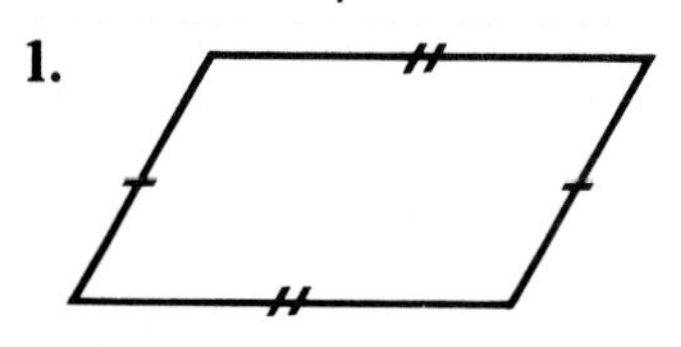

2. 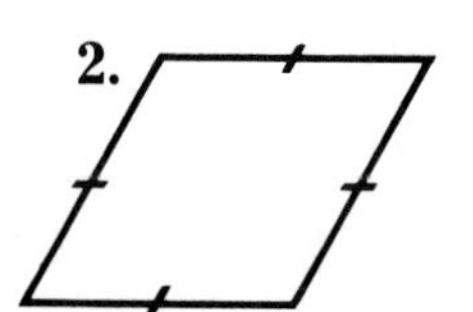

3. 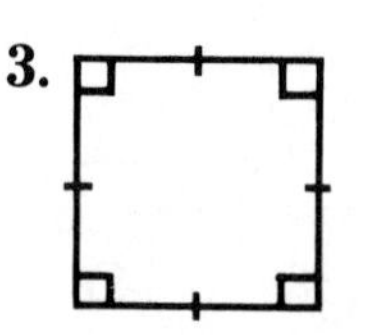

4. 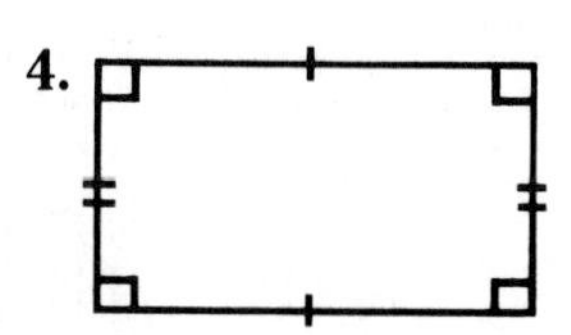

Problem Solving Applications

READ • PLAN • SOLVE • CHECK

The students in Mr. Martinez's woodshop class are working on their class projects. Some of Mr. Martinez's students are interested in careers in carpentry, while others enjoy woodworking as a hobby. They have learned that good math skills are necessary for making their projects.

AS YOU READ

- bits—replaceable parts of a tool used for drilling

Answer these questions about the students' woodworking projects.

1. Ned is cutting 6 shelves for a bookcase. He calculates that each shelf will be $3\frac{5}{8}$ ft long. The lumber comes in 12-ft lengths. How many 12-ft boards will he need?

2. Nora is building a utility box using a 4-ft by 8-ft sheet of plywood. If she cuts 5 pieces, each 4 ft by $1\frac{1}{4}$ ft, what is the measure of the remaining scrap piece of plywood? (Draw a picture.)

3. Annie is building a spice rack. Mr. Martinez recommends that finishing nails be no shorter than $1\frac{1}{2}$ times the thickness of the wood. The wood she is using is $\frac{3}{4}$ in. thick. How long should the finishing nails be?

4. Jack is making footstools for the student craft sale. The diameter of each top will be $1\frac{1}{8}$ ft. The board he plans to use for the tops is 12 ft long and $1\frac{1}{8}$ ft wide. At most, how many footstools can he make?

5. Chris is using a drill to make a pencil holder. The drill has *bits* with diameters of $\frac{13''}{32}$, $\frac{1''}{4}$, $\frac{19''}{64}$, $\frac{1''}{2}$, and $\frac{3''}{8}$. (The $''$ means inches.) If most pencils are about $\frac{5''}{16}$ in diameter, what is the smallest bit he could use?

6. Shirley needs two pieces of wood each $16\frac{3}{8}$ in. wide for a jewelry box. One piece she cut measures $16\frac{3}{4}$ in. How much should she plane off to get the correct width?

PERIMETER

To find the perimeter of a polygon

- Add the measures of the lengths of the sides.
- Use formulas when possible.

Perimeter (P) is the distance around a polygon. You can find the perimeter of any polygon by adding the measures of the lengths of the sides. The perimeter of certain polygons can be found by formulas.

EXAMPLE 1

Find the perimeter of the polygon.

P = sum of lengths of sides.
= **1.8 + 2.7 + 2.8 + 1.3**
P = **8.6**
So, the perimeter of the polygon is **8.6 cm.**

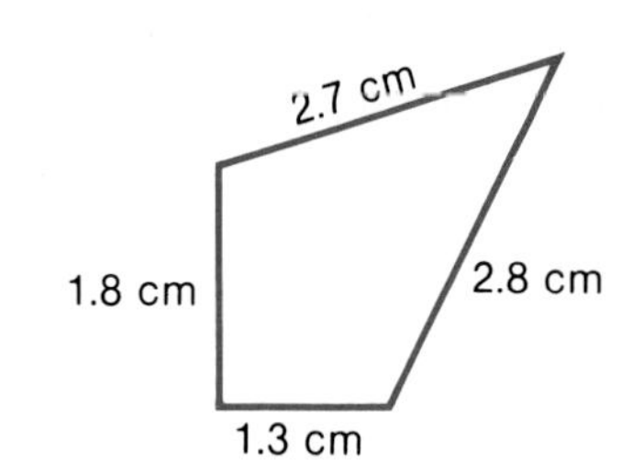

EXAMPLE 2

Find the perimeter of each quadrilateral.

P of a rectangle = 2 × length + 2 × width.
= **2 × 9 + 2 × 3**
P = **18 + 6 = 24**

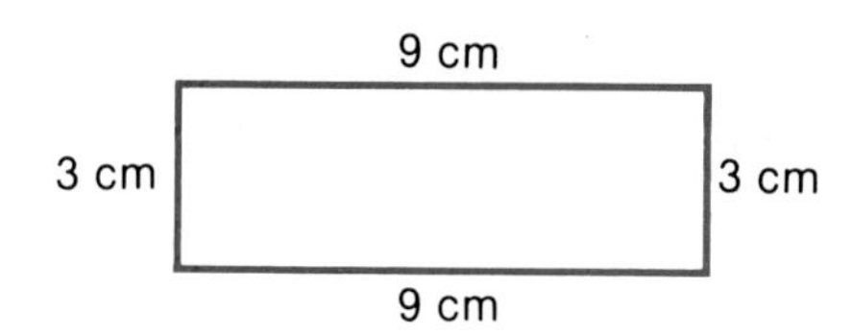

P of a square = 4 × length of a side.
= **4 × 2.7**
P = **10.8**
So, the perimeter of the rectangle is **24 cm,** and the perimeter of the square is **10.8 cm.**

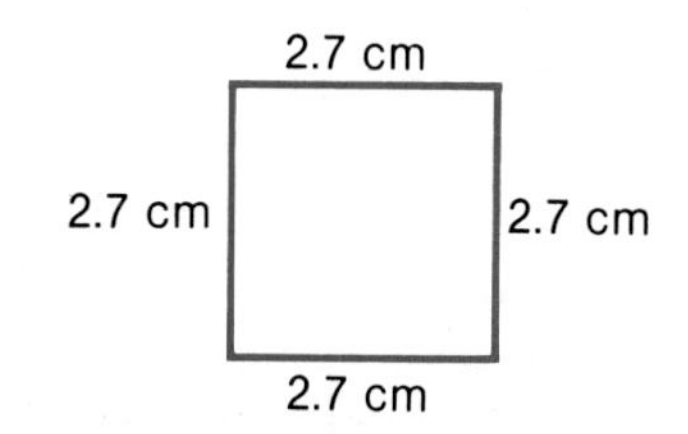

EXAMPLE 3

Find the perimeter of the triangle.

P of an equilateral triangle = 3 × length of a side.
= **3 × 5.8**
P = **17.4**
So, the perimeter of the triangle is **17.4 cm.**

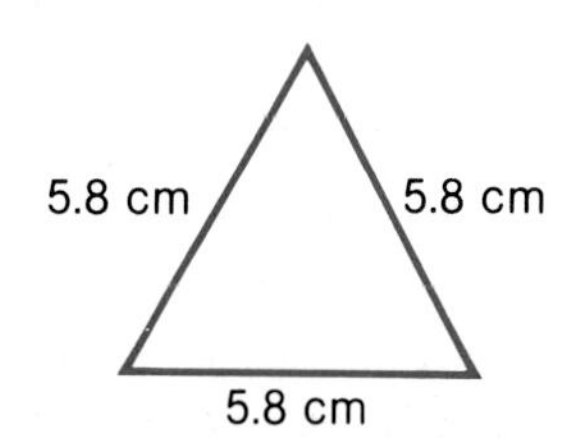

GETTING READY

Find the perimeter of each polygon.

1.

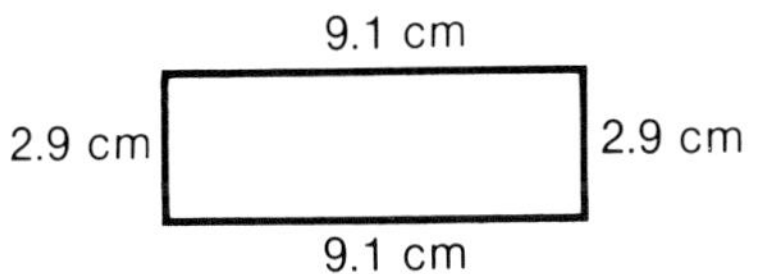

2.

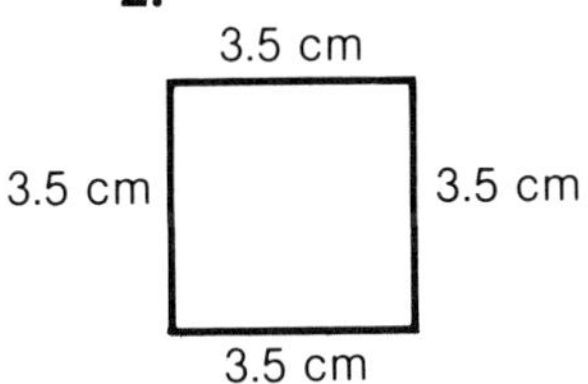

3.

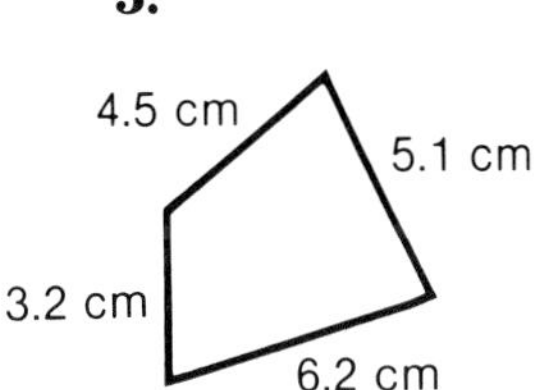

EXERCISES

Find the perimeter of each polygon.

1.

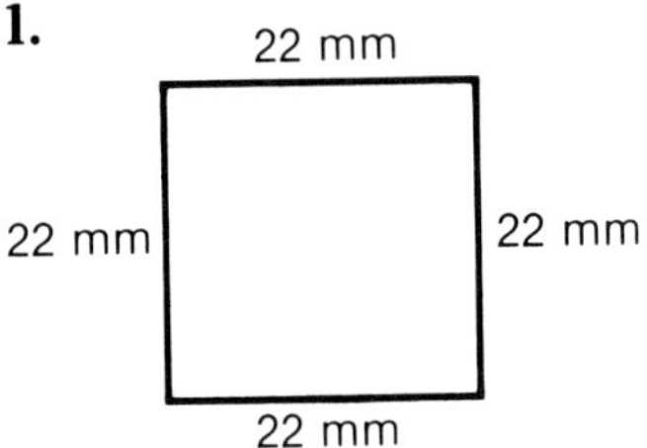

2.

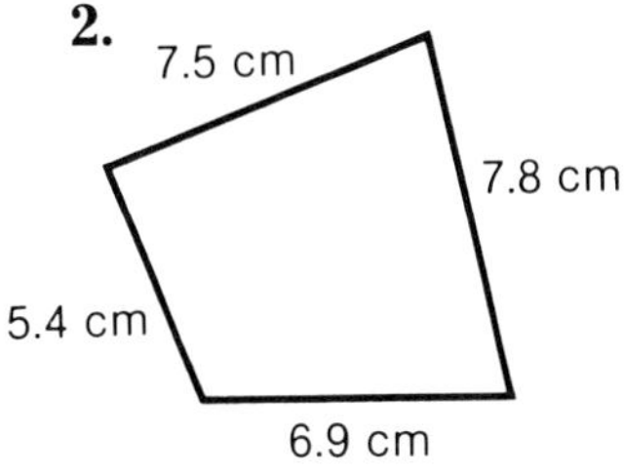

3.

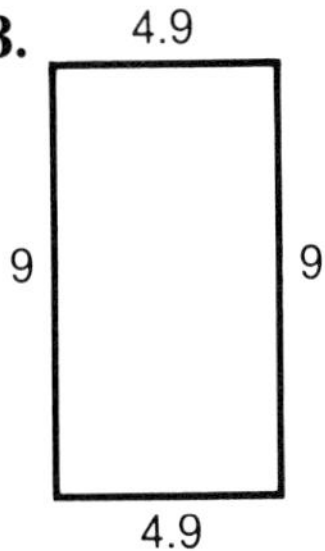

4.

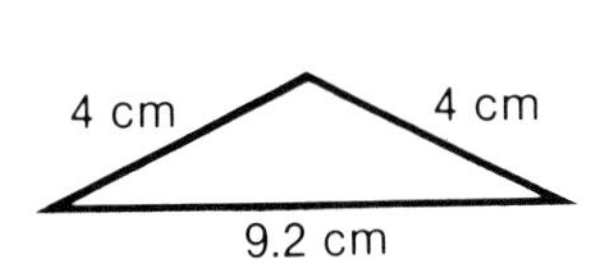

5.

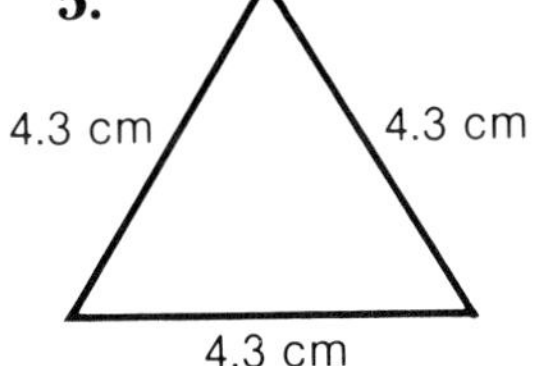

6.

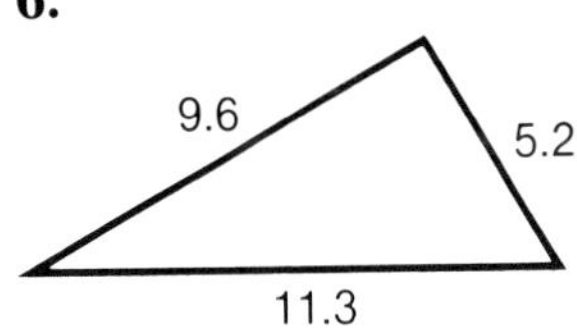

Solve.

7. A bulletin board is in the shape of a rectangle. The length is 125 cm and the width is 98 cm. What is the perimeter?

8. A baseball diamond is in the shape of a square with each side 90 ft long. What is the distance Pete must run if he hits a home run?

9. Jesse is making a tablecloth. He wants to trim the edges with braid. If the cloth is in the shape of a rectangle 2.5 m by 1.8 m, how much braid must he buy?

***10.** The library at York High School is square. The perimeter is 133.4 m. What is the length of each side of the library?

AREA OF A RECTANGLE AND A SQUARE

To find the area of a rectangle and of a square

Area (A) is the number of square units needed to cover a region. In this rectangle the length (l) is 6 units. The width (w) is 5 units. The area (A) is 30 square units.

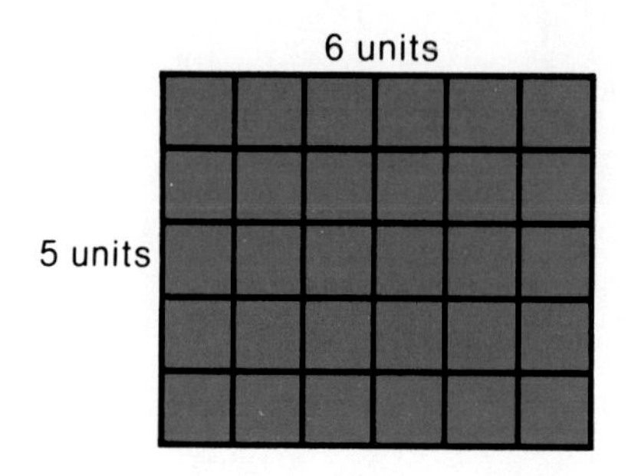

EXAMPLE 1

Find the area of a rectangle that measures 44 mm by 23 mm.

The formula for finding the area of a rectangle is:

$$\begin{aligned} \text{Area} &= \text{length} \times \text{width} \\ A &= l \times w \\ &= \mathbf{44 \times 23} \\ \mathbf{A} &= \mathbf{1{,}012\ mm^2} \end{aligned}$$

So, the area is **1,012 mm²**.

Area of rectangle = length × width, or $A = l \times w$.

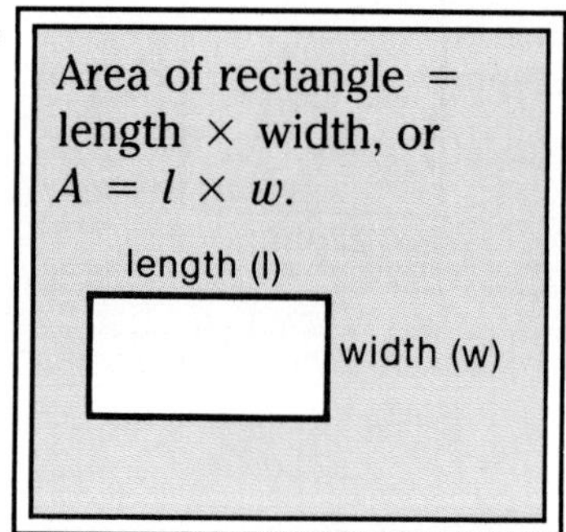

EXAMPLE 2

Find the area of a square with each side 6.5 cm long.

The formula for finding the area of a square is:

$$\begin{aligned} \text{Area} &= \text{length of side} \times \text{length of side} \\ A &= s \times s \\ &= \mathbf{6.5 \times 6.5} \\ \mathbf{A} &= \mathbf{42.25\ cm^2} \end{aligned}$$

So, the area of the square is **42.25 cm²**.

Area of square = length of side × length of side, or $A = s \times s$.

side (s)

side (s)

GETTING READY

Find the area of each rectangle or square.

1. **2.**

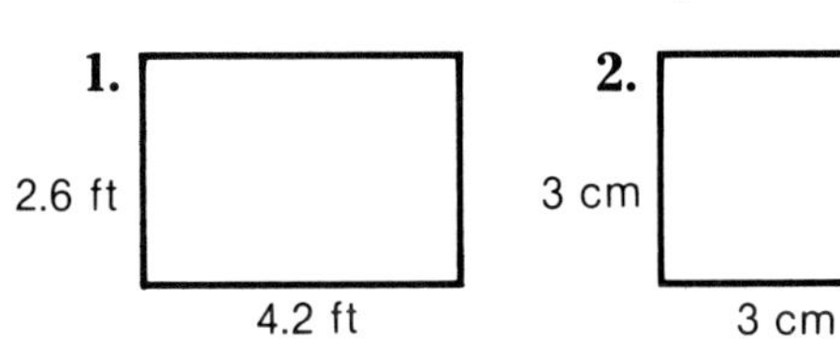

3. length = 8.2 m
width = 7 m

4. l = 35 mm
w = 15 mm

EXERCISES

Find the area of each rectangle or square.

1.

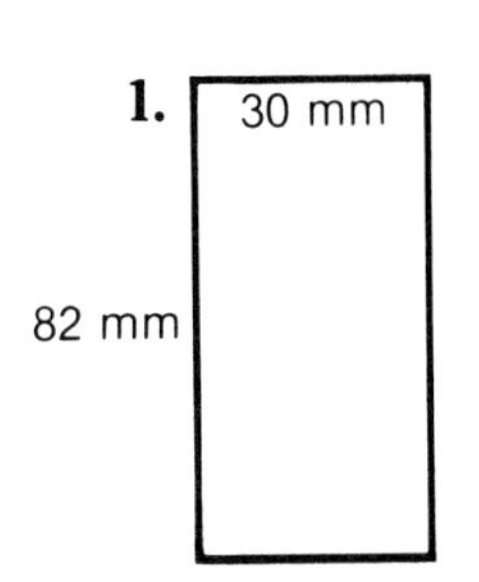

2. $5\frac{1}{2}$ in.
$2\frac{1}{2}$ in.

3.

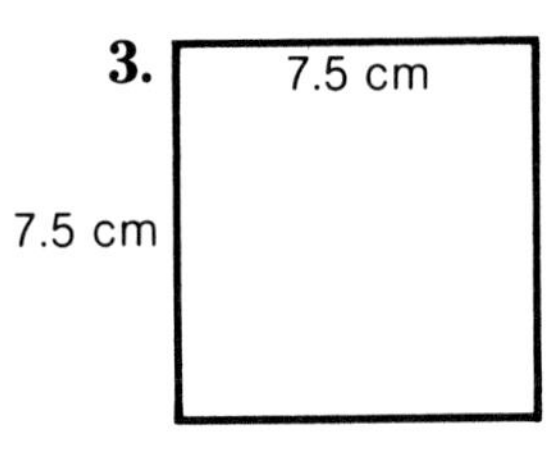

4. length = 8.5 in.
width = 6 in.

5. l = 28 ft
w = 12 ft

6. **7.**

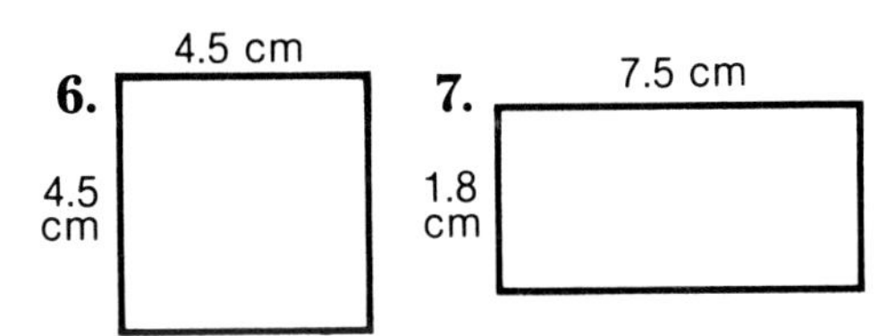

8. length = 8 yd
width = 6 yd

9. l = 108 cm
w = 0.5 cm

10. l = 88 in.
w = 20 in.

11. l = 42.4 cm
w = 25 cm

Solve.

12. What is the area of a square patio that measures 3.5 m on each side?

13. What is the area of the rectangular front cover of a book 24 cm by 19.5 cm?

***14.** Janice is covering her storeroom floor with linoleum squares. Each square is 1 ft by 1 ft. How many linoleum squares are needed if the room is a rectangle 10 ft by 12 ft?

***15.** The area of the school playground is 5,760 m^2. If the length of the playground is 12,800 cm, what is the width?

AREA OF A TRIANGLE

To find the area of a triangle

- Use the formula $A = \frac{1}{2} \times \text{base} \times \text{height}.$

EXAMPLE 1

Find the area of a right triangle with a height of 4 cm and a base of 6 cm.

A right triangle is half of a rectangle.
So, think of a rectangle that measures 6 cm by 4 cm.
Find the area of the rectangle.

$$\textbf{Area} = \mathbf{6 \times 4}$$
$$\textbf{Area} = \mathbf{24\ cm^2}$$

Now divide by 2 to get half the area.

$$\mathbf{24 \div 2 = 12}$$

So, the area of the triangle is **12 cm^2.**

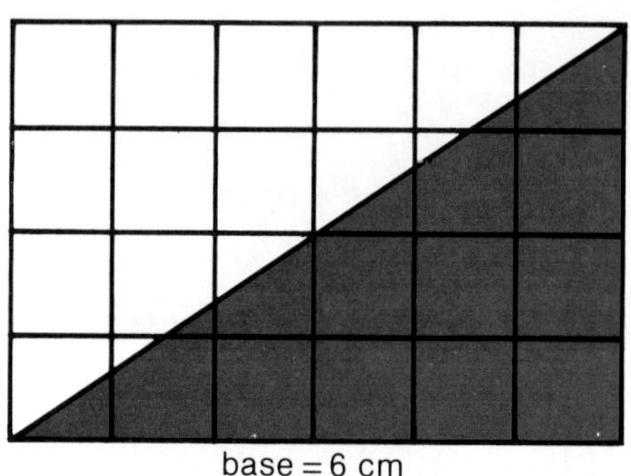

EXAMPLE 2

Find the area of the triangle.
The formula for finding the area of a triangle is:

$$\text{Area} = \frac{1}{2} \times \text{base} \times \text{height}.$$
$$= \frac{1}{2} \times 9 \times 7$$
$$A = \frac{63}{2}, \text{ or } 31.5$$

So, the area of the triangle is **31.5** square units.

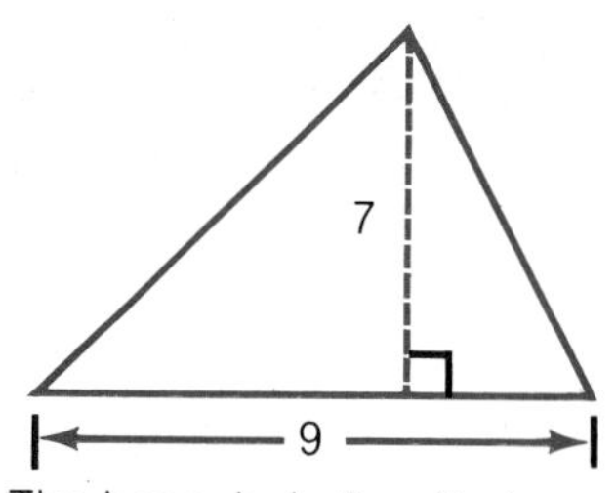

The base, b, is 9 units long.
The height, h, is 7 units long.

GETTING READY

Find the area of each triangle.

1.

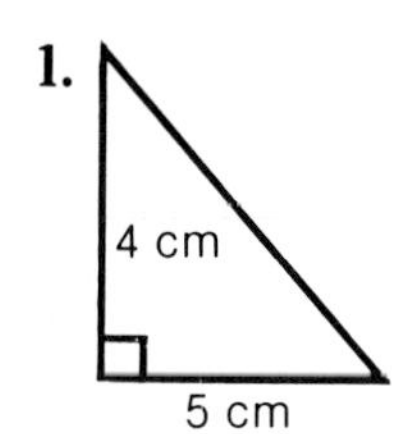

2.

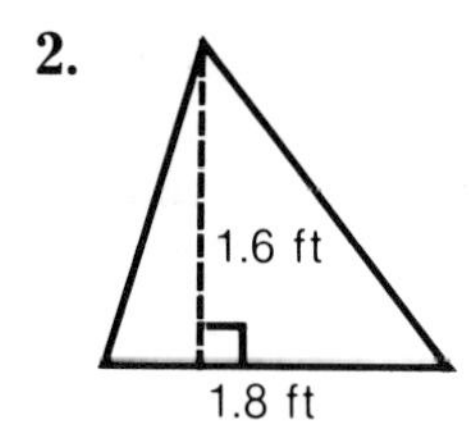

3. base = 3.7 cm
height = 6 cm

4. b = 42 mm
h = 25 mm

EXERCISES

Find the area of each triangle.

1.

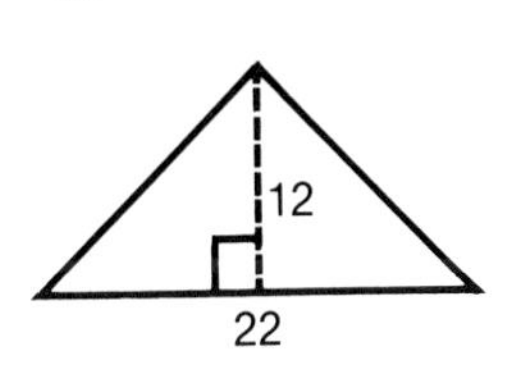

2.

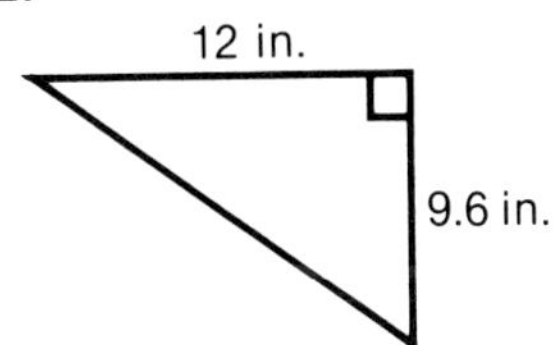

3.

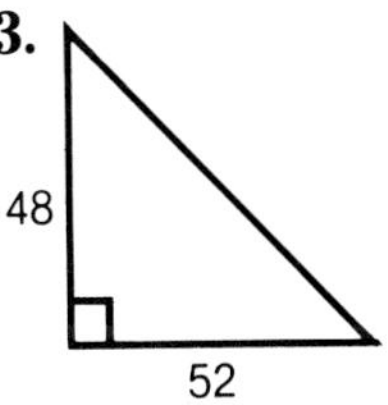

4.

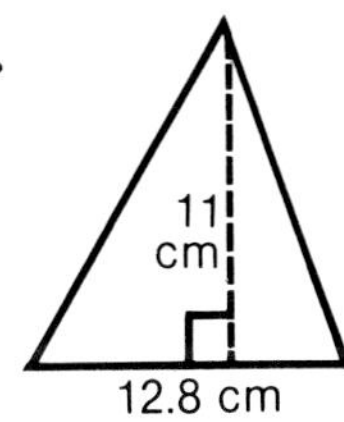

5.

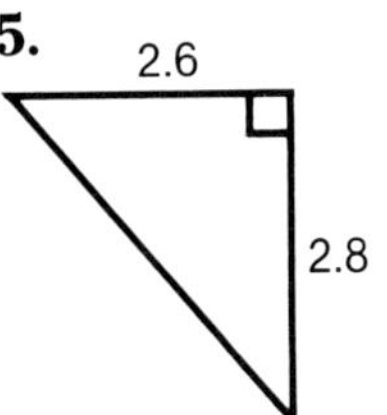

6.

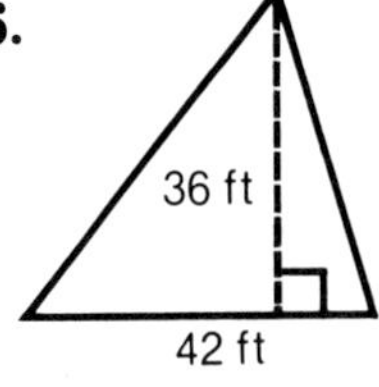

7. $b = 8$ mm
$h = 33$ mm

8. $b = 102$
$h = 12$

9. $b = 28$ in.
$h = 10$ in.

10. $b = 88$ cm
$h = 20$ cm

11. $b = 12.5$ m
$h = 10.2$ m

12. $b = 28$ yd
$h = 19$ yd

Solve.

13. One sail on Hugo's sailboat is in the shape of a right triangle. Find the area if the base is 3 m and the height is 5 m.

14. A pane of glass is in the shape of a right triangle. The base is 8 in. and the height is 6 in. What is the area?

***15.** The base of a right triangle is 5 cm and the height is 10 cm. Find its area in m^2.

***16.** The area of a right triangle is 24 m^2 and its base is 8 m. Find the height of the triangle.

ON YOUR OWN

Draw diagrams to help answer the following questions.

1. What happens to the area of a square when the length of the sides is doubled?

2. What happens to the perimeter of an equilateral triangle when the length of the sides is doubled?

3. Do figures with equal perimeters have equal areas?

AREAS OF IRREGULAR FIGURES

To find the area of an irregular figure

- Divide the figure into rectangles, triangles, or squares.

EXAMPLE 1

Find the area.

(1) Divide the figure into two rectangles.
(2) Find the area of the smaller rectangle.

$$A = l \times w$$
$$= \mathbf{3 \times 1}$$
$$\mathbf{A = 3\ cm^2}$$

(3) Find the area of the larger rectangle.

$$A = l \times w$$
$$= \mathbf{4 \times 2}$$
$$\mathbf{A = 8\ cm^2}$$

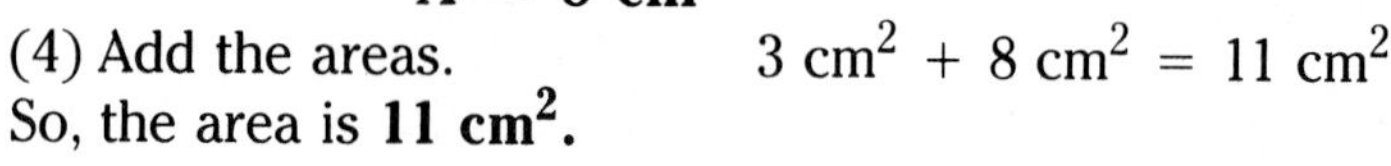

(4) Add the areas. $3\ cm^2 + 8\ cm^2 = 11\ cm^2$
So, the area is **11 cm²**.

EXAMPLE 2

Find the area of the parallelogram.

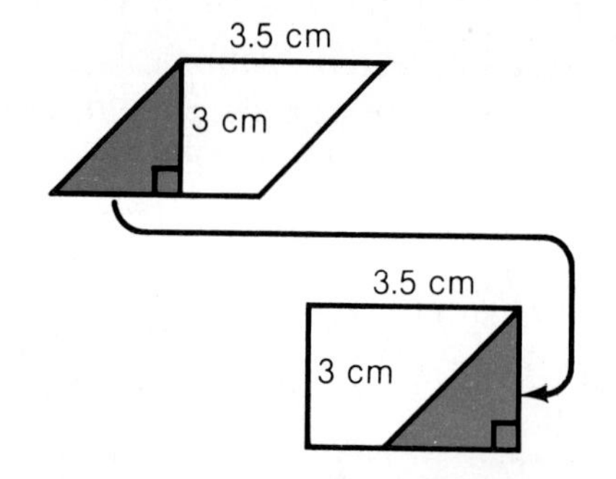

Move the triangle as shown to form a rectangle. The length is 3.5 cm and the width is 3 cm.
Find the area of the rectangle.

$$A = l \times w$$
$$= \mathbf{3.5 \times 3}$$
$$\mathbf{A = 10.5\ cm^2}$$

So, the area of the parallelogram is **10.5 cm²**.

EXAMPLE 3

Find the area of the shaded region.

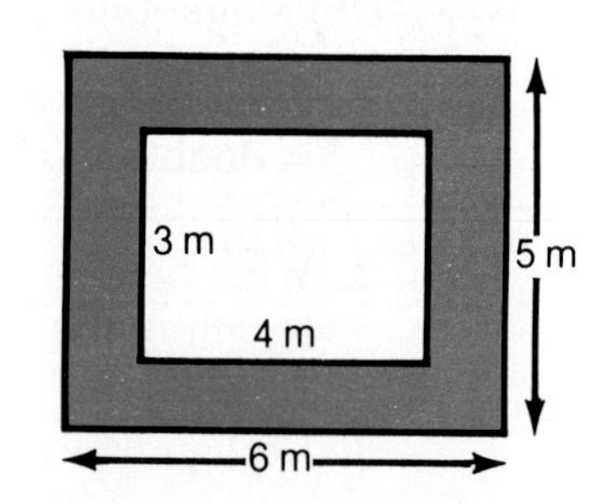

(1) Find the area of the larger rectangle.

$$\mathbf{A = 6 \times 5}$$
$$\mathbf{A = 30\ m^2}$$

(2) Find the area of the smaller rectangle.

$$\mathbf{A = 4 \times 3}$$
$$\mathbf{A = 12\ m^2}$$

(3) Subtract the areas. $30\ m^2 - 12\ m^2 = 18\ m^2$
So, the area of the shaded region is **18 m²**.

GETTING READY

Find the area of each figure.

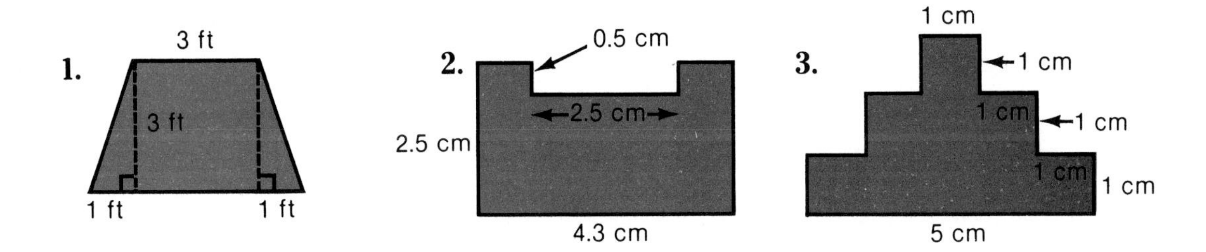

EXERCISES

Find the area of each figure or shaded region.

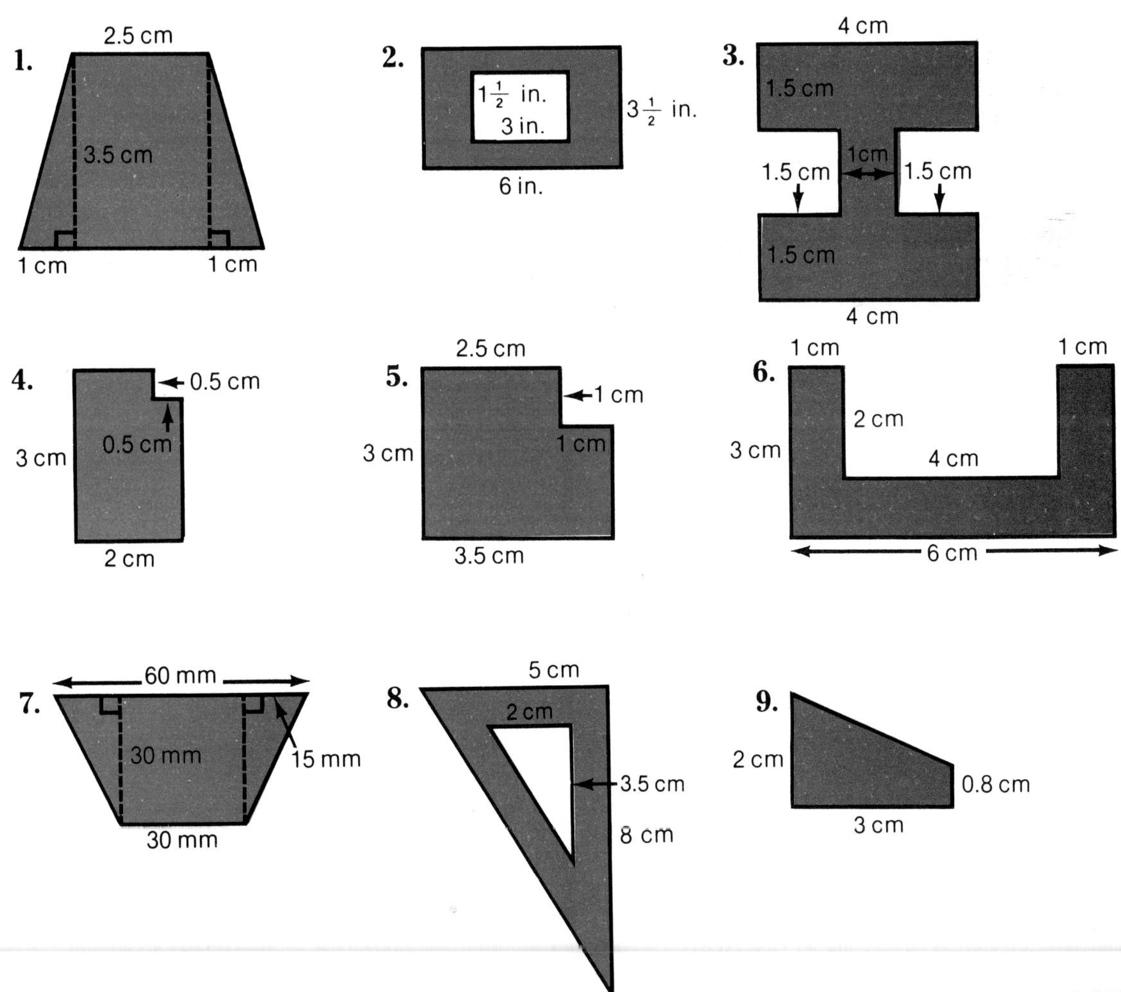

CIRCUMFERENCE OF A CIRCLE

To find the circumference of a circle

- Use the formula $C = \pi \times$ diameter.

A circle is named by its center. Each point on a circle is the same distance from the center of the circle.

In this circle, O is the center. The distance from P to R, called the diameter (d), is 6 units. The distance from O to S, called the radius (r), is 3 units. Notice that the diameter is 2 times the radius.

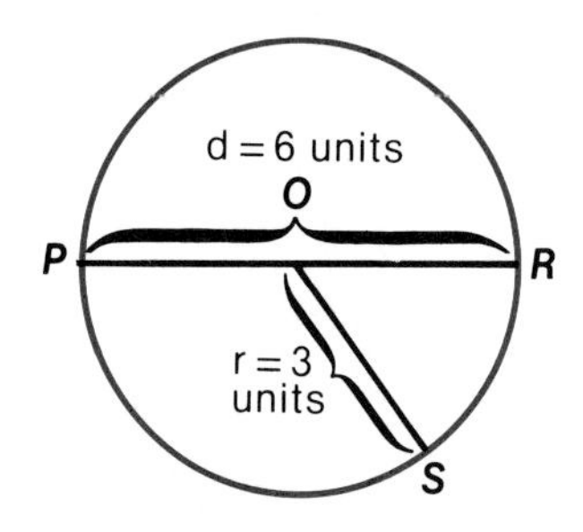

EXAMPLE 1

Find the circumference of the circle.

The diameter is 5 cm.
The formula for finding the circumference (C) of a circle is $C = \pi \times$ diameter.

$\mathbf{C = \pi \times d}$
$\mathbf{= 3.14 \times 5}$ ⟵ Use 3.14 for π.
$\mathbf{C = 15.7}$

So, the circumference is **15.7 cm.**

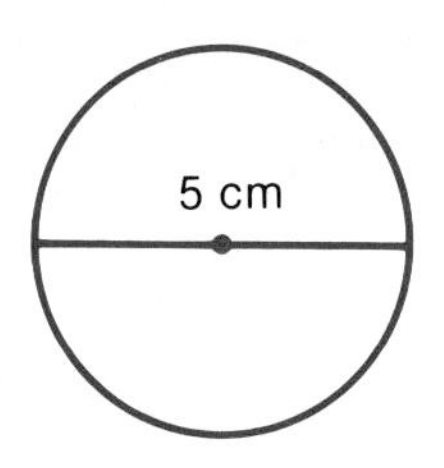

The distance around a circle is the **circumference.**

π is the Greek letter pi. It stands for a number very close to 3.14.

EXAMPLE 2

Find the circumference.

Think: diameter $= 2 \times$ radius
$d = 2 \times 14$, or 28
$C = \pi \times d$
$\mathbf{= 3.14 \times 28}$
$\mathbf{C = 87.92}$

So, the circumference is **87.92 mm.**

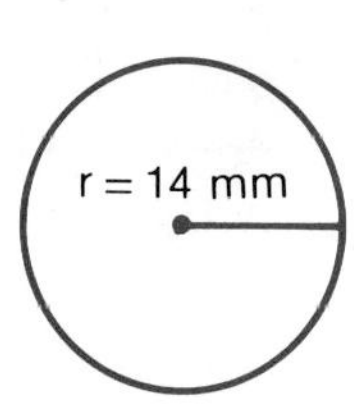

GETTING READY

Find the circumference. Use 3.14 for π.

1.
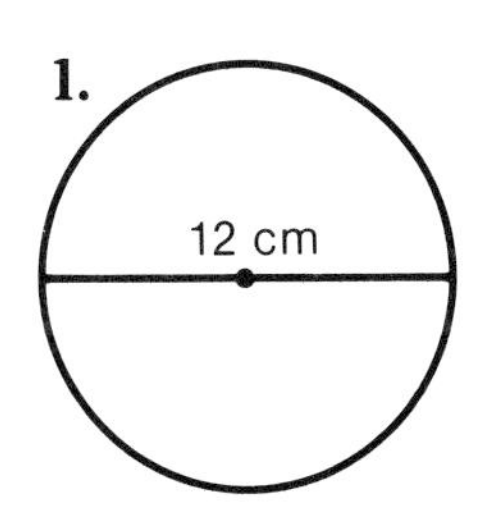

2.
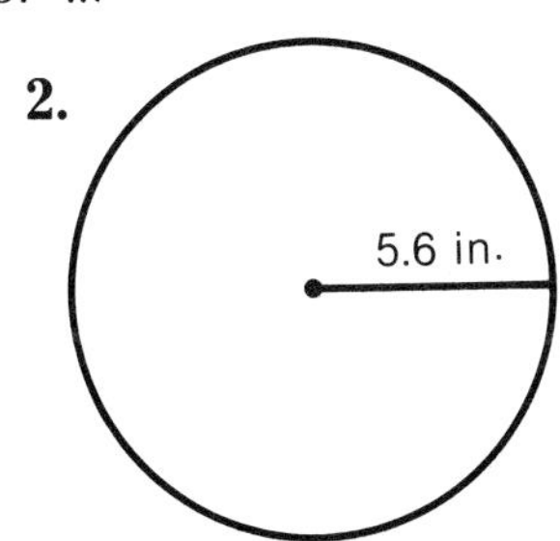

3.
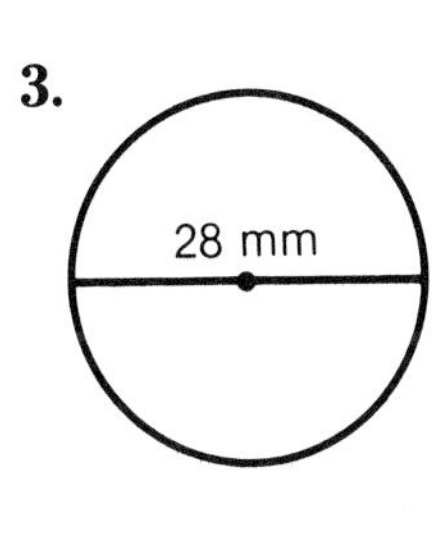

4. $r = 7$ m

5. $r = 3.5$ cm

6. $d = 119$ km

EXERCISES

Find the circumference. Use 3.14 for π.

1.
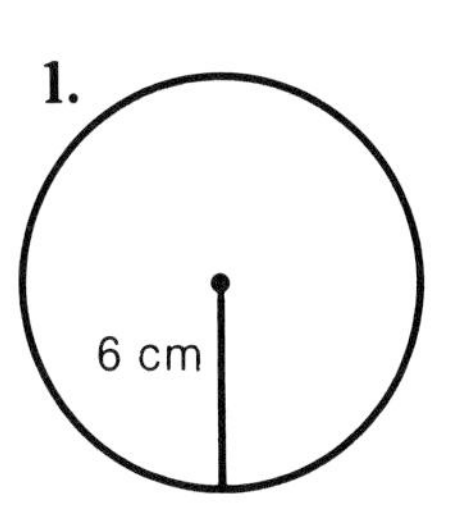

2.
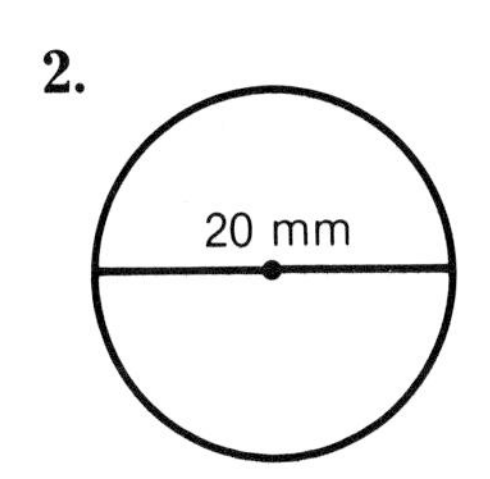

3.
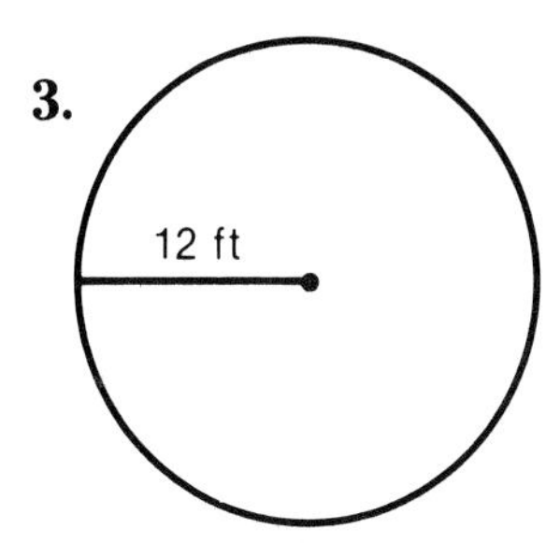

4. $r = 420$ cm

5. $r = 56$ in.

6. $r = 4.9$ km

7. $d = 140$ mm

8. A coin has a diameter of 2.5 cm. What is its circumference?

9. The diameter of a circular running track is 100 m. How far would you run in one lap?

10. A circular garden has a diameter of 5 m. Find its circumference.

11. A circular pool has a radius of 5 m. Find its circumference.

12. A bicycle tire has a diameter of 70 cm. About how many meters will it travel in 1,000 turns? (Hint: For one turn it travels a distance equal to its circumference.).

*13. One Ferris wheel is 15 m in diameter. Another is 17 m in diameter. How much farther will you ride in one complete turn of the larger Ferris wheel?

AREA OF A CIRCLE

To find the area of a circle

- Use the formula $A = \pi \times \text{radius} \times \text{radius}$.

EXAMPLE 1

Find the area of the circle.

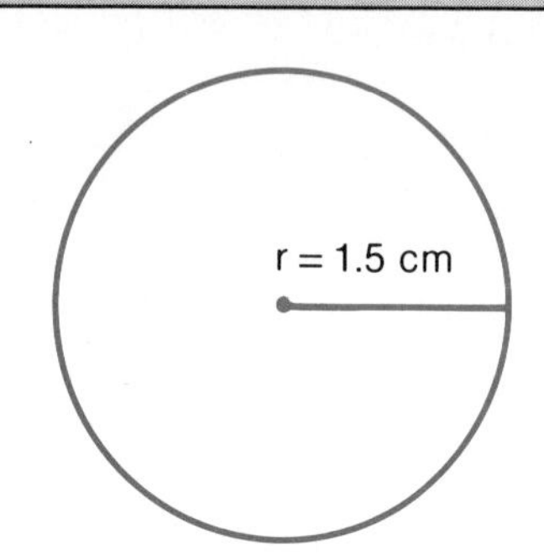

The formula for finding the area of a circle is:
Area = $\pi \times$ radius $\times$ radius.

$$A = \pi \times r \times r$$
$$= 3.14 \times 1.5 \times 1.5 \longleftarrow \textbf{Use 3.14 for } \pi.$$
$$= 3.14 \times 2.25 \longleftarrow 1.5 \times 1.5 = 2.25$$
$$A = 7.065 \text{ cm}^2$$

EXAMPLE 2

A round swimming pool has a diameter of 14 m. Find the area of its surface.

Think: radius = diameter $\div$ 2 $\qquad r = 14 \div 2$, or 7

$$A = \pi \times r \times r$$
$$= 3.14 \times 7 \times 7$$
$$A = 153.86 \text{ m}^2$$

So, the area of its surface is **153.86 m²**.

GETTING READY

Find the area of each circle. Use 3.14 for π.

1.

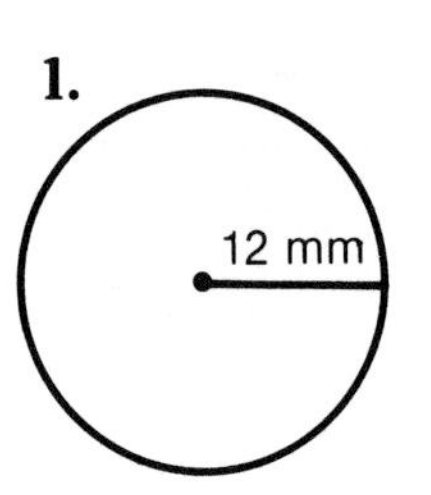

2.

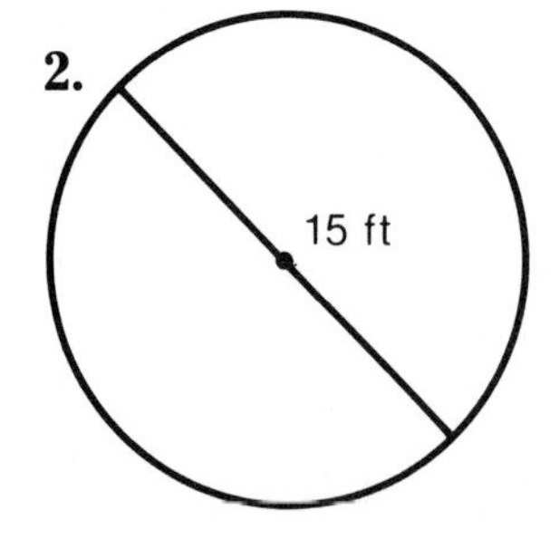

3.

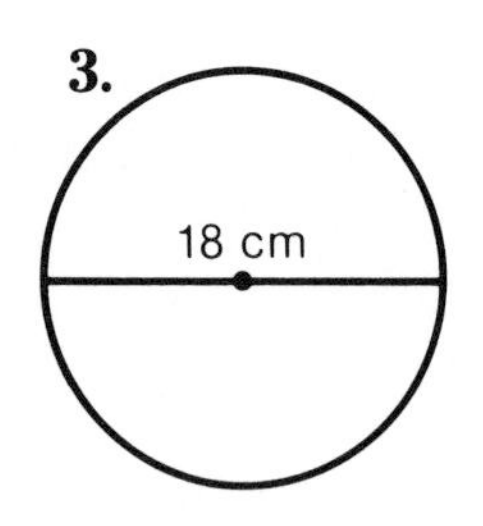

4. $r = 21$ m

5. $d = 32$ in.

6. $r = 6.3$ cm

EXERCISES

Find the area. Use 3.14 for π.

1.

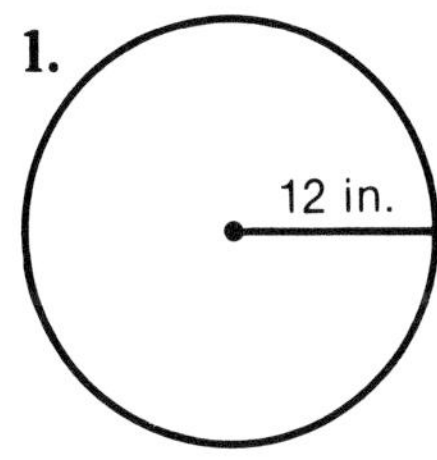

2.

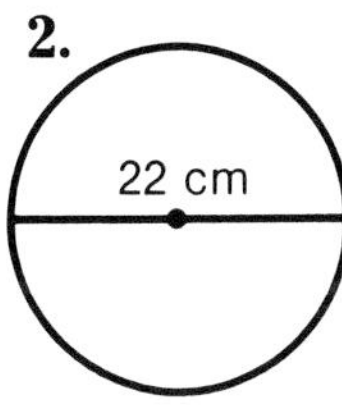

3.

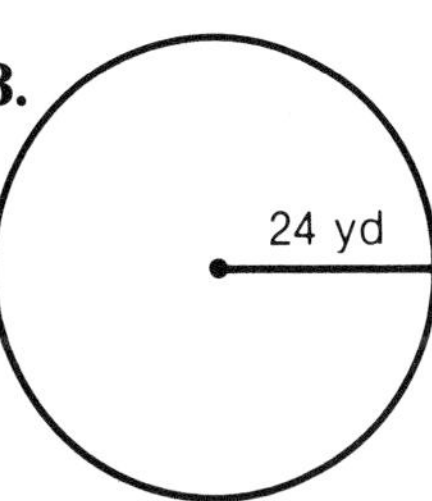

4. $r = 3$ cm **5.** $r = 9$ m **6.** $d = 16$ km **7.** $r = 2.4$ cm

8. $r = 10.2$ mm **9.** $r = 0.2$ m **10.** $r = 10$ cm **11.** $d = 13$ mm

12. $r = 28$ cm **13.** $r = 9.8$ m **14.** $d = 7$ m **15.** $d = 4.2$ m

Solve.

16. A record has a 15-cm radius. Find its area.

17. The diameter of a dime is 18 mm. Find its area.

18. A salad plate has a diameter of 20 cm. Find its area.

19. The radius of a quarter is 12 mm. Find its area.

Find the area of the shaded region.

***20.**

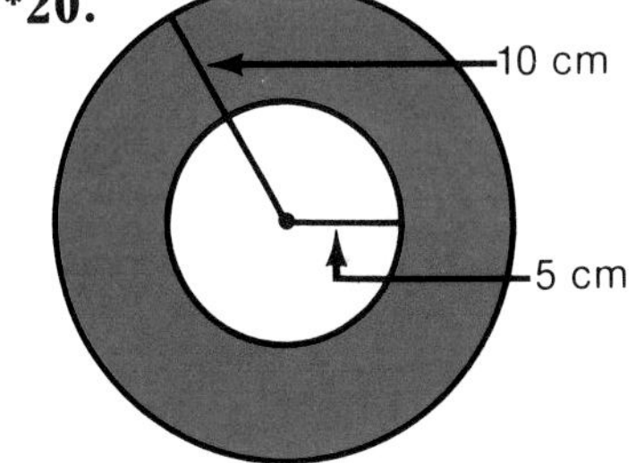

***21.**

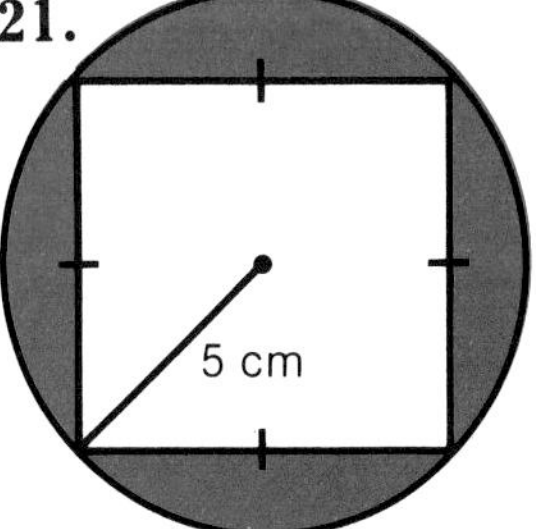

CALCULATOR

Use your calculator to find the area of each circle. Use 3.14 for π.

1. $r = 8.6$ cm **2.** $d = 3.4$ km **3.** $d = 2.3$ m

VOLUME

To find the volume of a rectangular solid

- Use the formula V = length × width × height.

Volume is the amount of space inside a solid figure. A cube 1 cm by 1 cm by 1 cm has a volume of 1 cubic centimeter, or 1 cm^3.

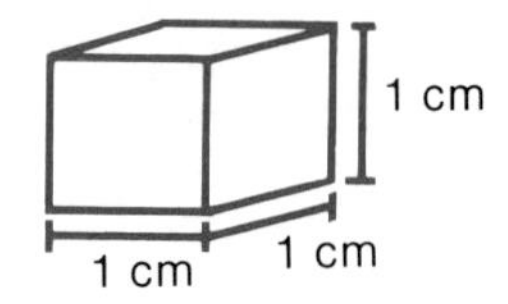

To find the volume of a solid figure in cubic centimeters, find the number of cubic centimeters needed to fill it.

EXAMPLE 1

Find the volume of the box.

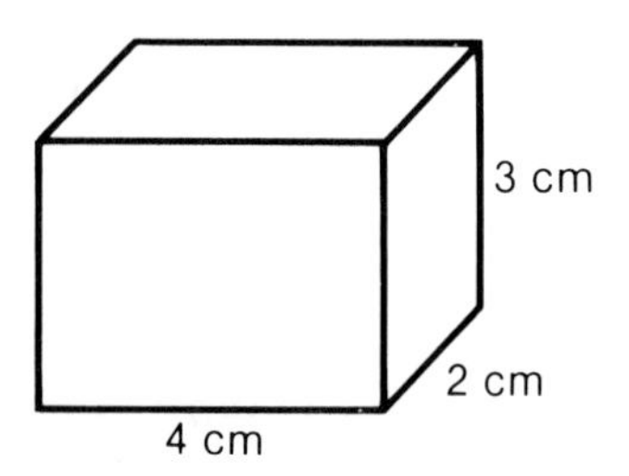

Think of cubes covering the bottom.

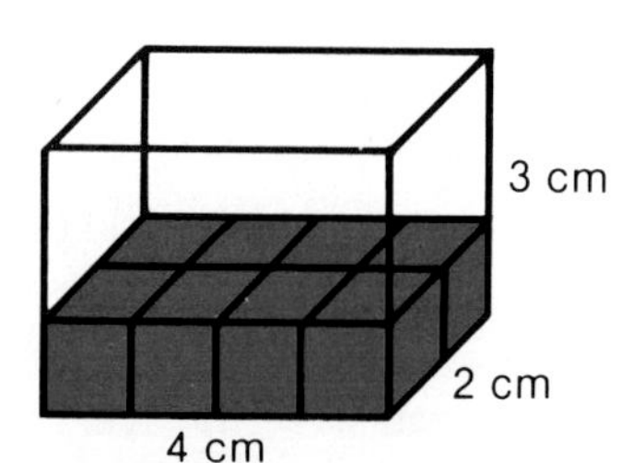

There are 2 rows of 4 cubes in the bottom layer. So, 2 × 4 = 8.

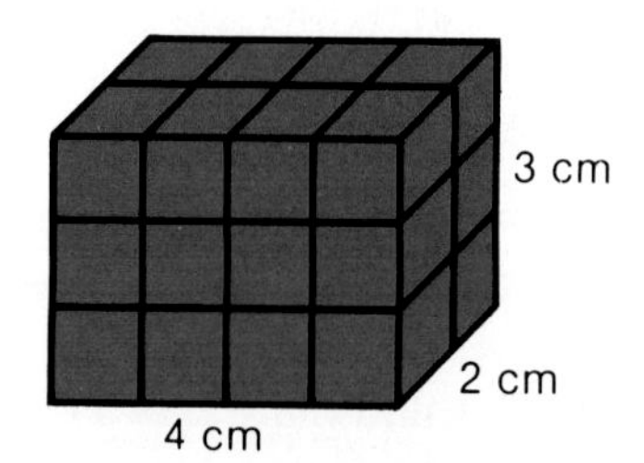

To fill the box, 3 layers are needed. So, 3 × 8 = 24.

So, the volume is 4 × 2 × 3, or **24 cm^3.**

EXAMPLE 2

Find the volume of the rectangular solid.

The formula for finding the volume of a rectangular solid is:
Volume = length × width × height.

$V = l \times w \times h$
$= \mathbf{5 \times 4 \times 3.5}$
$= \mathbf{20 \times 3.5} \longleftarrow \mathbf{5 \times 4 = 20}$
$\mathbf{V = 70\ cm^3} \longleftarrow \mathbf{20 \times 3.5 = 70}$

So, the volume is **70 cm^3.**

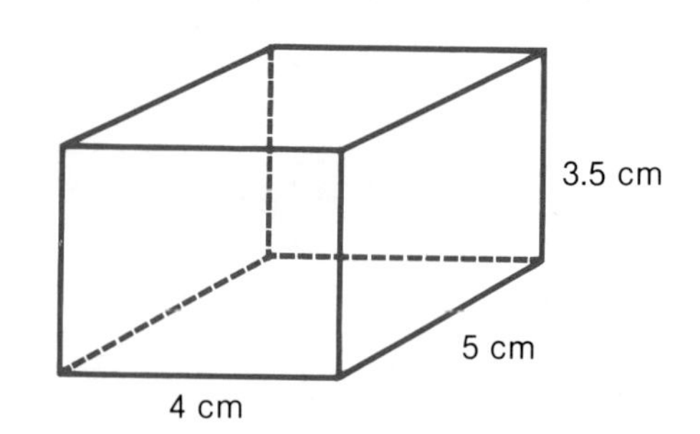

GETTING READY

Find the volume of the rectangular solid.

1.

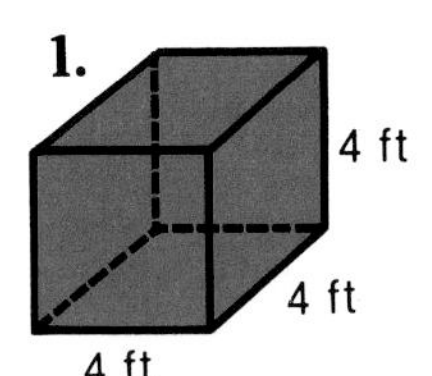

2.

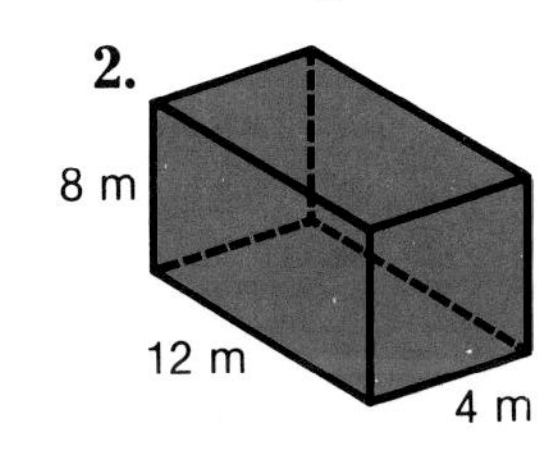

3. $l = 1.8$ m
$w = 1.2$ m
$h = 2.4$ m

4. $l = 7$ cm
$w = 3$ cm
$h = 4$ cm

EXERCISES

Find the volume of the rectangular solid.

1.

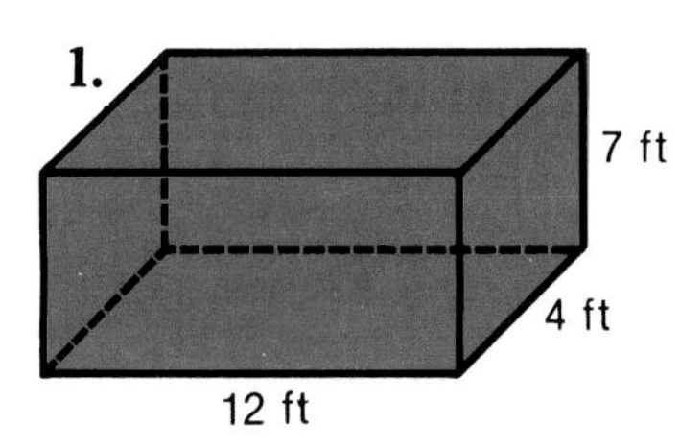

2.

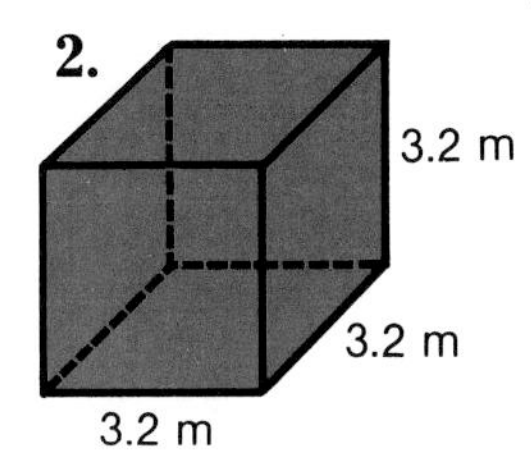

3.

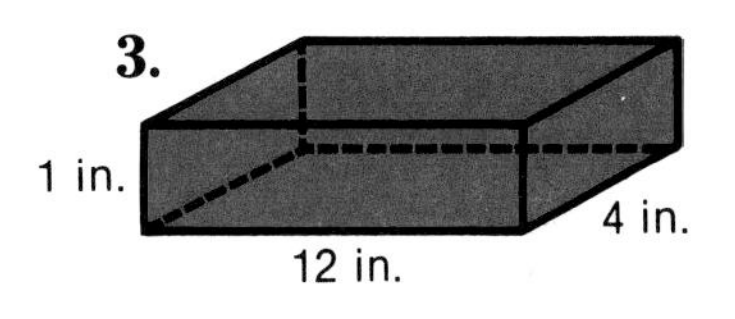

4. $l = 8$ cm
$w = 7$ cm
$h = 10$ cm

5. $l = 11$ in.
$w = 9$ in.
$h = 12$ in.

6. $l = 5$ cm
$w = 5$ cm
$h = 8$ cm

7. $l = 12$ m
$w = 6$ m
$h = 4$ m

8. $l = 20$ cm
$w = 10$ cm
$h = 50$ cm

9. $l = 4.2$ m
$w = 3.6$ m
$h = 1.6$ m

Solve each problem.

10. A rectangular-shaped tool chest is 1.5 m long, 0.8 m wide, and 0.8 m high. How much space is in the tool chest?

11. A truck bed is 3.2 m long, 2.2 m wide, and 1.2 m high. How many cubic meters of sand will the truck hold?

12. Which box will hold more?

A: 18 in. × 12 in. × 16 in.
B: 15 in. × 15 in. × 15 in.

13. Which box will hold more?

A: 20 cm × 15 cm × 12 cm
B: 30 cm × 14 cm × 9 cm

***14.** Mrs. Taylor is storing 1-cm cubes in boxes that are 15 cm by 12 cm by 10 cm. If she has 10,000 cubes, how many boxes will she need?

***15.** The volume of a rectangular box is 216 cm^3. It is 9 cm long and 4 cm wide. What is the height of the box?

Career: Carpenters

READ • PLAN • SOLVE • CHECK

Carpenters work for contractors, homebuilders, government agencies, schools, utility companies, and manufacturing firms. Large contractors employ carpenters, carpenters' helpers, apprentices, supervisors, and general construction supervisors. An apprentice program usually consists of 4 years of on-the-job training and at least 144 hours of classroom instruction each year.

AS YOU READ

- rough work—building frameworks, scaffolds and wooden forms

Solve.

1. About $\frac{1}{4}$ of all carpenters are self-employed. There are about 1,253,000 carpenters working today. About how many are self-employed?

2. Hubert King works for a small contractor. He is remodeling a kitchen. He cuts an $8\frac{3}{4}$-ft board into 4 equal pieces. How long is each piece?

3. John Gregg owns Gregg Fence Company. He is planning to order supplies to fence in a playground for toddlers. It measures $20' \times 15'$. How many feet of fencing should he order?

4. Andrea Palmer works for Benson Homebuilders. The height of a beam that she is using is $2\frac{1}{2}$ times its width. The beam is $16\frac{1}{4}$ in. wide. What is its height?

5. Elmer Richmond is a carpenter for a manufacturing firm. A recent flood damaged the first floor. To order supplies, Elmer must know the area of the first floor, which measures $130' \times 97'$. What is the area of the first floor?

6. Rita Davis is supervisor for the *rough work* done on the construction of two new homes. To complete the job on time, 5 carpenters need to work 16 h overtime each this weekend. How many overtime hours are needed altogether?

CHAPTER TEST 15

Classify, then measure these angles.

1.

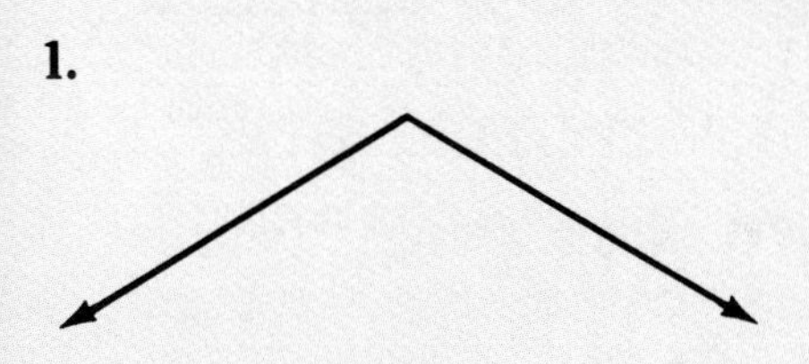

2.

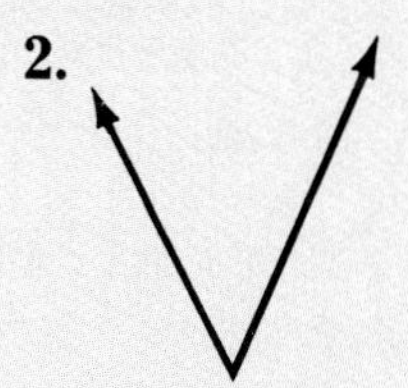

3.

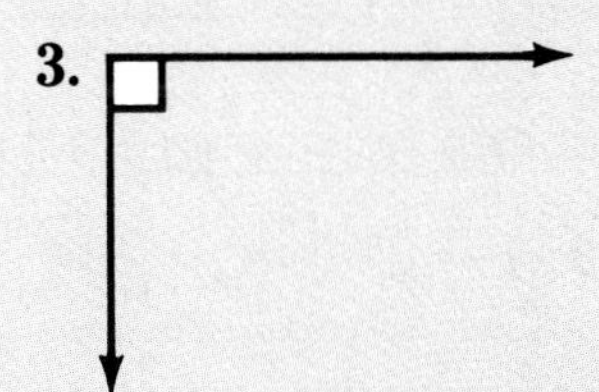

Name each polygon.

4.

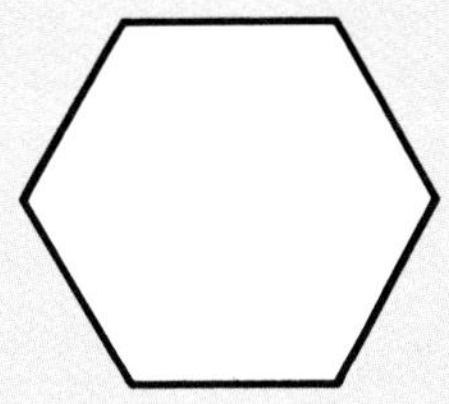

5.

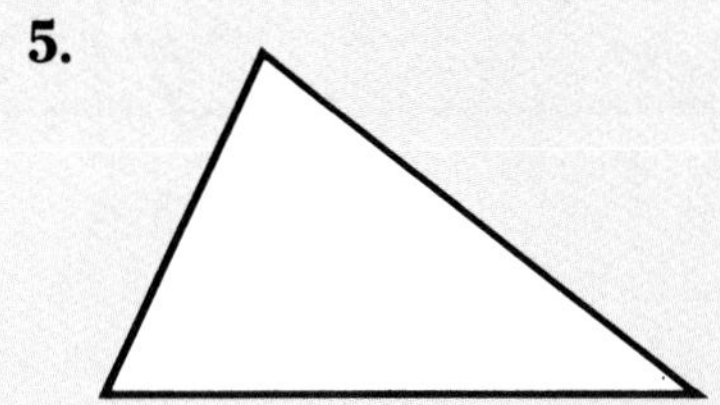

6.

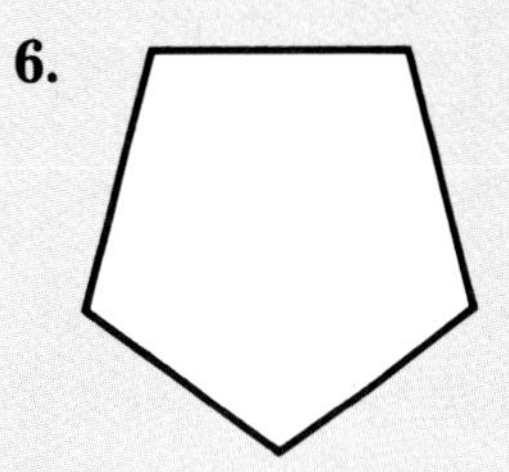

Name each quadrilateral.

7.

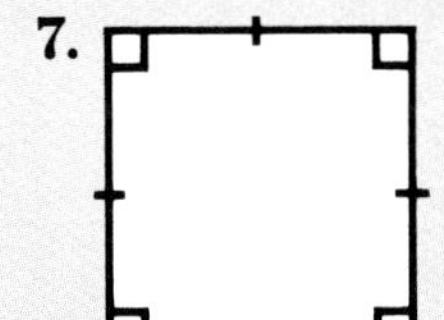

8.

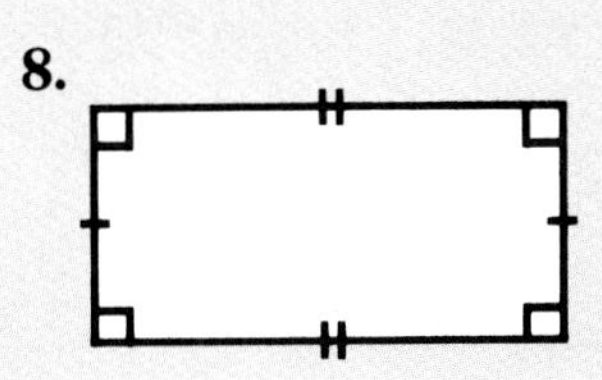

9.

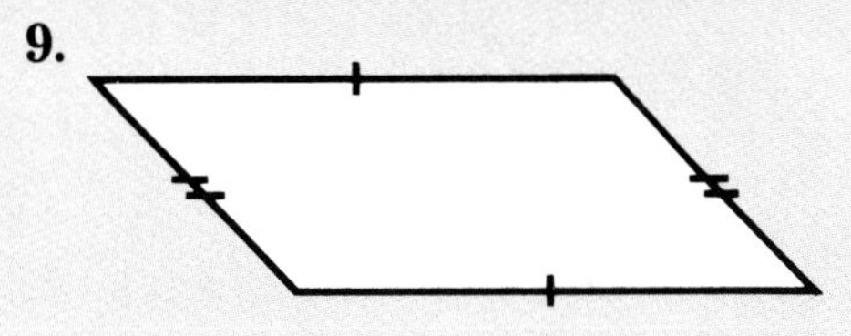

Classify the triangles by the measures of their angles.

10.

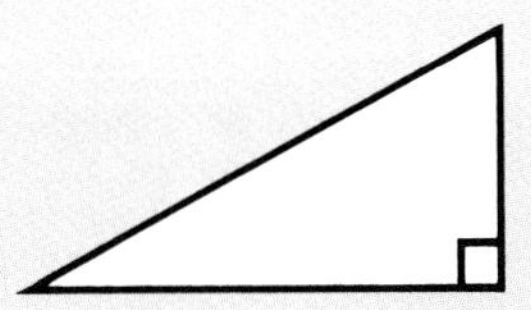

11.

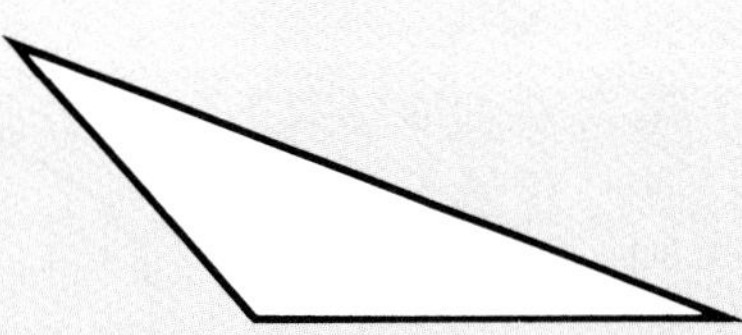

12.

Classify the triangles by the number of equal sides.

13.

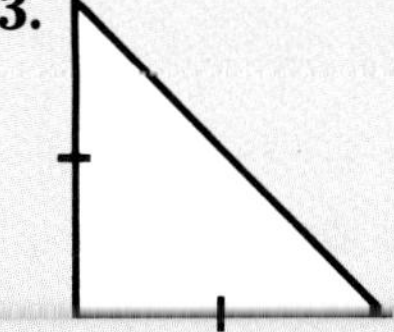

14.

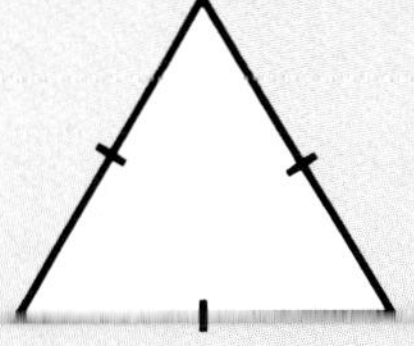

15.

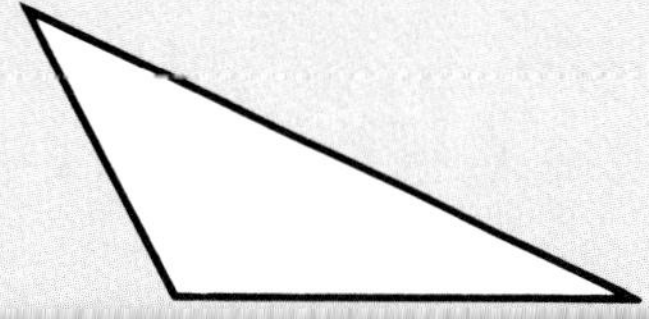

Find the perimeter of each polygon.

16. 4.5 cm, 4.5 cm, 4.5 cm, 4.5 cm

17. 25, 20, 14, 16

18. 17, 17, 15

Find the area of each rectangle or square.

19. $l = 22$ ft, $w = 10$ ft

20. square: length of side = 8 cm

Find the area of each triangle.

21.

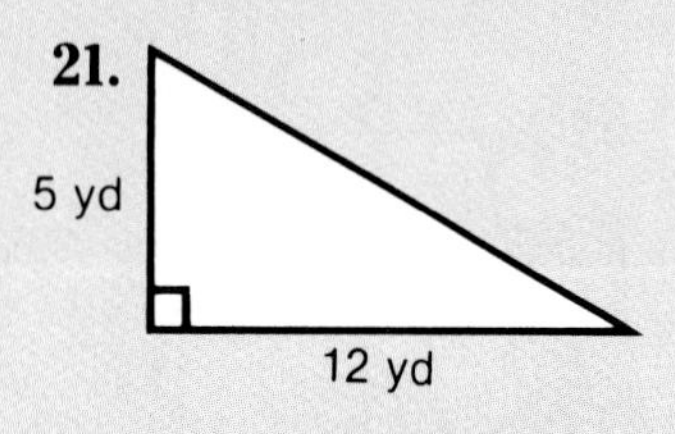

22. 14 cm, 24 cm

Find the area of the shaded region.

23.

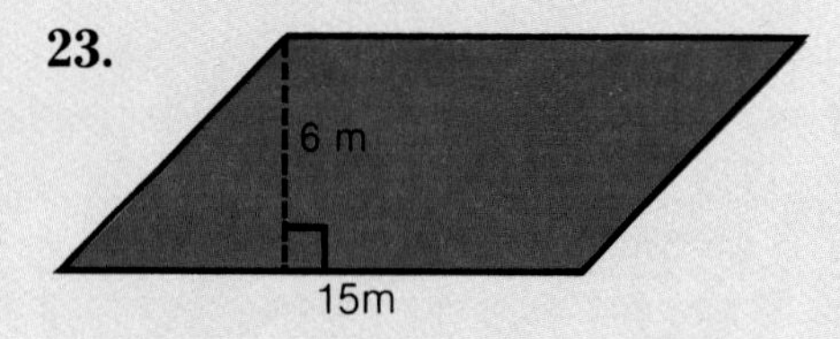

24. 10 cm, 6 cm, 2 cm, 4 cm

Find the circumference. Use 3.14 for π.

25. $d = 10$ in.

26. $r = 4$ m

Find the area. Use $\pi = 3.14$.

27.

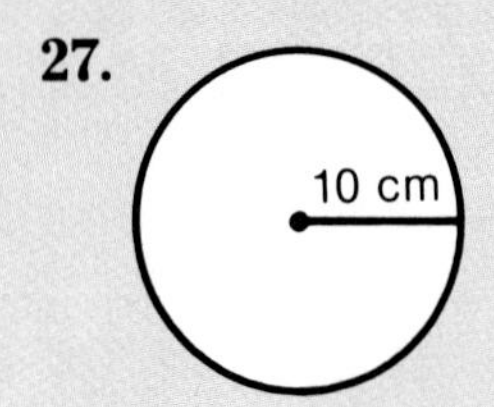

28.

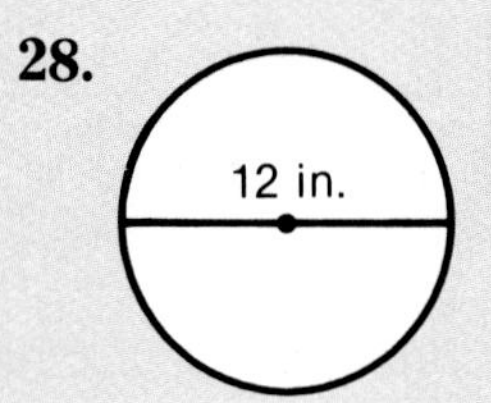

Solve.

29. A rectangular box has a length of 6 cm, a width of 3 cm, and a height of 5 cm. What is its volume?

30. On a floor plan, 2 cm = 3 m. What is the actual length of a room if the length in the floor plan is 6 cm?

COMPUTER LITERACY

A computer can be programmed to use all kinds of formulas. Then when the computer is given values or numbers for the variables in a formula, it can calculate the answer. A computer can store formulas in its memory so that different values can be entered whenever the programmer needs to use them.

AS YOU READ

- memory—the place in the computer where instructions and data are stored
- *(asterisk)—the symbol used in BASIC for multiplication.

A computer places given values in its *memory* before it finds the answer to a calculation. You can think of memory locations like the mailboxes in a post office. Each mailbox has an address and each can be filled. In a BASIC program, the variable name is the "address" of a memory location. The value of the variable fills the memory location. For example, in LET S = 4, S is the address and 4 fills the location. This program finds the area of a square and the volume of a cube.

```
10  LET S = 4
20  LET A = S * S
30  LET V = S * S * S
40  PRINT
50  PRINT "THE AREA OF A SQUARE
    WITH SIDE LENGTH ";S;" IS ";A

60  PRINT "THE VOLUME OF A CUBE
    WITH EDGE LENGTH ";S;" IS ";V


]RUN

THE AREA OF A SQUARE WITH SIDE LENGTH 4 IS 16
THE VOLUME OF A CUBE WITH EDGE LENGTH 4 IS 64
```

The program on the previous page produces the area of a square whose side is 4 units long and the volume of a cube whose edge is 4 units long. The area formula is found in line 20: A = S * S. The volume formula is found in line 30: V = S * S * S. Remember that * means multiply.

The instructions in a BASIC program must be numbered to indicate to the computer the order in which they are to be performed. The computer will carry out the instructions starting with the lowest line number. The following program uses the simple interest formula.

```
10  LET P = 100
20  LET R = .05
30  LET T = 2
40  LET I = P * R * T
50  PRINT
60  PRINT "INTEREST =  ";I
70  END

]RUN

INTEREST =  10
```

This program finds the simple interest on a principal of $100, at a rate of 5% per year, for a 2 year time period. The simple interest formula is shown in line number 40: I=P*R*T.

Because a computer can store programs in its memory, the programs can be used over and over. Answers to problems can be labeled in the program, which make them easier to understand.

EXERCISES

1. Enter and run the AREA and VOLUME program on the previous page. What change would you have to make in the AREA program if you wanted to find the area of a square whose side is 9 units long?

2. Enter and run the INTEREST program above. What change would you have to make in the INTEREST program if you wanted to find the interest on a principal of $250 at a rate of 5% for a 4 year time period?

3. Choose another formula and write a program in BASIC. Write and run the program and check it with your calculator.

Statistics

Problem Solving Skill

To organize data in a table

- Choose an appropriate interval or group.
- Use tally marks to indicate the number of times each item occurs.
- Count the number of tally marks in each interval or group.

To make a set of numbers or data easier to read, it helps to organize it in table form. The data can be listed individually or grouped in equal intervals.

EXAMPLE

The prices of 15 different styles of sports shoes are given below. Make a table to show the data.

$18.95	$24.50	$33.95	$21.00	$34.95	$41.50	$29.95	$27.95
$19.50	$22.95	$25.95	$24.95	$32.50	$31.95	$20.75	

Since each price is different, choose an appropriate interval. Make a tally mark for each of the 15 prices. Then count the number of tallies.

Price	Tally	Number
Under $20.00	II	2
$20.00 to $24.99	卌	5
$25.00 to $29.99	III	3
$30.00 to $34.99	IIII	4
$35.00 and over	I	1

EXERCISES

Make a table for each set of data.

1. scores of 20 Bowling League Members (List scores individually.)

200	243	230	201	199
210	210	200	200	199
230	225	200	230	200
199	199	225	210	230

2. prices of Selected Wristwatches on Sale (Use equal intervals.)

$29.95	$79.95	$60.00
$59.95	$295.00	$89.00
$150.00	$145.00	$100.00
$140.00	$75.00	$50.00

3. raffle Tickets Sold by 15 Club Members (Use equal intervals.)

22	14	35	11	17
6	12	27	15	28
9	30	22	10	25

4. typing Speeds of 15 Typing Students (Use equal intervals.)

34	51	80	50	37
47	48	45	40	55
52	60	66	70	73

FINDING RANGE AND MODE

To find the range and the mode of a given set of numbers

- Subtract the smallest from the largest to find the range.
- Select the number that appears most often to find the mode.

EXAMPLE

Find the range and the mode of temperatures for 15 U.S. cities one summer day.

88, 96, 83, 86, 96, 63, 54, 83,
83, 76, 74, 80, 76, 83, 80

Arranging the temperatures in order can help you find the range and the mode more easily.

54, 63, 74, 76, 76, 80, 80, 83, 83, 83, 83, 86, 88, 96, 96

96 − 54 = 42 So, the **range is 42.**

83 appears four times, or most often.
So, **83 is the mode** for the set of temperatures.

The **range** is the difference between the largest and the smallest number in a set of data.

The **mode** is the number that occurs most often in a set of data.

GETTING READY

Find the range and the mode.

1. 45, 47, 48, 36, 32, 23, 54, 48, 63, 45, 45, 48, 56, 37, 45, 54, 56, 45, 73, 45

2. 65, 67, 64, 53, 28, 56, 75, 46, 53, 43, 53, 57, 53, 53, 34, 34, 53, 46, 64, 53

EXERCISES

Find the range and the mode.

1. 39, 38, 35, 18, 26, 37, 28, 35, 39, 26, 25, 39

2. 40, 45, 44, 40, 40, 41, 41, 39, 44, 42, 38, 42

3. 108, 112, 120, 222, 222, 234, 212, 108, 210, 120, 222, 222

4. 14, 15, 16, 24, 17, 26, 22, 14, 26, 25, 20, 12, 14, 17, 25, 15

***5.** Make up real-life applications for **1–4.**

FINDING THE MEDIAN

To find the median of a given set of numbers

- Arrange the numbers in order from smallest to largest.
- If the number of items is odd, select the middle number.
- If the number of items is even, add the two middle numbers and divide by 2.

EXAMPLE 1

Find the median price.

\$77, \$78, \$78, \$78, \$79, \$80, \$80, \$82, \$83

The prices are already in numerical order.
There are nine prices.
Nine is odd, so select the middle number.
\$77, \$78, \$78, \$78, \$79, \$80, \$80, \$82, \$83

↑ middle number

So, the median price is **\$79.**

> The **median** is the middle number in a set of data arranged in order.

EXAMPLE 2

Find the median.

13, 15, 11, 30, 16, 14

Arrange the numbers in order.

11, 13, 14, 15, 16, 30

↑ middle numbers

There are six numbers.
Six is even, so add the two middle numbers and divide by 2.
So, the median is **14.5.**

$$\frac{(14 + 15)}{2} = \frac{29}{2} = \mathbf{14.5}$$

GETTING READY

Find the median.

1. 95, 87, 80, 72, 75,
75, 75, 75, 35, 32,
32, 41, 56, 76, 43

2. 1.95, 1.87, 1.80, 1.72,
1.57, 1.57, 1.53, 1.22,
1.13, 1.14, 1.87, 1.76

EXERCISES

Find the median.

1. 245, 250, 255, 250, 260, 230, 210, 270

2. 82, 30, 92, 40, 30, 63, 24, 44, 50

3. 73.5, 39.9, 1.3, 169.3, 28.7

4. 2,640; 3,800; 5,180; 3,900; 4,200; 1,620

5. $37.50, $28.50, $29.00, $31.00, $19.98, $36.29, $15.00

6. $8.25, $11.00, $16.95, $13.00, $15.50, $6.95

Find the range, median, and mode of each set of data.

7. Tigers

Player	Height in meters
Deana	1.86
Missy	1.76
Lisa	1.72
Jenny	1.66
Anne	1.59
Jill	1.64
Leslie	1.64
Karol	1.77

8. Rangers

Player	Height in meters
Marty	1.87
John	1.76
Benji	1.96
José	1.89
Frank	1.84
Tommy	1.77
Kenny	1.89
Eric	1.89

9.

Player	Points scored
Tracey	109
Maria	95
Pam	94
Debra	102
Laura	79
Meg	107
Tara	79

10.

Player	Points scored
Art	93
Danny	97
Jake	102
Joe	109
Louis	87
Juan	105
Phil	93

ON YOUR OWN

Find the median and average of each set of scores. How close are the median and the average?

1. 2, 3, 51, 99, 100

2. 50, 50, 51, 100, 100

Here are four scores of a dart game: 350, 700, 620, 529

3. Choose a fifth score so that the average of all five is 556.

FINDING THE MEAN

To find the mean of a given set of numbers

- Add the numbers to find their sum.
- Divide the sum by the number of items in the set.

EXAMPLE

In five games a basketball team scored 40, 37, 51, 44, and 60 points respectively. Find the mean.

Add the numbers. **40 + 37 + 51 + 44 + 60 = 232**

Divide the sum by the number of games.

$$\begin{array}{r} 46.4 \\ 5\overline{)232.0} \\ \underline{20} \\ 32 \\ \underline{30} \\ 2\,0 \\ \underline{2\,0} \end{array}$$

number of games → 5

So, the mean is **46.4**.

> The **mean** is the average of the numbers in a set of data.

GETTING READY

Find the mean.

1. 120, 94, 111, 139, 99

2. \$3.23, \$2.90, \$3.80

3. 2.63, 0.94, 1.06, 10.37

EXERCISES

Find the mean.

1. 95, 87, 88, 90, 79, 74

2. 14.8, 1.1, 2.7

3. \$23.10, \$37.50, \$28.30, \$39.50

4. 6,142; 9,020; 4,893

5. 100, 2, 0.6, 1.8

6. \$220, \$390, \$281, \$312, \$360

ON YOUR OWN

Tell what happens to the mean in a set of data if you do the following:

1. increase each number by 7

2. multiply each number by 5

MAKING A LINE GRAPH

To make a line graph

- Look at the data to choose a scale.
- Mark off the scales on the vertical and horizontal axes.
- Make a dot for each item.
- Connect the dots with straight lines.

EXAMPLE

Make a line graph to show the information in the table.

PEG'S VAN SALES

Jan	60	Mar	90
Feb	120	Apr	180

Choose a scale.
The largest number of vans is 180 and the smallest number is 60. You can use a scale with ticks at every 30 units.

Label the horizontal axis: Month.

Label the vertical axis: Number of Vans.

Make a dot for each month and connect them.

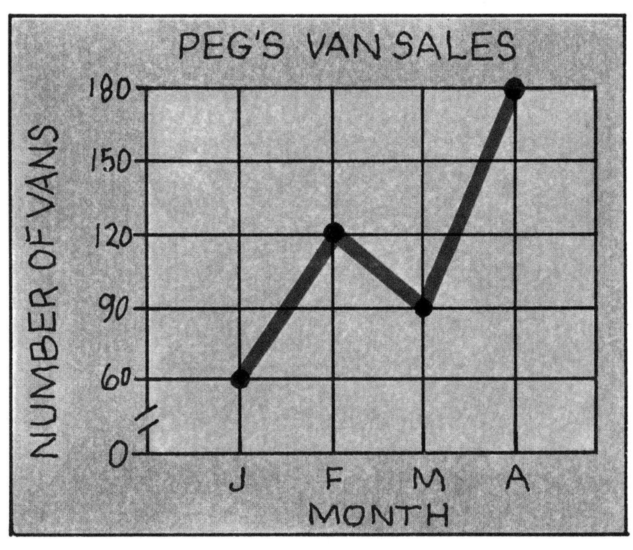

EXERCISES

Make a line graph to show the information in each table.

1. Air Conditioner Sales

Month	Value of Sales
June	$400,000
July	$600,000
Aug.	$500,000
Sept.	$100,000

***2.** Annual Circulation, Memorial Library

Year	Number of Books Circulated
1979	175,000
1980	225,000
1981	300,000
1982	325,000

MAKING A BAR GRAPH

To make a bar graph

- Draw and name a vertical axis and a horizontal axis.
- Mark off a convenient scale.
- Draw and label each bar.
- Give the graph a title.

EXAMPLE

Make a bar graph based on the table at the right.

Since the enrollment numbers range from 700 to 1,100, the vertical scale should include this range. Tick marks for every 100 students will make a convenient scale.

Enrollment at Central High

Year	Students	Year	Students
1960	750	1975	1,050
1965	950	1980	900
1970	1,100	1985*	700

*Estimated

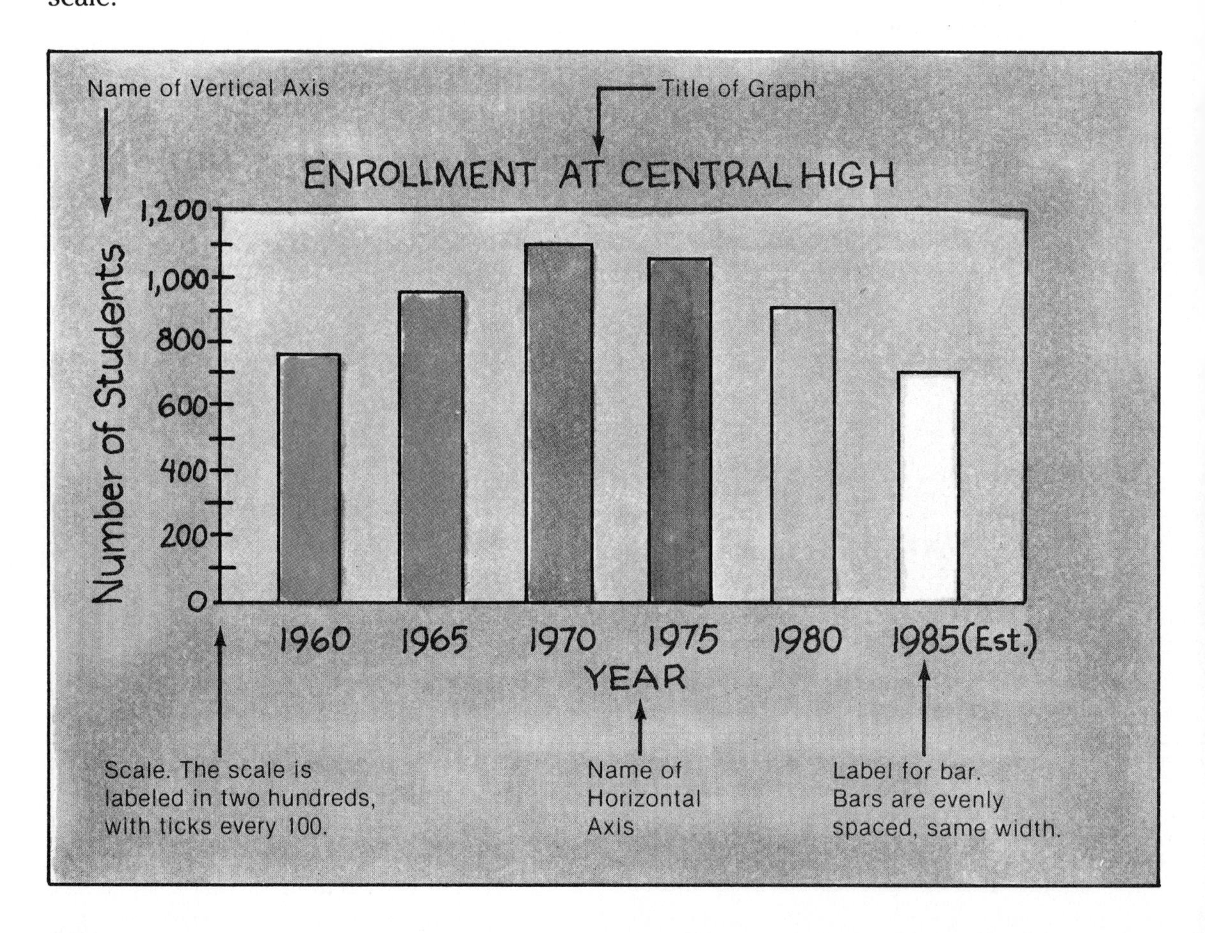

GETTING READY

Make a bar graph to show the information in each table.

1. Average Number of Hours Worked After School

Jimmy	4.5
Fran	4
Alex	3.5
Mona	5.5
Carl	4.5

2. Amounts Spent for Lunch

Monday	\$3.50
Tuesday	\$2.75
Wednesday	\$3.00
Thursday	\$1.80
Friday	\$4.00

EXERCISES

Make a bar graph to show the information in each table.

1. Maximum Temperatures in May

Juneau, Alaska	14°C
St. Paul, Minnesota	10°C
Charleston, S. Carolina	27°C
Baltimore, Maryland	19°C
Key West, Florida	28°C
Raleigh, N. Carolina	23°C
St. Louis, Missouri	18°C

2. Number of Voters per Election District

District 1	10,500
District 2	9,750
District 3	8,750
District 4	11,000
District 5	10,250
District 6	9,000
District 7	12,500

3. Number of Tickets Sold for the Week

Tuesday	650
Wednesday	560
Thursday	420
Friday	600
Saturday Matinee	590
Saturday Evening	650
Sunday Matinee	630

***4.** Highest Mountains by Continent

Asia	29,028 ft
Europe	18,481 ft
South America	23,035 ft
North America	20,320 ft
Africa	19,340 ft
Antarctica	15,100 ft
Australia	7,316 ft

Problem Solving Applications

READ • PLAN • SOLVE • CHECK

Some of the students in the business department at South High School belong to the business services club. Club members help with business-related tasks in the school and in the community.

AS YOU READ

- tallying—making a count of

Answer these questions about the business services club.

1. Vincent is *tallying* the checks from ad sales for the yearbook committee. What is the total amount of the sales? What is the mean of the sales? What is the mode of the sales?

Ads	Tally
\$20	𝍸 𝍸 𝍸
\$30	𝍸 |||
\$50	||

2. The committee plans to raise \$3,500 from ad sales. How much more money needs to be raised by the club members?

3. If there are 8 students selling the ads, how much more in sales is needed from each person in order to reach the goal of \$3,500?

4. Club members help the attendance clerk in the main office. What is the mean number of absences for the week? What is the mean number of students who were late during the week?

5. On Friday, if $\frac{1}{3}$ of the late students were also marked absent, how many were actually absent?

Week of Nov. 13		
Day	No. Absent	No. Late
Mon.	87	39
Tues.	93	52
Wed.	92	41
Thurs.	126	53
Fri.	152	45

Career: Clerical Workers

READ • PLAN • SOLVE • CHECK

Clerical workers are employed throughout business, manufacturing, and government. Some examples of clerical workers are bookkeepers, clerks, cashiers, office machine operators, secretaries, and stenographers. Training varies from high school preparation to 1 or 2 years at business schools.

AS YOU READ

- inventory—the amount of goods on hand
- receipts—money received

Solve.

1. Mary Jenner is head cashier at Wright's Wholesale Distributors. She supervises 15 cashiers. If the total *receipts* for one day were $73,145.70, what were the average receipts for each cashier that day?

2. Bart Flemming is a typist at World Wide Insurance. It takes him about 9 min to type a letter and the copier 30 s to make 20 copies. How long will it take Bart to type 2 letters and make 200 copies of each?

3. Arthur Faulkens is an *inventory* clerk at a large appliance service center. At the beginning of the month, he had 4,568 parts in inventory. He received 1,354 new parts and dispensed 2,438 parts. What is his present inventory?

4. A rate clerk computes charges on shipments. The charge for Zone 1 for 1 metric ton is $49.85 and $15.75 for each additional ton or part of a ton. How much should be charged for a shipment weighing 3.75 metric tons to Zone 1?

***5.** Arnold Peterson is a statistical clerk for the health department of Platt City. The population is 525,000. This winter it is predicted that 52,000 persons will get the flu. What is the flu rate for Platt City? (Hint: 1 flu case per how many people?)

6. Ed Lopez is a shipping clerk at Sunnyside Produce. Ed computes the mean weight of an item in order to plan for shipping. Five sample crates of lettuce weighed 25.25 lb, 23.5 lb, 21.75 lb, 24.5 lb, and 22.75 lb. What is the mean weight of a crate of lettuce?

CHAPTER TEST 16

Here are the test scores of 25 students.

90 80 85 80 75 70 65 65 75 85 80 80 90
80 95 75 75 70 80 60 85 80 75 80 85

1. Find the mode.
2. Find the range.
3. Find the median.
4. Find the mean.
5. Make a bar graph.

Population of Allentown

1950	3,000	1980	4,200
1960	4,500	1990 (est.)	3,500
1970	4,000		

6. Make a line graph.

Attendance at Football Games

Game 1	850	Game 4	1,000
Game 2	950	Game 5	700
Game 3	1,200	Game 6	800

7. Make a table to show the data. Use equal intervals.

Heights of 16 Students (in centimeters)

157 179 163 183 181 168 169 172
180 175 163 159 174 160 177 175

*8. The tally shows distribution of family size in a neighborhood.

Number of Children	Number of Families
0	IIII
1	I
2	卌 III
3	IIII
4	卌
5	III
6	II

Find the average number of children per family (the mean), the median number, and the most common number (the mode).

Percent

MEANING OF PERCENT

To write a percent as a fraction with denominator of 100

Percent means "per hundred" or "out of a hundred." You can read 75% as "75 percent" or "75 out of a 100." This can be written as $\frac{75}{100}$.

EXAMPLE 1

Write 36% as a fraction with 100 as the denominator.
36% means 36 out of 100, or $\mathbf{\frac{36}{100}}$.

The symbol % stands for percent. **Percent** means "per hundred."

EXAMPLE 2

Write 150.2% as a fraction with 100 as the denominator.
150.2% is 150.2 out of 100, or $\mathbf{\frac{150.2}{100}}$.

EXAMPLE 3

Write 304% as a fraction with 100 as the denominator.
304% is 304 out of 100, or $\mathbf{\frac{304}{100}}$.

GETTING READY

Write each percent as a fraction with a denominator of 100.

1. 28% **2.** 85.3% **3.** 205% **4.** 0.5%

EXERCISES

Write each percent as a fraction with a denominator of 100.

1. 45% **2.** 16% **3.** 100% **4.** 167%

5. 24.6% **6.** 10.5% **7.** 3% **8.** 0.4%

9. 1% **10.** 0.95% **11.** 1.8% **12.** 100.1%

13. A sales tax of 3% **14.** A 200% increase in sales

Problem Solving Applications

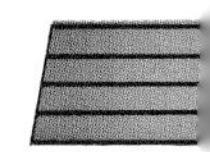

READ • PLAN • SOLVE • CHECK

In Floyd County, many students are active members of the 4-H Club and the Future Farmers of America (FFA). They raise livestock, plant crops, repair farm equipment, and participate in many other activities. They enter local and state competitions in such events as plant identification, livestock judging, quilting, canning, talent, and public speaking.

AS YOU READ

- auction—a public sale to the highest bidder
- profit—amount earned less expenses

Answer these questions about the students.

1. Wayne bought a calf for $250. Four months later he sold the calf for $720. If feed cost $310, how much *profit* did he make?

2. Seventeen high school teams entered the plant identification contest. If each team has 4 members, how many high school students were in the contest?

3. A calf gains an average of $2\frac{1}{2}$ lb per day. If Luis feeds his calf for 145 days, how many pounds will the calf gain?

4. Sandra sold her calf at *auction* for $547.50. If the calf weighed 750 lb, what was the price per pound?

5. In the county livestock judging contest, Ron earned 48 points for his ranking in group I, 35 points in group II, and 50 points in group III. Find Ron's total points earned in the contest.

***6.** Charlotte helps with the plowing after school. If the plow path is 12 ft wide, about how many paths will she make in a field that measures 208 ft by 208 ft? (Round to the nearest whole number.)

PERCENTS AS FRACTIONS

To write a percent as a fraction in simplest form

- Write the percent as a fraction with a denominator of 100.
- Simplify the fraction.

EXAMPLE 1

Write 40% as a fraction in simplest form.

Write as a fraction with a denominator of 100. $\mathbf{40\% = \frac{40}{100}}$

Simplify the fraction. $\mathbf{\frac{40}{100} = \frac{40 \div 20}{100 \div 20} = \frac{2}{5}}$

So, $40\% = \mathbf{\frac{2}{5}}$.

EXAMPLE 2

Write $33\frac{1}{3}\%$ as a fraction in simplest form.

Write as a fraction with a denominator of 100. $\mathbf{33\frac{1}{3}\% = \frac{33\frac{1}{3}}{100}}$

Simplify. $\mathbf{\frac{33\frac{1}{3}}{100} = 33\frac{1}{3} \div 100}$

$$\mathbf{\frac{100}{3} \div \frac{100}{1} = \frac{\overset{1}{\cancel{100}}}{3} \times \frac{1}{\underset{1}{\cancel{100}}} = \frac{1}{3}}$$

So, $33\frac{1}{3}\% = \mathbf{\frac{1}{3}}$.

EXAMPLE 3

Write 12.5% as a fraction in simplest form.

Write as a fraction with a denominator of 100. $\mathbf{12.5\% = \frac{12.5}{100}}$

Multiply numerator and denominator by 10 to get rid of the decimal. $\mathbf{\frac{12.5 \times 10}{100 \times 10} = \frac{12.5}{100.0} = \frac{125}{1{,}000}}$

Simplify. $\mathbf{\frac{125}{1{,}000} = \frac{125 \div 125}{1{,}000 \div 125} = \frac{1}{8}}$

So, $12.5\% = \mathbf{\frac{1}{8}}$.

EXAMPLE 4

Write 375% as a mixed number.

Write as a fraction with a denominator of 100. $\mathbf{375\% = \frac{375}{100}}$

Change to a mixed number. $\mathbf{\frac{375}{100} = \frac{300}{100} + \frac{75}{100} = 3 + \frac{3}{4} = 3\frac{3}{4}}$

So, $375\% = \mathbf{3\frac{3}{4}}$.

GETTING READY

Write as a fraction, as a mixed number, or as a whole number.

1. 17% **2.** 50% **3.** $16\frac{2}{3}\%$ **4.** 9.4% **5.** 220% **6.** 400%

EXERCISES

Write as a fraction, as a mixed number, or as a whole number.

1. 25% **2.** 75% **3.** $62\frac{1}{2}\%$ **4.** 9% **5.** 60% **6.** 2%

7. 35% **8.** $87\frac{1}{2}\%$ **9.** 48% **10.** 5.6% **11.** $83\frac{1}{3}\%$ **12.** $6\frac{1}{4}\%$

13. 125% **14.** 53% **15.** 1% **16.** 250% **17.** 20% **18.** 200%

19. 100% **20.** 85% **21.** 5% **22.** 10% **23.** 30.5% **24.** 140.8%

ON YOUR OWN

Copy and complete the chart below by filling in the missing fractions in simplest form. The percents for these fractions are commonly used, so you will find the chart helpful.

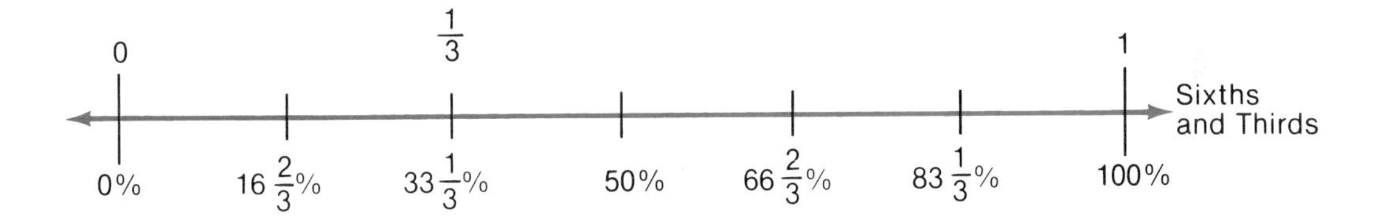

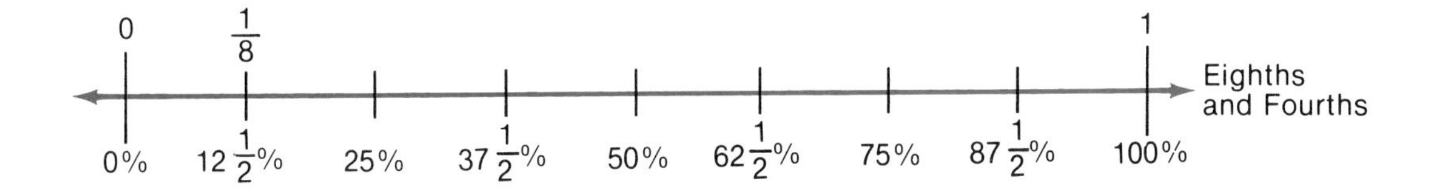

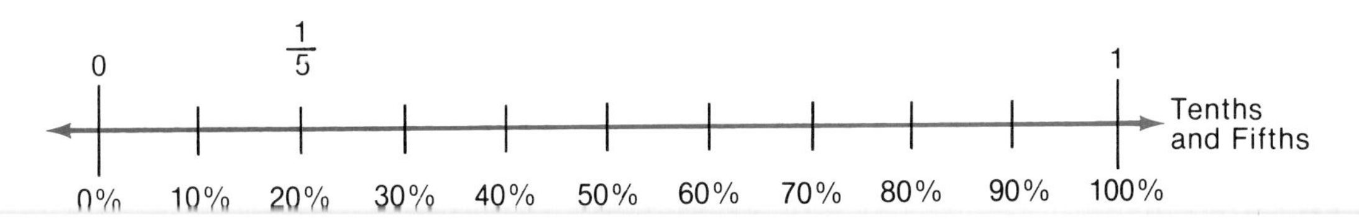

PERCENTS AS DECIMALS

To write a percent as a decimal

- Drop the percent symbol.
- Move the decimal point two places to the left.

EXAMPLE 1

Write 78% as a decimal.

Think: 78% means $\frac{78}{100}$.

Move the decimal point two places to the left.

78 ÷ 100 = 7 8.

So, 78% = **0.78.**

A whole number can be written with a decimal point.
Example: 83 is 83.

EXAMPLE 2

Write 350% as a decimal.

Think: 350% means $\frac{350}{100}$.

Move the decimal point.

350 ÷ 100 = 3 5 0.

So, 350% = **3.5.**

Percents larger than 100% are decimals greater than 1.

EXAMPLE 3

Write 6% as a decimal.

Think: 6% means $\frac{6}{100}$.

Move the decimal point.

6 ÷ 100 = 0 6. ⟵ Add a zero in order to move the decimal point two places.

So, 6% = **0.06.**

EXAMPLE 4

Write 0.5% as a decimal.

Think: 0.5% means $\frac{0.5}{100}$.

Move the decimal point.

0.5 ÷ 100 = 0 0.5 ⟵ Add a zero.

So, 0.5% = **0.005.**

EXAMPLE 5

Write $43\frac{1}{2}\%$ as a decimal.

Think: $\frac{1}{2}$ is 0.5, so $43\frac{1}{2}\%$ is 43.5%.

$43.5\% = \frac{43.5}{100}$

Move the decimal point. **43.5 ÷ 100 = 4 3.5**

So, $43\frac{1}{2}\% =$ **0.435.**

GETTING READY

Write as a decimal.

1. 25%	**2.** 8%	**3.** 132%	**4.** 0.1%	**5.** $12\frac{1}{2}\%$	**6.** 86.3%

EXERCISES

Write as a decimal.

1. 10%	**2.** $1\frac{1}{2}\%$	**3.** 4%	**4.** 125%	**5.** $12\frac{1}{4}\%$	**6.** 0.8%
7. 500%	**8.** 93%	**9.** 2.0%	**10.** 83.4%	**11.** 375%	**12.** $5\frac{2}{3}\%$
13. 0.9%	**14.** 200%	**15.** $33\frac{1}{3}\%$	**16.** 0.03%	**17.** 8%	**18.** 62.5%
19. 39%	**20.** 2.9%	**21.** 390%	**22.** $12\frac{3}{4}\%$	**23.** 4.8%	**24.** 143%
25. 3.5%	**26.** 80.0%	**27.** 0.6%	**28.** $41\frac{1}{2}\%$	**29.** 325%	**30.** 0.09%

ON YOUR OWN

Four cats and 3 kittens weigh 18.5 kg.

Three cats and 4 kittens weigh 16.5 kg.

How much do a cat and a kitten weigh together?

DECIMALS AS PERCENTS

To write a decimal as a percent

- Move the decimal point two places to the right.
- Add the percent symbol.

EXAMPLE 1

Write 0.37 as a percent.

Move the decimal point two places to the right. 0.37 ⟶ **0.37**
So, 0.37 is **37%.**

EXAMPLE 2

Write 0.8 as a percent.

Move the decimal point. Add one zero. 0.8 ⟶ **0.80**
So, 0.8 is **80%.**

EXAMPLE 3

Write 0.249 as a percent.

Move the decimal point. 0.249 ⟶ **0.249**
So, 0.249 is **24.9%.**

EXAMPLE 4

Write 1.45 as a percent.

Move the decimal point. 1.45 ⟶ **1.45**
So, 1.45 is **145%.**

Decimals larger than 1 are percents greater than 100%.

GETTING READY

Write as a percent.

1. 0.78 **2.** 0.3 **3.** 4.53 **4.** 1.8 **5.** 0.09

EXERCISES

Write as a percent.

1. 0.07 **2.** 0.13 **3.** 0.38 **4.** 0.009 **5.** 0.7 **6.** 3.00

7. 0.100 **8.** 0.02 **9.** 0.176 **10.** 0.381 **11.** 1.5 **12.** 4.85

Career: Agricultural Workers

READ • PLAN • SOLVE • CHECK

Agricultural workers include such jobs as county agents, farmers, ranchers, soil analysts, feed store clerks, commodity graders, and storage bin operators. Some jobs require only a minimal education. Others require many hours of training or a college degree.

AS YOU READ

- crop dusting—spraying crops with pesticides from an airplane

Solve.

1. Ten years ago the total value of farm products sold in Franklin County was \$5,967,000. This year the total value increased to \$6,718,800. Find the amount of increase.

2. A report by the county agent of Asheland County shows that over a 5-y period the total acreage in farms decreased from 118,500 to 109,000 acres. How many acres was the decrease?

3. Alan Smithey has 160 acres of his cropland planted in soybeans. If he has 240 acres in all, what part of his cropland is planted in soybeans?

4. Sharon Richards is a pilot for a *crop-dusting* company. If her plane's 160-horsepower engine is running at 0.85 efficiency, what horsepower is it providing?

5. A rancher's 600-lb calf eats 20 lb of feed a day. If it can be expected to gain 1 lb for every 6 lb of feed eaten, how many pounds will it gain per day?

*6. A farmer finds that 18 oz of spray will cover 100 ft^2 of cropland. How many gallons of spray are needed to cover an acre of land? (1 acre = 43,560 ft^2; 1 gal = 128 oz)

FRACTIONS AS PERCENTS

To write a fraction as a percent

- Write as a fraction with a denominator of 100 or as a decimal.
- Change to a percent.

EXAMPLE 1

Write $\frac{3}{5}$ as a percent.

One method: Write $\frac{3}{5}$ as a fraction with a denominator of 100.

$$\frac{3}{5} = \frac{3 \times 20}{5 \times 20} = \frac{60}{100} = 60\%$$

Another method: Write $\frac{3}{5}$ as a decimal to hundredths by dividing.

Divide 3 by 5.

$$\begin{array}{r} 0.60 \\ 5\overline{)3.00} \\ \underline{30} \end{array}$$

$$0.60 = 60\%$$

So, $\frac{3}{5}$ = **60%.**

EXAMPLE 2

Write $2\frac{3}{4}$ as a percent.

$$2\frac{3}{4} = 2 + \frac{3}{4}$$

Write each number as a fraction with a denominator of 100.

$$= \frac{200}{100} + \frac{75}{100}$$

$$= \frac{275}{100} = 275\%$$

So, $2\frac{3}{4}$ = **275%.**

EXAMPLE 3

Write $\frac{5}{8}$ as a percent.

Write as a decimal.

$$\begin{array}{r} 0.625 \\ 8\overline{)5.000} \\ \underline{4\ 8} \\ 20 \\ \underline{16} \\ 40 \\ \underline{40} \end{array}$$

Change to a percent.

$$0.625 = 62.5\%$$

So, $\frac{5}{8}$ = **62.5%.**

EXAMPLE 4

Write $\frac{1}{3}$ as a percent. Round the answer to the nearest tenth of a percent.

Write as a decimal.

$$\begin{array}{r} 0.3333 \\ 3\overline{)1.0000} \\ \underline{9} \\ 10 \\ \underline{9} \\ 10 \\ \underline{9} \\ 10 \\ \underline{9} \end{array}$$

Change to a percent. $0.3333 = \mathbf{33.33\%}$

Round to the nearest tenth. $33.33\% \longrightarrow \mathbf{33.3\%}$

So, $\frac{1}{3} = \mathbf{33.3\%}$.

GETTING READY

Write as a percent. Round to the nearest tenth of a percent when necessary.

1. $\frac{3}{4}$ **2.** $\frac{2}{10}$ **3.** $\frac{7}{8}$ **4.** $\frac{1}{9}$ **5.** $3\frac{4}{5}$

EXERCISES

Write as a percent. Round to the nearest tenth of a percent when necessary.

1. $\frac{2}{5}$ **2.** $\frac{2}{3}$ **3.** $\frac{3}{10}$ **4.** $\frac{1}{4}$ **5.** $\frac{9}{10}$

6. $\frac{1}{8}$ **7.** $\frac{49}{100}$ **8.** $\frac{3}{8}$ **9.** $1\frac{1}{4}$ **10.** $\frac{11}{10}$

11. $4\frac{1}{5}$ **12.** $5\frac{1}{2}$ **13.** $3\frac{3}{10}$ **14.** $\frac{4}{3}$ **15.** $\frac{25}{10}$

16. $\frac{325}{100}$ **17.** $\frac{7}{5}$ **18.** $\frac{1}{20}$ **19.** $\frac{2}{25}$ **20.** $\frac{9}{5}$

21. $\frac{7}{4}$ **22.** $\frac{400}{100}$ **23.** $\frac{5}{200}$ **24.** $\frac{5}{6}$ **25.** 2

FINDING A PERCENT OF A NUMBER

To find a percent of a number

- Use the formula $\frac{\text{part}}{\text{base}} = \frac{\text{percent}}{100}$ to write a proportion.
- Solve the proportion.

The statement 18 is 50% of 36 gives the proportion:

part ⟶ $\frac{18}{36} = \frac{50}{100}$ ⟵ percent
base ⟶

The percent symbol (%) indicates the percent, 50.
The word "of" leads you to the base, 36.

EXAMPLE 1

What amount is 15% of $940?

The percent is 15. The base is $940.
You must find the missing part.

Write a proportion. part ⟶ $\frac{n}{940} = \frac{15}{100}$ ⟵ percent
Let n equal **part.** base ⟶

Solve for n.

$$100 \times n = 15 \times 940$$
$$100n = 14{,}100$$
$$n = 14{,}100 \div 100 = 141$$

So, 15% of $940 is **$141.**

EXAMPLE 2

Forty percent of the 1,380 Hillcrest High School seniors will go to college after graduation. How many will go to college?

Write a proportion. part ⟶ $\frac{n}{1{,}380} = \frac{40}{100}$ ⟵ percent
Let n equal **part.** base ⟶

Solve for n.

$$100 \times n = 40 \times 1{,}380$$
$$100n = 55{,}200$$
$$n = 55{,}200 \div 100 = 552$$
$$40\% \text{ of } 1{,}380 = 552$$

Another method: Multiply base times percent.

40% = 0.40

$$\begin{array}{r} 1{,}380 \leftarrow \text{base} \\ \times 0.40 \leftarrow \text{percent} \\ \hline 552.00 \end{array}$$

So, **552** seniors will go to college.

EXAMPLE 3

Find $6\frac{1}{2}\%$ of \$3,211.

Write a proportion. Let n equal **part.**

part → $\frac{n}{3,211} = \frac{6\frac{1}{2}}{100}$ ← percent
base →

Solve for n.

$$100n = 6\tfrac{1}{2} \times 3,211$$
$$= 6.5 \times 3,211 = 20,871.5$$
$$n = 20,871.5 \div 100 = 208.715$$

Round to the nearest cent. \$208.715 **\$208.72** So, the answer is **\$208.72.**

GETTING READY

1. In "20% of 25 is 5," what is the **base?** The **part?** The **percent?**

Compute. Round to the nearest tenth or nearest cent when necessary.

2. 25% of 84 **3.** $10\frac{1}{2}\%$ of \$326 **4.** 8% of 120 **5.** 33.3% of \$2,496

EXERCISES

Compute. Round to the nearest tenth or nearest cent when necessary.

1. 25% of 96 **2.** $37\frac{1}{2}\%$ of 64 **3.** 80% of 44 **4.** 15% of \$78.30

5. 10% of \$240 **6.** 60% of \$17.50 **7.** 2% of \$345 **8.** 400% of 25

Solve. Round to the nearest tenth or nearest cent when necessary.

9. A newspaper contained 1,860 ads. If 40% of the ads produced sales, how many sales were produced by the ads?

10. Mary Belle gets a 3% commission on all sales she makes. She sold \$1,800 worth of merchandise. How much commission did she earn?

11. Leo saved 10% of \$342.50. How much did he save?

12. Alex paid a $1\frac{1}{2}\%$ service charge on his overdue account. He owed \$58. How much was his service charge?

***13.** The Chiu family spends 31% of their income on food. The Chius earned \$1,525.80 last month. How much did they spend on food?

***14.** Mr. Contreras earned \$14,728 last year. If 22.4% of his income was deducted for taxes, how much was deducted for taxes?

FINDING PERCENTS

To find what percent one number is of another

- Use the formula $\frac{\text{part}}{\text{base}} = \frac{\text{percent}}{100}$ to write a proportion.
- Solve the proportion.

EXAMPLE 1

22 is what percent of 110?

One method: Write a proportion.
Let n equal **percent.**
Solve for n.

part → $\frac{22}{110} = \frac{n}{100}$ ← percent
base →

$$110 \times n = 22 \times 100$$
$$110n = 2{,}200$$
$$n = 2{,}200 \div 110 = 20$$

22 is 20% of 110.

Another method: Divide part by base.

part → $\frac{22}{110}$
base →

$$\begin{array}{r} .2 \\ 110\overline{)22.0} \\ \underline{22\ 0} \end{array} = 0.20 = 20\%$$

So, the answer is **20%.**

EXAMPLE 2

Jana has 30 boxes of cookies to sell. She has sold 14 boxes. What percent has she sold? Round the answer to the nearest tenth of a percent.

Restate the question. 14 is what percent of 30?

Write a proportion.
Let n equal **percent.**
Solve for n.

part → $\frac{14}{30} = \frac{n}{100}$ ← percent
base →

$$30 \times n = 14 \times 100$$
$$30n = 1{,}400$$
$$= 1{,}400 \div 30$$
$$n = 46.66$$

$$\begin{array}{r} 46.66 \\ 30\overline{)1{,}400.00} \\ \underline{1\ 20} \\ 200 \\ \underline{180} \\ 20\ 0 \\ \underline{18\ 0} \\ 2\ 00 \\ \underline{1\ 80} \end{array}$$

Round to the nearest tenth. 46.66 → **46.7**

So, Jana has sold about **46.7%** of her boxes of cookies.

GETTING READY

Compute. Round to the nearest tenth of a percent when necessary.

1. 55 is what percent of 75?

2. What percent of $280 is $84?

3. 30 is what percent of 70?

4. What percent of 150 is 300?

EXERCISES

Compute. Round to the nearest tenth of a percent when necessary.

1. 19 is what percent of 38?

2. 91 is what percent of 728?

3. What percent of 108 is 27?

4. What percent of 32 is 8?

5. 48 is what percent of 60?

6. 12 is what percent of 120?

7. What percent of 72 is 18?

8. What percent of 300 is 450?

9. 19 is what percent of 57?

10. 8 is what percent of 12?

Solve. Round to the nearest tenth of a percent when necessary.

11. Tanya saves $10 a week from her salary. Her weekly salary is $250. What percent of her salary does Tanya save each week?

12. Joli borrowed $300 from Fred. Fred charged her $24 interest. What percent interest did Fred charge?

13. A new car, which costs $5,280, depreciates $1,531.20 in value during the first year. What percent of the original cost is the depreciation?

***14.** The enrollment at Northside High School is 1,480. On Tuesday, 1,375 students were present. What percent of the students enrolled were absent on Tuesday?

CALCULATOR

Try these with your calculator.

1. Multiply 5,144 by 24 and see what your calculator does.

2. Multiply 218,107 by 3.

3. Multiply your age by 7 and that total by 1,443.

FINDING THE BASE NUMBER

To find a number when a percent of it is known

- Use the formula $\frac{\text{part}}{\text{base}} = \frac{\text{percent}}{100}$ to write a proportion.
- Solve the proportion.

EXAMPLE 1

9 is 20% of what number?

Write a proportion. part $\longrightarrow$ $\frac{9}{n} = \frac{20}{100}$ $\longleftarrow$ percent

Let n equal **base.** base $\longrightarrow$

Solve for n.

$$\mathbf{20 \times n = 9 \times 100}$$
$$\mathbf{20n = 900}$$
$$\mathbf{n = 900 \div 20 = 45}$$

So, 9 is 20% of **45.**

EXAMPLE 2

Louise has a collection of 42 stamps. This is 60% of the number of stamps in Edna's collection. How many stamps does Edna have in her collection?

Restate the question. 42 stamps are 60% of how many stamps?

Write a proportion. part $\longrightarrow$ $\frac{42}{n} = \frac{60}{100}$ $\longleftarrow$ percent

Let n equal **base.** base $\longrightarrow$

Solve for n.

$$\mathbf{60 \times n = 42 \times 100}$$
$$\mathbf{60n = 4{,}200}$$
$$\mathbf{= 4{,}200 \div 60}$$
$$\mathbf{n = 70}$$

$$\begin{array}{r} 70 \\ 60\overline{)4{,}200} \\ \underline{4\ 20} \end{array}$$

So, Edna has **70** stamps in her collection.

GETTING READY

Compute. Round to the nearest whole number when necessary.

1. 13 is 25% of what number?

2. \$5.80 is 10% of what amount?

3. 165 is 150% of what number?

4. 6 is 11% of what number?

5. Ms. Weygandt spent \$78 for ads. This was 15% of her receipts. How much were her receipts?

EXERCISES

Compute. Round to the nearest whole number when necessary.

1. 16 is 80% of what number?

2. 15 is 5% of what number?

3. 12 is 10% of what number?

4. 3 is 20% of what number?

5. 50% of what number is 100?

6. 15% of what number is 45?

7. 7 is 35% of what number?

8. 80 is 40% of what number?

9. 25% of what number is 16?

10. 125% of what number is 50?

11. 3 is 12.5% of what number?

12. 21 is 73% of what number?

Solve. Round to the nearest whole number or nearest cent when necessary.

13. Last year, $1,100 in taxes was deducted from Tom's pay. This was 20% of his pay. How much did he earn that year?

14. Ms. Halpern saves $30 every week. This is 12% of her pay. How much does she earn per week?

15. All the employees at Voight's received a 5% raise. Jane received a $6.50-a-week increase. What was her weekly pay before that raise?

16. One year Pete's savings account earned $27.50 in interest. This was 5% of his original deposit. How much was his original deposit?

17. Ivan received a commission of $54.75. This was 3% of his total sales. What were his total sales?

18. Corinne sold 120 all-occasion cards one week. This was about 68% of her supply. How many cards did she have at first?

19. Josh won 57 table tennis games last year. This was 75% of the total number of games he played. How many games did he play? Can you find how many games he lost?

20. Debra received a $4.50 dividend for one share of stock. The dividend is 4% of the value of the stock. What is the value of the stock?

PERCENT CHANGE

To find the percent increase or decrease

- Find the amount of increase or decrease.
- Write a proportion to find the percent.

You can use the proportion $\frac{\text{part}}{\text{base}} = \frac{\text{percent}}{100}$ to find a percent of increase or a percent of decrease.

EXAMPLE 1

A $23 dress was marked down to $18.40. What was the percent decrease in price?

The decrease is the **part.** $23 − $18.40 = $4.60 decrease
The original price is the **base.** $23 is the original price.
Write a proportion.
Let n equal **percent decrease.**

part ⟶ base ⟶ $\frac{4.60}{23} = \frac{n}{100}$ ⟵ percent decrease

Solve for n.

$$23 \times n = 4.60 \times 100$$
$$23n = 460$$
$$n = 460 \div 23 = 20$$

So, the percent decrease in price is **20%.**
This is also called the **discount rate.**

EXAMPLE 2

Over a period of five years, the number of graduating seniors increased from 280 to 426 at Wilshire High School. What was the percent increase in graduating seniors? Round the answer to the nearest tenth of a percent.

The increase is the **part.** 426 − 280 = 146 increase
The original number of seniors is the **base.** 280 is the original number of seniors.

Write a proportion.
Let n equal **percent increase.**

part ⟶ base ⟶ $\frac{146}{280} = \frac{n}{100}$ ⟵ percent increase

Solve for n.

$$280 \times n = 146 \times 100$$
$$280n = 14{,}600$$
$$n = 14{,}600 \div 280$$

Divide to hundredths. Then round to tenths.

$280\overline{)14{,}600.00}$ = **52.14** ⟶ **52.1**

So, the percent increase was about **52.1%.**

GETTING READY

Solve. Round to the nearest tenth of a percent when necessary.

1. What is the percent increase in price when changing from \$2.40 to \$3?

2. What is the percent decrease in weight when going from 55 kg to 48 kg?

3. A warehouse increased its furniture stock from 250 to 500 pieces. What was the percent increase in the amount of furniture in stock?

4. What is the percent decrease in length when cutting 8 feet of rope from 16 feet of rope?

5. A tree used to be 2 m tall. Now it is 4.5 m tall. Find the percent increase in height.

EXERCISES

Solve. Round to the nearest tenth of a percent when necessary.

1. Heidi sold 40 brushes last week. This week she sold 62 brushes. What was the percent increase in sales?

2. The cash price of a used car is \$2,500. The installment price is \$3,000. What is the percent increase in price?

3. Peter made 15 errors on a typing test. On the next test he made 9 errors. What was the percent decrease in errors?

4. Last year Rob averaged 22.5 points per game. This year he averaged 16 points per game. What was the percent decrease in his average?

5. A \$120 suit was marked to sell for \$145. What was the percent increase in price?

6. Beth was earning \$4.30 per hour. After her raise, she was earning \$4.46 per hour. What was the percent increase in pay?

7. A year ago Jamie weighed 100 kg. He now weighs 80 kg. What is the percent decrease in weight?

8. A \$96 typewriter was on sale. Toni bought it for \$80 at the sale. What was the percent decrease in price?

9. Last year, Iron Fencing, Inc., had 600 employees. This year the company has 800 employees. What is the percent increase in the number of employees?

10. In 1939, a 2-L container of milk could be bought for 20¢. It now costs \$1.40. What is the percent increase in price?

ESTIMATING A PERCENT OF A NUMBER

To estimate a percent of a number

- Use a fraction for the percent.
- Round the number.

Certain percents occur quite often in daily life. The table below lists some of these percents and the corresponding fractions. You can use the fractions to estimate a percent of a number.

$\frac{1}{5} = 20\%$	$\frac{2}{5} = 40\%$	$\frac{2}{3} = 66\frac{2}{3}\%$	$1 = 100\%$	$\frac{1}{8} = 12\frac{1}{2}\%$	$\frac{1}{6} = 16\frac{2}{3}\%$
$\frac{1}{4} = 25\%$	$\frac{1}{2} = 50\%$	$\frac{3}{4} = 75\%$	$2 = 200\%$	$\frac{1}{10} = 10\%$	$\frac{7}{10} = 70\%$
$\frac{1}{3} = 33\frac{1}{3}\%$	$\frac{3}{5} = 60\%$	$\frac{4}{5} = 80\%$	$3 = 300\%$	$\frac{3}{10} = 30\%$	$\frac{9}{10} = 90\%$

EXAMPLE 1

Estimate a 10% discount on a suit priced at \$69.96.

Use a fraction. $10\% = \frac{1}{10}$

Round the price to the nearest dollar. \$69.95 ⟶ **\$70**

$\$70 \times \frac{1}{10}$ is the same as $\$70 \div 10$.

$\$70 \div 10 = \$7$

So, the estimated discount is **\$7.**

EXAMPLE 2

Estimate the 5% sales tax on a bill of \$16.49.

5% is $\frac{1}{2}$ of 10%, so estimate 10% first, then find half of it.

Round to the nearest dollar. \$16.49 ⟶ **\$16**

Divide mentally. **$\$16 \div 10 = \$1.60$**

Find half of \$1.60. **$\$1.60 \div 2 = \$0.80$**

So, the 5% tax on \$16.49 is about **\$0.80.**

EXAMPLE 3

The price of a \$47 item is to be increased by 50%.
Estimate the new price.

50% is $\frac{1}{2}$, so find half of \$47. **Half of \$47 is about \$24.**

Add to the original cost. **$\$24 + \$47 = \$71$**

So, the estimated new price is **\$71.**

EXAMPLE 4

The fare for a taxi ride is \$15.45. Estimate a 15% tip.
To find 15%, use 10% and 5% (half of 10%).

Round the total cost.	\$15.45 ⟶ **\$15**
Find 10% of \$15.	**\$15 ÷ 10 = \$1.50**
Find 5% of \$15. Divide \$1.50 by 2.	**\$1.50 ÷ 2 = \$0.75**
Add.	**\$1.50 + \$0.75 = \$2.25**

So, the 15% tip is about **\$2.25.**

GETTING READY

Estimate.

1. 10% of \$14.89
2. 5% of \$62.17
3. 50% of 36
4. 20% of 150
5. 15% of \$26.95
6. 25% of \$28.35

EXERCISES

Estimate.

1. 10% of \$5.69
2. 50% of 18
3. 100% of 97
4. 5% of \$18.33
5. 20% of \$86.14
6. 15% of \$7.55
7. 1% of \$9.15
8. 25% of 160
9. 5% of \$24.69

Solve by estimating.

10. Find the 15% tip for a \$9.85 taxi fare.

11. What is the 5% sales tax on a \$39.95 dress?

12. If a \$40 item is marked up 10%, what is the new price?

13. If an \$8 shirt is reduced in price by 50%, what is the new sale price?

14. What is the 10% discount on a \$7.88 bath towel?

15. Find the 15% tip for a \$12.15 taxi fare.

Problem Solving Skill

To read and interpret a circle graph

- Find the total amount represented by the full circle.
- Identify the part of the whole represented by each section.

EXAMPLE

Use the circle graph at the right to answer each question.

What is the total amount of all monthly expenses?

The title shows that the total amount is **$1,600.**

What percent of monthly income is budgeted for clothing?

The full circle represents 100%. The total of the sections except clothing is **25 + 30 + 5 + 10 + 10, or 80%.**
100% − 80% is 20%.
So, **20%** is budgeted for clothing.

What amount is budgeted for car expenses?

Car expenses take 10% of the total amount.
10% of $1,600 is $160.
So, **$160** is budgeted for car expenses.

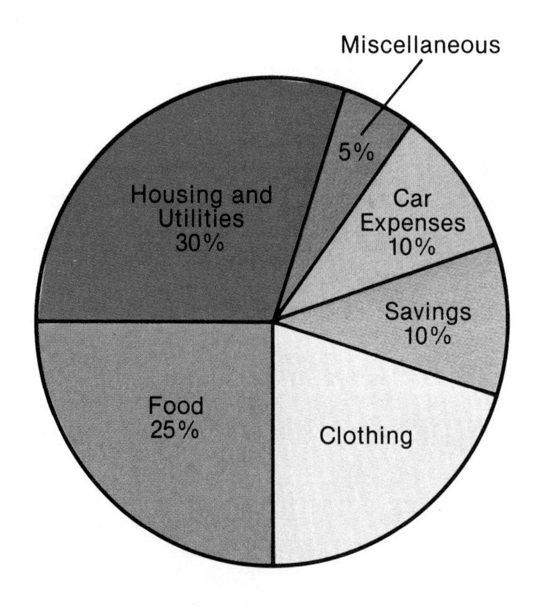

EXERCISES

Use the graph above to answer **1–4.**

1. What amount is budgeted for food?

2. What amount is budgeted for miscellaneous expenses?

3. What percent of monthly income is budgeted for housing expenses, food, and clothing?

4. According to the budget, how much will be spent on housing and utilities in one year?

Use the circle graph at the right to answer **5–10.**

5. What housing need requires the most energy each year?

6. Do cooling and water heating combined take more or less energy than space heating?

7. What percent of the annual amount of energy is used for heating or cooling the house?

8. If the stove section and the washer and dryer section are the same size, what percent does each represent?

9. If the all-electric house uses 18,000 kW·h of electricity in a year, how many kilowatt-hours are used by the refrigerator?

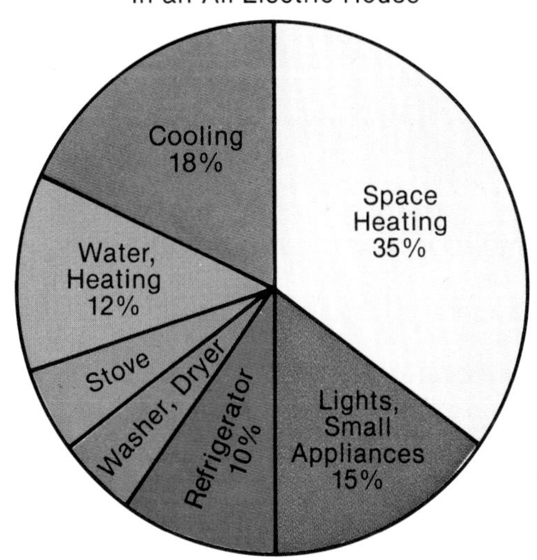

***10.** If the house uses 3,960 kW·h in a year for cooling, what is the total amount of electric energy used?

ON YOUR OWN

Here are estimated populations of major regions of the world, in millions:

North America	260	USSR	280
Latin America	360	Asia	2,600
Europe	500	Africa	500

1. Find the total population of all these regions.

2. Use a calculator to find what percent of the total population lives in each region. Round to the nearest percent.

3. Draw a circle graph showing the percent living in each region.

CHAPTER TEST 17

Write as a fraction with a denominator of 100.

1. 27% **2.** 6.7% **3.** 320%

Write as a decimal.

4. 35% **5.** $10\frac{1}{2}$% **6.** 140%

Write as a fraction in simplest form.

7. 85% **8.** 4% **9.** 160%

Write as a percent.

10. 0.01 **11.** 0.875 **12.** 3.29 **13.** $\frac{7}{10}$ **14.** $\frac{3}{4}$ **15.** $2\frac{1}{8}$

Compute.

16. 25% of 480 **17.** 7 is what percent of 28? **18.** 35 is 20% of what number?

Estimate.

19. 15% of $18.37 **20.** 50% of $689.87

Solve. Round to the nearest tenth or cent when necessary.

21. Anna won 32 of 40 tennis sets. What percent of the sets did she win?

22. David saved 20% of $735.80. How much did he save?

23. Barry saved $5 on a shirt that was marked down 20%. How much was the shirt before it was marked down?

24. A $250 TV set was marked down to sell for $212.50. What was the percent decrease in price?

Use the circle graph to answer **25–26.**

25. What is Mike's greatest expense for his car?

26. If Mike spends $4,000 on his car, how much will he spend on tune-ups and oil changes?

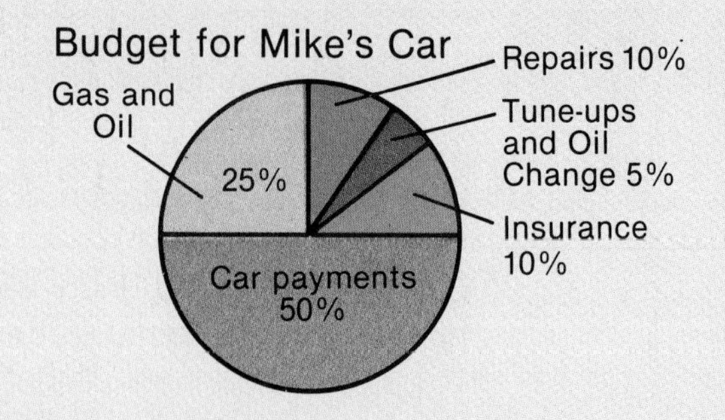

Earning Money

EARNINGS

To compute the amount earned for a given number of hours of work

- Multiply the rate per hour by the total hours.

Many workers are paid by the hour and receive overtime pay for the hours they work over 40 h per week.

EXAMPLE 1

Jerry works in an automobile assembly plant. He receives time and a half for all hours worked over 40 h per week. How many hours' pay is received for 47 h?

> **Time and a half** means 1.5 hours' pay for each hour worked.

Think: 47 − 40 = 7 h overtime

7 h overtime is the same as 10.5 h regular time. **7 × 1.5 = 10.5**

Add to find the total.

$$\begin{array}{r} \mathbf{40.0} \\ \mathbf{+10.5} \\ \hline \mathbf{50.5} \end{array}$$

So, **50.5** hours' pay is received for 47 h.

EXAMPLE 2

Sandra works as a line installer for a telephone company. She earns $9.50 per hour. Last week she worked 45 h. She earns time and a half for all hours over 40. Find Sandra's gross pay.

> **Gross pay** is the total amount earned before deductions for such things as taxes and social security are made.

Think: 45 − 40 = 5 h overtime

Multiply by 1.5 to find the equivalent in regular hours. **5 × 1.5 = 7.5**

Add to find the total. **40.0 + 7.5 = 47.5**

Multiply the rate per hour by the total hours.

$$\begin{array}{r} \mathbf{\$9.50} \\ \mathbf{\times 47.5} \\ \hline \mathbf{4\ 750} \\ \mathbf{66\ 50} \\ \mathbf{380\ 0} \\ \hline \mathbf{\$451.250} \end{array}$$

So, Sandra's gross pay is **$451.25.**

GETTING READY

Tell how many hours' pay is received. Use time and a half for all hours over 40.

1. 52 h **2.** 48 h **3.** 51 h **4.** 35 h

Find the gross pay. Use time and a half for all hours over 40.

5. 54 h at $3.80 per hour **6.** 60 h at $10.50 per hour

EXERCISES

Tell how many hours' pay is received. Use time and a half for all hours over 40.

1. 38 h **2.** 53 h **3.** 46 h **4.** 39 h **5.** 62 h

6. 58 h **7.** 49 h **8.** 59 h **9.** 55 h **10.** 57 h

Find the gross pay. Use time and a half for all hours over 40.

11. 38 h at $3.25 per hour **12.** 30 h at $4.75 per hour

13. 35 h at $6.90 per hour **14.** 39 h at $10.25 per hour

15. 44 h at $3.76 per hour **16.** 54 h at $4.80 per hour

17. 42 h at $8.90 per hour **18.** 60 h at $7.60 per hour

19. 55 h at $10.50 per hour **20.** 49 h at $11.25 per hour

Find the amount earned based on the time cards below. Use time and a half for all hours over 40.

21.

Emma Farber	Rate $4.50/h
Mon	$7\frac{1}{2}$ h
Tues	8 h
Wed	9 h
Thurs	8 h
Fri	$10\frac{1}{2}$ h

22.

Craig Adams	Rate $4.20/h
Mon	$8\frac{1}{2}$ h
Tues	$8\frac{1}{2}$ h
Wed	$9\frac{1}{2}$ h
Thurs	8 h
Fri	$10\frac{1}{2}$ h

COMMISSION

To compute commission

- Multiply the sales by the rate of commission.

EXAMPLE 1

Carla works in an appliance store. She gets a 12% commission on all sales. How much commission does she earn on sales of $3,200?

The rate of commission is 12%. Multiply.

12% = 0.12

$3,200	⟵	sales
× 0.12	⟵	rate of commission
64 00		
320 0		
$384.00		

So, Carla's commission is **$384.**

EXAMPLE 2

Peter works for an automobile dealer. He earns $120 a week plus commission. His commission is 3% of his sales. Last week he sold two cars. One car sold for $5,540 and the other sold for $7,650. How much money did he earn last week?

Add to find the total sales.

$5,540
+7,650
$13,190

Find the commission.

$13,190	⟵	sales
× 0.03	⟵	rate of commission
$395.70		

Add to find total earnings.

$120.00	⟵	salary
+395.70	⟵	commission
$515.70		

So, Peter earned **$515.70** last week.

GETTING READY

Compute the commission.

1. 13% commission on sales of $4,500
2. 10% commission on sales of $3,682
3. Mr. Dunsay earns $100 per week plus a 4% commission on his sales. Last week his total sales were $9,000. How much did he earn?
4. Ms. Cato earns $125 per week plus a 5% commission on her sales. She made two sales last week for $800 and $1,500. How much did she earn?

EXERCISES

Compute the commission.

1. 15% commission on sales of $520

2. 17% commission on sales of $960

3. 11% commission on sales of $1,250

4. 12% commission on sales of $2,380

Solve.

5. Beverly Goodson earns $120 per week plus a 4% commission on her sales. Last week she sold a car for $5,795. How much did she earn?

6. Jim Martin earns $175 per week plus a 3% commission on all sales. He made two sales last week, for $12,380 and $8,550. How much did he earn?

7. Joanna is a travel agent. She earns $210 a week plus a commission of 7% on all trips that she arranges. Last week she arranged two trips, one for $1,850 and the other for $2,720. How much did she earn last week?

8. Elliot is a salesperson for the Everyday Equipment Company. He earns $750 per month plus a 5% commission on all sales. Last month his sales were $14,754. How much did he earn last month?

***9.** Ernesto received $520 in commission. His rate of commission is 8%. What were his total sales?

***10.** Ms. Palmer received $3,870 commission on total sales of $64,500. What was her rate of commission?

CALCULATOR

Solve each problem on the calculator. The correct solution, when read from the calculator turned upside down, will form a word. A clue is given in each exercise.

1. 8,586 − 3,972 opposite of low

2. 105,065 − 97,351 small mountain

3. 5.1 + 87.2 + 983.3 + 2,731.4 part of the ear

4. 8 × 4.63 part of a doughnut

COMMISSION ON A GRADUATED SCALE

To compute commission on a graduated scale

- Use the scale to find the commissions for the sales.

Some workers receive one rate of commission on part of their sales and a different rate on the rest of their sales.

EXAMPLE

Don's commission is based on this scale.

Sales	Rate
\$0–\$25,000	8%
over \$25,000	12%

How much did Don earn on sales of \$34,200?

Subtract to find Don's sales over \$25,000. **\$34,200 − \$25,000 = \$9,200**

The rate of commission is 8% on \$25,000 and 12% on \$9,200.

Multiply sales by rate of commission.

\$25,000	← sales →	**\$9,200**
×0.08	← rate of commission →	**×0.12**
\$2,000.000		**\$1,104.00**

Add to find the total commission. **\$2,000 + \$1,104 = \$3,104**

So, Don earned **\$3,104** on the sales.

GETTING READY

Solve.

1. Victoria's commission is based on this scale.

Sales	Rate
\$0–\$8,000	9%
over \$8,000	13%

Victoria's sales were \$15,500. What was her commission?

2. Carlos's commission is based on this scale.

Sales	Rate
\$0–\$90,000	3%
over \$90,000	5%

Carlos's sales were \$85,000. What was his commission?

3. Mark receives a salary of \$1,200 per month plus commission. His commission is 6% on sales up to \$10,000 and 12% on sales over \$10,000. How much did he earn last month if his sales were \$16,500?

EXERCISES

Use the information below to find the commission.

	Sales	Rate	Total sales
1.	$0–$6,000 over $6,000	6% 7%	$ 5,800
3.	$0–$40,000 over $40,000	3% 5%	$45,000
5.	$0–$12,000 over $12,000	7% 9%	$12,580

	Sales	Rate	Total sales
2.	$0–$5,500 over $5,500	4% 6%	$ 5,400
4.	$0–$3,500 over $3,500	12% 14%	$ 6,000
6.	$0–$18,000 over $18,000	6% 9%	$22,570

7. Cindy's earnings are based on this scale.

Sales	Rate
$0–$1,100 over $1,100	12% 15%

How much did she earn on sales of $1,470?

8. Manuel's commission is based on this scale.

Sales	Rate
$0–$500 over $500	20% 25%

How much did he earn on sales of $2,470?

9. Judy sells books. She is paid a 6% commission on the first $14,000 in sales each month and 8% on all sales over $14,000. In November her sales were $17,500. What was her commission?

10. Carl sells cosmetics. He is paid a 7% commission on the first $8,000 in sales each month and 10% on all monthly sales over $8,000. How much will he receive if his sales last month were $9,025?

11. Jethro works in a motorcycle shop. His commission is 10% on monthly sales of $6,500 or less and 12% for any sales over $6,500. Last month his sales were $7,200. What was Jethro's commission?

***12.** Ben receives a salary of $1,400 per month plus commission. His commission is 4% on sales up to $20,000, 5% on sales from $20,000 to $30,000, and 7% on sales over $30,000. His sales last month were $35,000. How much did he earn last month?

Problem Solving Skill

To understand paycheck stub amounts

- Recognize the meanings of the entries on the paycheck stub.

Federal With. Tax, State, and **City** are the withholding taxes.

Net is the amount of the paycheck: gross pay minus the total of all deductions.

F.I.C.A. (Federal Insurance Contributions Act) is the amount paid to Social Security.

Year-to-Date Totals show how much has been earned and deducted to date.

PENINSULA SHIPBUILDING COMPANY — Check No. 039885

Social Security No.	Pay End Date	Hourly Rate	Reg. Hours	O.T. Hours	Reg. Earnings	O.T. Earnings	Misc. Earnings	Gross
325 48 1008	5-25-84	7 50	40		300 00			300 00

Deductions

Federal with. Tax	F.I.C.A.	State	City	Retirement	Union Dues	Credit Union		Net
52 50	20 10	13 15		18 58	6 50	25 00		164 17

Employee Name: HELEN AMOS   Employee No: 605940

Year-To-Date Totals

Gross	Federal with. Tax	F.I.C.A.	State	City	Retirement	Union Dues	Credit Union	Date of Check
6522 00	1003 08	436 97	248 77		371 60	130 00	500 00	6-1-84

EXERCISES

Answer these questions about the paycheck stub.

1. Whose paycheck stub is shown?
2. What is her Social Security number?
3. How many hours did she work this week?
4. What are her total (gross) earnings this week?
5. What was deducted for city tax?
6. What was deducted for Social Security?
7. What were her total deductions this week?
8. What is her net take-home pay this week?
9. What are her total (gross) earnings so far this year?
10. What are her total deductions so far this year?

Problem Solving Applications

READ • PLAN • SOLVE • CHECK

Some of the students at Lakeland Regional High School have after-school businesses. Four of the students have a tackle-and-bait shop in a boathouse. They make many of the lures, flies, sinkers, and floats they sell.

AS YOU READ

- flies—fishhooks decorated to look like insects
- lures—artificial bait for catching fish

Answer these questions about the tackle-and-bait shop.

1. The shop is open from 3:00 pm until 8:30 pm Monday through Friday and 6:00 am to 5:00 pm on Saturday. How many hours is it open each week?

2. Louis buys three floats for 20¢ each, eight *flies* for 79¢ each, and three *lures* for 75¢ each. If the sales tax is 55¢ and he gives Miguel $10, what is his change?

3. Maria is making flies. She can make 100 flies from a skein of yarn for 89¢, a spool of thread for 59¢, and hooks for 8¢ each. If she sells the flies for 79¢ each, how much profit will she make on 100 flies?

4. Saturday morning there were $105 in bills and $25.50 in coins in the cash drawer. At the end of the day, Miguel counted $422.74. What was the amount of sales on Saturday?

5. Olivia is pouring lead into molds for sinkers. Lead that costs $24 will make 200 sinkers. What is the cost per sinker?

6. Paula is buying a rod and reel costing $120. Don tells her she can pay 25% down and $18 a week for six weeks. How much will she pay altogether?

COMPUTING NET PAY

To compute net pay

- Subtract the total deductions from the gross pay.

EXAMPLE 1

Gregory Haynes works as an appliance repair trainee. He earns \$4.25 an hour. Last week he worked 30 h. His deductions were Federal Withholding Tax, \$9.60; F.I.C.A., \$8.54; and State Withholding Tax, \$2.80. Find his net pay.

> **Net pay** equals gross pay minus total deductions.

Multiply to find the gross pay.

$$\begin{array}{rl} \$4.25 & \leftarrow \text{hourly rate} \\ \times 30 & \leftarrow \text{number of hours} \\ \hline \$127.50 & \leftarrow \text{gross pay} \end{array}$$

Add to find the total deductions. **\$9.60 + \$8.54 + \$2.80 = \$20.94**

Subtract to find the net pay.

$$\begin{array}{rl} \$127.50 & \leftarrow \text{gross pay} \\ -20.94 & \leftarrow \text{total deductions} \\ \hline \$106.56 & \leftarrow \text{net pay} \end{array}$$

So, Gregory's net pay is **\$106.56.**

EXAMPLE 2

Cynthia worked a total of 48 h last week. She earns \$5.75 per hour, with time and a half for all hours over 40. Her deductions were Federal Withholding Tax, \$39.70; F.I.C.A., \$20.03; and State Withholding Tax, \$11.80. Find her net pay.

Find the total hour's pay.
8 h overtime is the same
as 12 h regular time.

$$\begin{array}{r} 48 \\ -40 \\ \hline 8 \end{array} \quad \begin{array}{r} 1.5 \\ \times 8 \\ \hline 12.0 \end{array} \quad \begin{array}{rl} 40 & \\ +12 & \\ \hline 52 & \leftarrow \text{total hours' pay} \end{array}$$

Multiply to find the gross pay.

$$\begin{array}{rl} \$5.75 & \leftarrow \text{hourly rate} \\ \times 52 & \leftarrow \text{number of hours} \\ \hline 11\ 50 & \\ 287\ 5 & \\ \hline \$299.00 & \leftarrow \text{gross pay} \end{array}$$

Add to find the total deductions. **\$39.70 + \$20.03 + \$11.80 = \$71.53**

Subtract to find the net pay.

$$\begin{array}{rl} \$299.00 & \leftarrow \text{gross pay} \\ -\ \ 71.53 & \leftarrow \text{total deductions} \\ \hline \$227.47 & \leftarrow \text{net pay} \end{array}$$

So, Cynthia's net pay is **\$227.47.**

GETTING READY

Solve.

1. Betsy works part-time as a photographer. Last week she worked 25 h at $3.90 per hour. Her deductions were Federal Withholding Tax, $14.90; State Withholding Tax, $3.80; and F.I.C.A., $6.53. Find her net pay.

2. Van worked 44 h at $6.25 an hour, with time and a half for overtime over 40 h. His deductions were Federal and State Withholding Taxes, $37.50; F.I.C.A., $19.26; and $6.75 for the credit union. Find his net pay.

EXERCISES

Solve. Use time and a half for all hours over 40.

1. David works a 40-h week in an assembly plant and earns $5.10 an hour. What is his take-home pay if his deductions are Federal Withholding Tax, $28.56; State Withholding Tax, $5.71; and F.I.C.A., $13.67?

2. Cheryl earns $15.50 per hour as a model. Last week she worked 20 h. Her deductions were Federal Withholding Tax, $49.60; State Withholding Tax, $8.25; and F.I.C.A., $20.77. What was her net pay?

3. Amy Clark worked 46 h last week. Her hourly pay is $4.75. Her deductions were Federal Withholding Tax, $34.85; F.I.C.A., $15.59; and State Withholding Tax, $8.70. Find her net pay.

4. Fred Arnold worked 48 h last week. His hourly pay is $9.50. His deductions were Federal Withholding Tax, $92.50; F.I.C.A., $33.10; and State Withholding Tax, $25.00. Find his net pay.

5. Gary Garett worked 40 h last week. He earns $4.50 per hour. His deductions are $27.80 for Federal Withholding Tax, $12.06 for F.I.C.A., and $4.56 for State Withholding Tax. He also pays $15.00 to the credit union. Find his net pay.

6. Maria is paid a salary of $850 per month plus 8% commission on all sales. Last month she had sales of $12,000. What was her gross pay? Find her net pay if she had deductions of $362.00, $72.40, and $121.27.

EXAMINING A PAYCHECK

To determine whether a paycheck is correct

- Check gross pay, total deductions, and net pay on the paycheck stub.

Record of Wages and Deductions Name: KIP MARCOS	
Hours Worked: 40	Rate: $7.55/h
Gross Pay	$302.00
Retirement Federal Tax F.I.C.A. State Tax Credit Union	18.12 20.23 45.30 15.10 14.19
Total Deductions	$112.94
Net Pay	$189.06

READY CUSTODIAL SERVICE

APRIL 6 1984 No. 3278

Pay to the Order of KIP MARCOS $189.06

ONE HUNDRED EIGHTY-NINE AND 06/100 Dollars

bank west

110123307041 President Chris Young

EXAMPLE

Is Mr. Marcos's paycheck correct? Find the errors, if any, on the paycheck stub.

Check the gross pay.

$$\begin{array}{rl} \mathbf{\$7.55} & \longleftarrow \text{hourly rate} \\ \underline{\mathbf{\times 40}} & \longleftarrow \text{number of hours} \\ \mathbf{\$302.00} & \longleftarrow \text{gross pay} \end{array}$$

The gross pay is correct.
Check the total deductions.

$$\mathbf{\$18.12 + \$20.23 + \$45.30 + \$15.10 + \$14.19 = \$112.94}$$

The total deductions are correct.

Check the net pay.

$$\begin{array}{rl} \mathbf{\$302.00} & \longleftarrow \text{gross pay} \\ \underline{\mathbf{-112.94}} & \longleftarrow \text{total deductions} \\ \mathbf{\$189.06} & \longleftarrow \text{net pay} \end{array}$$

The net pay is correct.

So. Mr. Marcos's paycheck **is correct.**

GETTING READY

Find and correct the errors, if any.

1.

Record of Wages and Deductions	
Name: KATRINA BOOTH	
Hours Worked: 40	Rate: $4.25/h
Gross Pay	$170.00
Retirement	11.90
Federal Tax	20.40
F.I.C.A.	11.39
State Tax	10.50
Credit Union	15.00
Total Deductions	$ 58.74
Net Pay	$111.26

2.

Record of Wages and Deductions	
Name: MANDY REIS	
Hours Worked: 40	Rate: $3.98/h
Gross Pay	$159.20
Retirement	7.90
Federal Tax	28.38
F.I.C.A.	10.67
State Tax	6.20
Credit Union	32.00
Total Deductions	$ 85.15
Net Pay	$ 75.05

EXERCISES

Find and correct the errors, if any.

1.

Record of Wages and Deductions	
Name: J.J. OLINSKY	
Hours Worked: 40	Rate: $5.15/h
Gross Pay	$206.00
Retirement	9.96
Federal Tax	22.66
F.I.C.A.	13.80
State Tax	5.66
Credit Union	23.00
Total Deductions	$ 76.08
Net Pay	$129.92

2.

Record of Wages and Deductions	
Name: KAREN ROGERS	
Hours Worked: 40	Rate: $6.10/h
Gross Pay	$244.00
Retirement	17.08
Federal Tax	43.92
F.I.C.A.	19.52
State Tax	10.98
Credit Union	16.35
Total Deductions	$107.85
Net Pay	$136.15

3.

Record of Wages and Deductions	
Name: SAMUEL ARTHUR	
Hours Worked: 40	Rate: $5.60/h
Gross Pay	$214.00
Retirement	24.64
Federal Tax	35.84
F.I.C.A.	15.00
State Tax	4.48
Credit Union	2.50
Total Deductions	$ 82.46
Net Pay	$131.54

4.

Record of Wages and Deductions	
Name: SHEILA GOLDMAN	
Hours Worked: 40	Rate: $4.90/h
Gross Pay	$196.00
Retirement	11.76
Federal Tax	24.40
F.I.C.A.	13.13
State Tax	13.72
Credit Union	4.61
Total Deductions	$ 57.62
Net Pay	$138.38

Career: Conservationists

READ • PLAN • SOLVE • CHECK

Conservationists protect, develop, and manage natural resources such as forests, rangelands, wildlife, soil, and water. Jobs range from tour guides and animal caretakers in large zoos to forest technicians, rangers, and range managers in national forests.

AS YOU READ

- replant—to plant again

Solve.

1. Michael Isaacs is a forestry technician trainee and is paid $8,428 a year. Anna Martin, a forestry technician, earns $14,680 a year. How much more does Anna earn than Michael?

2. Sue Stockman is *replanting* trees after a forest fire. She plants 45 trees per acre for each of 68 acres. If she has planted 1,530 trees, how many trees are still to be planted?

3. Sarah Hawkins is a tour guide at a zoo. She is paid $4.75 an hour. Saturday she worked from 8:30 am to 6:00 pm, with a half hour for lunch. How much did she earn?

4. Marilyn Valquez earns $125 a week in a national park's gift shop. She also earns a 7% commission on all sales. Last week her sales were $1,528. How much was Marilyn paid last week?

5. Gene O'Connell sells fish for tourists to feed to the seals. He charges 50¢ for each packet. Yesterday he sold 325 packets. How much money did he take in?

6. William White is a caretaker at a zoo. Last week he worked 36 h. If he is paid $4.25 an hour, how much did he earn last week?

CHAPTER TEST 18

Solve.

1. How many hours' pay is received for 50 h if time and a half is paid for all hours worked over 40 h?

2. Lucille worked 48 h last week. She is paid $4.70 per hour, with time and a half for all hours worked over 40 h. What was her gross pay?

3. How much did Dale earn this week? He receives time and a half for all hours worked over 40 h per week.

Dale Greene	$5.40/h
Mon	10 h
Tues	$10\frac{1}{2}$ h
Wed	$12\frac{1}{2}$ h
Thurs	$9\frac{1}{2}$ h
Fri	$11\frac{1}{2}$ h

4. Amy receives time and a half for all hours worked over 40 h per week. How much did she earn this week?

Amy Cantrell	$6.20/h
Mon	9 h
Tues	$10\frac{1}{2}$ h
Wed	10 h
Thurs	9 h
Fri	$8\frac{1}{2}$ h

5. Mr. Epps sold a house for $55,000. He receives a 7% commission on all sales. How much commission did he earn?

6. Dolores earns $250 a week plus an 8% commission on customer sales. For a week in which her sales were $2,400, how much did she earn?

7. Joan's commission is based on this scale. Her sales are $8,500. Find her commission.

Sales	Rate
$0–$6,000	7%
over $6,000	11%

8. Use this scale to find the commission on sales of $32,000.

Sales	Rate
$0–$9,999	3%
$10,000–$20,000	5%
over $20,000	7%

9. Nick's deductions are Federal Withholding Tax, $21.88; F.I.C.A., $13.12; State Withholding Tax, $4.38; and credit union, $24.00. What are his total deductions?

10. Marna Hill worked 40 h last week. She is paid $3.82 per hour. Her deductions were Federal Withholding Tax, $25.65; State Withholding Tax, $8.13; and F.I.C.A., $10.24. Find her net pay.

11. Find Ellen's net pay by using her paycheck stub below.

Record of Wages and Deductions Name: ELLEN MARTINEZ		
Hours Worked: 40	Rate: $6.50/h	
Gross Pay	$260	00
Retirement Federal Tax F.I.C.A. State Tax Credit Union	 41 17 6 15	 60 42 90 00
Total Deductions		
Net Pay		

12. Find and correct the errors, if any, on the paycheck stub below.

Record of Wages and Deductions Name: BILL BROWN		
Hours Worked: 40	Rate: $5.35/h	
Gross Pay	$216	00
Retirement Federal Tax F.I.C.A. State Tax Credit Union	17 32 14 5	12 10 34 35
Total Deductions	$ 82	25
Net Pay	$131	75

***13.** What are the meanings of the following paycheck stub entries: F.I.C.A.; State; Net?

ON YOUR OWN

Your employer at the car dealership offers you a choice of two wage plans:

Plan A: You earn $13 an hour for a 35-h week and $20 for each hour you work over 35 h.

Plan B: Your salary is a flat $200 a week, but you also get a 5% commission on sales over $5,000.

You are willing to work 40 h a week and figure you can make sales of $13,000 in an average week.

Which plan earns you more?

Managing Money

BANK DEPOSITS

To read and interpret information from a bank deposit slip

Money that you put into your account in a bank is called a **deposit.** Banks provide deposit slips containing the depositor's name, address, and account number, the bank's name, and a place to show the date and the amount of the deposit. On some deposit slips, cash is separated into bills (currency) and coins.

EXAMPLE

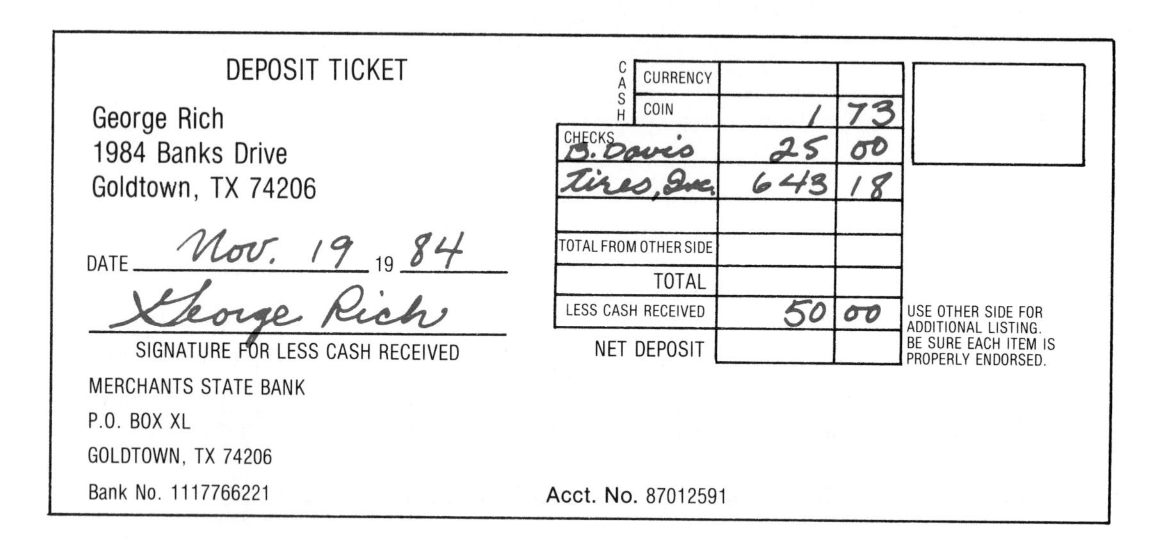
DEPOSIT TICKET

George Rich
1984 Banks Drive
Goldtown, TX 74206

DATE Nov. 19 19 84

George Rich
SIGNATURE FOR LESS CASH RECEIVED

MERCHANTS STATE BANK
P.O. BOX XL
GOLDTOWN, TX 74206
Bank No. 1117766221

CASH	CURRENCY		
	COIN	1	73
CHECKS	B. Davis	25	00
	Tires, Inc.	643	18
	TOTAL FROM OTHER SIDE		
	TOTAL		
	LESS CASH RECEIVED	50	00
	NET DEPOSIT		

USE OTHER SIDE FOR ADDITIONAL LISTING. BE SURE EACH ITEM IS PROPERLY ENDORSED.

Acct. No. 87012591

What is George's account number?	**87012591**
How much cash did he deposit?	**$1.73**
What was the total amount of checks deposited?	**$25.00 + $643.18 = $668.18**
Find the total of checks and cash.	**$668.18 + $1.73 = $669.91**
George needed some cash, so he kept $50 of his total. Find the net deposit.	**$669.91 − $50 = $619.91** So, **$619.91** is the net deposit.

GETTING READY

Use the information below to find the net deposit.

1. Aug. 7: Cash $ 93.70
Checks 124.50
26.34
423.00

2. Sept. 28: Checks $ 25.30
389.01
Less cash received 230.00

Bank No. 200534177

Date *Mar. 7* 19*84*

Deposit to the account of:

Margaret McBroom

2317 Durkee St.

Deansville, Pa. 19411

Merrill State Bank
Alma, PA 19419

Acct. No. *304090504*

	Dollars	Cents
Cash		
Checks 1	20	16
List 2	239	05
Singly 3	587	20
4		
Total		
Less cash received	225	00
Net Deposit		

EXERCISES

Answer the following questions about the bank deposit slip above.

1. What is Margaret's account number?

2. What is the name of her bank?

3. How many checks were deposited?

4. How much did she deposit in checks?

5. How much cash did she receive?

6. Find the net deposit.

Use the information below to find the net deposit for each date.

7. Jan. 1:

Cash	$ 90.00
Checks	29.58
	222.61
	43.51

8. Mar. 3:

Checks	$ 82.50
	123.16
	36.14
	50.88
Less cash received	100.00

9. June 14:

Checks	$ 81.62
	72.11
	52.86
	36.96
	17.00
Less cash received	250.00

10. Apr. 26:

Checks	$22.61
	19.28
	21.00
	66.12
	36.11
	65.62

***11.** Make out a deposit slip from a local bank with this information: June 15, Cash $60.00 and Checks $47.18, $510.52, $53.20, $92.51. What is the net deposit?

CHECKING ACCOUNTS

To write checks on a checking account
To complete records of checking account transactions

A check that is written on your checking account acts as a receipt. The person receiving the money endorses the check by signing his or her name on the back. This is proof that the money was received.

Check records are kept in a check register. The register shows the amounts for checks (withdrawals), deposits, and fees for any services.

EXAMPLE

Examine the check and check register below.

Marian Davies
2514 Elm Blvd.
Cedartown, KY 42807

Number 276

May 4 19 84

Pay to the Order of Styles Unlimited $ 48.98

Forty-eight and 98/100 Dollars

Alma State Bank
Cedartown, KY 42811

Marian Davies

BE SURE TO DEDUCT FEES OR SERVICE CHARGES — DEBITS — CREDITS

ITEM NO. OR TRANSACTION CODE	DATE	DESCRIPTION OF TRANSACTION	AMOUNT OF PAYMENT/WITHDRAWAL (−)		✓ TAX	FEE OR OTHER DEDUCT	AMOUNT OF DEPOSIT OR INTEREST (+)			BALANCE FORWARD	
										203	64
276	5/4	TO Styles Unlimited FOR	48	98					+ −	48	98
									BAL	154	66
	5/7	TO (paycheck) FOR					681	03	+ −	681	03
									BAL	835	69
277	5/8	TO Unico Gas FOR	12	80					+ −	12	80
									BAL		
278	5/11	TO Dr. Bruce Hooper FOR	115	—					+ −	115	—
									BAL		

To whom is check #276 made out? **Styles Unlimited**

What is the balance after check #276 is written? $203.64 – 48.98 = **$154.66**

GETTING READY

Use the check and check register on page 362 for **1–6**.

1. What is the date of check #276?

2. What is the amount of check #276?

3. How much was paid to Dr. Bruce Hooper?

4. What was the balance in the account after the paycheck was deposited?

5. On what day was the check to Unico Gas Station written?

6. Complete the check register by finding the two missing balances.

EXERCISES

Make out a series of checks with the information below. Use the current year.

1. March 3: Check #15 to J. L. Pope Co. for $19.25

2. March 5: Check #16 to City Utilities for $35

3. March 6: Check #17 to Dr. A. J. Holmesly for $18.00

4. March 8: Check #18 to Waisman's Dept. Store for $97.51

5. March 11: Check #19 to Cash for $50

Make out a check register record. Use the information below with the current year and find all necessary balances.

Begin with an old balance of $216.83.

6. May 1: Check #47 to Johnson TV Cable for $37.50

7. May 4: Check #48 to Acme Furniture for $18.50

8. May 5: Deposited $500.00 paycheck

9. May 6: Check #49 to Perez Dry Cleaning for $7.98

10. May 9: Check #50 to Cash for $75

BANK STATEMENTS

To reconcile a bank statement

A bank statement shows the checks that have cleared and the deposits that were recorded during the period covered by the statement. Any checks written during this period but not shown on the statement are called **outstanding** checks. These must be subtracted from the final balance on the statement. Any recent deposits not shown must be added to this balance. The process of checking and correcting the balance is called **reconciling** a bank statement.

EXAMPLE

BANK STATEMENT

Date of Statement: March 28, 1984

DATE	WITHDRAWALS	DEPOSITS	BALANCE
March 9			183.00
March 12	8.00		175.00
March 16	35.00		140.00
March 29		55.60	195.60
March 26	64.50		131.10

A check for $47.80 was written on March 29, and a deposit of $200 was made on March 30. Find the actual balance on March 30.

Subtract the outstanding check from the final balance.

$$\begin{array}{r} \$131.10 \\ -47.80 \\ \hline \$\ 83.30 \end{array}$$

Now add the deposit not shown on the statement.

$$\begin{array}{r} \$\ 83.30 \\ +200.00 \\ \hline \$283.30 \end{array}$$

So, the actual balance is **$283.30.**

GETTING READY

Solve.

1. Myrna's bank statement balance is $540.00. She has an outstanding check for $55. What is her actual balance?

2. Delmar's bank statement balance is $364.89. He has a $210.00 deposit that is not shown on the bank statement. What is his actual balance?

EXERCISES

Solve.

1. Mrs. Duff's bank statement shows a balance of $464.38. She has an outstanding check for $50.00 and a deposit of $64.80 that are not shown. Find the actual balance.

2. Mr. Culvert's bank statement shows a balance of $249.89. He has outstanding checks for $20.00 and $35.16. Find his actual balance.

3. Claude's bank statement balance is $174.63. He has outstanding checks for $27.00 and $51.99 and deposits of $165.25 and $86.77 that are not shown. What is the actual balance?

4. As a service fee, Luis's bank charges him 75¢ each month plus 10¢ for each check he writes. In March, Luis wrote 21 checks. What was his service fee?

Complete the balance column of the bank statement below. Then answer the questions about the statement.

BANK STATEMENT

Date of Statement: June 25, 1984

DATE	WITHDRAWALS	DEPOSITS	BALANCE
June 4			416.00
June 6	15.95		**5.** ?
June 7	40.00		**6.** ?
June 21	127.80	85.15	**7.** ?
June 22	3.00s		**8.** ?

Note: s = monthly service charge

9. In reconciling his bank statement, Josh found two outstanding checks: June 23, $240; and June 24, $76.39. Has he overdrawn his account?

10. Suppose Josh has no outstanding checks and on June 26 he deposits $200. What will be his actual balance?

11. Josh's bank charges a flat $3 monthly service charge. How much does the service charge cost for a year?

12. North State Bank charges a $6 service fee only if the balance falls below $200 during the month. If Josh had had his account at this bank, would he have paid the $6 service fee?

SIMPLE INTEREST

To compute simple interest
To find the total amount to be paid after simple interest is computed

When you borrow money, you have to pay interest for the use of the money. If you pay simple interest, you make only one payment at the end of the loan period. To compute simple interest, use this formula: Interest = principal × rate (per year) × time (in years), or $I = p \times r \times t$.

EXAMPLE 1

Martha loaned Jerry $1,400. She charged him $12\frac{1}{2}\%$ simple interest for 3 y. How much interest did she receive?

> **Principal** is the amount of money borrowed.
>
> **Rate** is the percent interest per year.
>
> **Time** is the number of years for the loan.

Use the formula $I = p \times r \times t$.

$p = \$1{,}400 \quad r = 12\frac{1}{2}\% \quad t = 3 \text{ years}$

$$\mathbf{r = 12.5\%}$$
$$\mathbf{= 0\,1\,2.5}$$
$$\mathbf{r = 0.125}$$

$$I = p \times r \times t$$
$$= \mathbf{\$1{,}400 \times 0.125 \times 3}$$
$$\mathbf{I = \$525}$$

So, Martha received **$525** in interest.

EXAMPLE 2

Find the total amount (principal plus interest) to be paid back in 9 months on a loan of $800 at 10%.

Use the formula $I = p \times r \times t$ to find the simple interest.

$p = \$800 \quad r = 10\% \quad t = 9 \text{ months or } \frac{3}{4} \text{ of a year}$

$$\mathbf{r = 010.} \qquad \mathbf{t = 0.75}$$
$$\mathbf{r = 0.10}$$

$$I = p \times r \times t$$
$$= \mathbf{\$800 \times 0.10 \times 0.75}$$
$$\mathbf{I = \$60}$$

Add to find the total amount to be paid back.

$800	←	principal
+ 60	←	interest
$860	←	total amount

So, the total amount to be paid back is **$860.**

GETTING READY

Find the simple interest.

1. $p = \$500$, $r = 10\%$, $t = 3$ y
2. $p = \$750$, $r = 8\frac{1}{2}\%$, $t = 1$ y
3. $p = \$325$, $r = 12\%$, $t = 6$ mo

Find the total amount to be repaid.

4. $p = \$1{,}200$, $r = 9.75\%$, $t = 10$ y
5. $p = \$400$, $r = 14\%$, $t = 2\frac{1}{2}$ y
6. $p = \$650$, $r = 8\%$, $t = 3$ mo

EXERCISES

Find the simple interest.

1. $p = \$875$, $r = 12\%$, $t = 1$ y
2. $p = \$1{,}500$, $r = 10.5\%$, $t = 6$ y
3. $p = \$760$, $r = 7\%$, $t = 9$ mo
4. $p = \$1{,}200$, $r = 9\frac{1}{4}\%$, $t = 4$ y

Find the total amount to be repaid.

5. $p = \$1{,}000$, $r = 6\frac{1}{2}\%$, $t = 1$ y
6. $p = \$2{,}400$, $r = 8\%$, $t = 18$ mo
7. $p = \$900$, $r = 11.75\%$, $t = 5$ y
8. $p = \$1{,}300$, $r = 15\%$, $t = 3$ y

Solve.

9. Lois Salivar borrowed \$500 and repaid the loan 18 mo later at 10% interest. What was the total amount that she paid?

10. The amount due on a loan is \$380.16. This includes \$28.16 in simple interest. How much was borrowed originally?

Kent Moss borrowed \$2,500 at 7% for 10 y. At the end of 10 y:

11. How much principal did Kent owe?

12. How much interest did Kent owe?

After a year, Alison Jones paid a total amount of \$570 on a \$500 loan.

13. How much simple interest did she pay?

*14. What was the rate of Alison's loan?

*15. Interest of \$206.63 is paid on a loan of \$725 for 3 y. What was the rate of interest used?

*16. Al paid \$210 interest on a \$500 loan. He paid the total amount in 10 equal payments. How much was each payment?

COMPOUND INTEREST

To compute compound interest

Banks and credit unions usually pay compound interest on savings accounts. Each time the interest is computed, it is added to the existing principal to get a new principal. The next time interest is computed, the new principal is used, and the process is repeated.

Times per year interest is computed:

compounded annually	1 time
compounded semiannually	2 times
compounded quarterly	4 times
compounded monthly	12 times

EXAMPLE

A savings account earns 6% interest compounded quarterly (4 times a year). If \$50 is deposited, what is the amount in savings after 9 mo? How much interest has been earned in 9 mo?

> Use the simple interest formula each time compound interest is computed.

Use $I = p \times r \times t$.

$p = \$50 \qquad r = 6\% \qquad t = 3$ months or $\frac{1}{4}$ of a year

$\qquad\qquad = 006. \qquad t = 0.25$

$\qquad r = 0.06$

Find the interest after 3 months. **\$50 × 0.06 × 0.25 = \$0.75**

Add the interest to the principal to find the new principal. **\$50 + \$0.75 = \$50.75** principal after 3 months

Find the interest after 6 months. **\$50.75 × 0.06 × 0.25 = \$0.76125**

Round. **= \$0.76**

Add. **\$50.75 + \$.76 = \$51.51** principal after 6 months

Find the interest after 9 months. **\$51.51 × 0.06 × 0.25 = \$0.77265**

Round. **= \$0.77**

Add. **\$51.51 + \$0.77 = \$52.28** principal after 9 months

Subtract to find the total interest earned.

\$52.28 ←	principal after 9 months
−50.00 ←	original deposit
\$ 2.28 ←	interest earned after 9 months

So, after 9 months the amount in savings is **\$52.28** and the interest earned is **\$2.28.**

GETTING READY

Solve.

1. How many times over a 3-year period is interest computed for an account that earns interest compounded quarterly?

2. Find the amount in savings if $100 is deposited at 7% per year compounded annually for 2 y.

3. Find the amount in savings if $60 is deposited at $6\frac{1}{2}$% per year compounded semiannually for 18 mo.

4. Find the interest earned for Exercises 2 and 3.

EXERCISES

Find the amount in savings.

1. $500 is deposited at 6% per year compounded annually for 2 y.

2. $200 is deposited at 8% per year compounded semiannually for 1 y.

Find the total interest earned.

3. $200 is deposited at 8% per year compounded quarterly for 9 mo.

4. $100 is deposited at $7\frac{1}{2}$% compounded semiannually for 18 mo.

5. Ian deposited $50 in a savings account that earns interest at 5% compounded semiannually. How much interest did the account earn after 1 y?

6. How much interest will Lizette earn after 9 mo if she deposits $125 in a savings account that pays 6% interest compounded quarterly?

7. Nai Chung deposited $100 in a savings account that earns interest at $7\frac{1}{2}$% per year compounded monthly. What is the amount in savings after 2 mo?

*8. A $1 investment at 5% interest compounded semiannually is worth $1.95 after $13\frac{1}{2}$ y. Find the value of a $4,500 investment under the same conditions.

CALCULATOR

Use your calculator to find how much money you would have to deposit, at 10% interest compounded quarterly, to earn $100 in interest by the end of 1 y. If the interest was not compounded, you would have to deposit $1,000. With compounding, you need a bit less. Try different amounts until you find one that works.

Problem Solving Skill

To plan effective use of money

- Make a budget.
- Use the budget to plan expenses.

EXAMPLE

Mr. Ohlhausen's take-home pay after deductions is $1,200 each month. He budgets his pay as follows:

Transportation	10%
Food	25%
Housing, utilities	30%
Clothing	20%
Savings	10%
Miscellaneous expenses	5%
	100%

Find the amount of money Mr. Ohlhausen has budgeted in each area.

Multiply take-home pay by the budget percent.

Transportation:	$1,200 × 0.10 = **$120**
Food:	$1,200 × 0.25 = **$300**
Housing, utilities:	$1,200 × 0.30 = **$360**
Clothing:	$1,200 × 0.20 = **$240**
Savings:	$1,200 × 0.10 = **$120**
Miscellaneous:	$1,200 × 0.05 = **$60**

Does Mr. Ohlhausen have enough money in the budget to buy a $135 suit?

Find the area which includes this purchase.
The suit will come out of the Clothing budget of $240.

Is the budgeted amount enough?
$135 is less than $240.

So, **yes** he has enough to buy the suit.

He decides to put an additional 5% of his salary into savings each month and reduce his clothing budget to 15%. How much more will he deposit in his savings account each month?

Find 5% of his take-home pay.
$1,200 × 0.05 = $60

So, he will deposit **$60** more each month.

EXERCISES

Stanley has a part-time job. He earns $80 a week. He lives with his parents and plans to attend a nearby technical school. His weekly budget is shown below.

Tuition for school	30%
Clothing	15%
Food	20%
Transportation	10%
Entertainment	5%
School supplies	10%
Miscellaneous expenses	10%
	100%

1. How much does Stanley save each week for his school tuition?

2. Can he afford to go to a movie for $3.50 and roller skating for $1.50 each week?

3. One week Stanley did not buy any clothes. How much money did he save?

4. He wants to buy a $65 jacket. How many weeks will he have to save his clothing allowance to buy the jacket?

Ada earns $240 a month take-home pay from a part-time job. Each week she spends $5.50 for a school lunch ticket and puts $4.00 in savings. She decides to buy a used car, making payments of $70 per month. Ada estimates that gas for the car will be $50 a month.

5. Prepare a monthly budget for Ada that will allow her to make the car payment each month as well as take care of her other expenses.

6. Car insurance is $600 a year. Does Ada have enough money in her monthly budget to cover this expense?

***7.** Make a list of your own expenses last week. Select budget areas and make a chart, listing each expense under the proper area. Find the total amount you spent for each area.

***8.** In which area did you spend the most money? Do you expect your budget amounts to change next week? Explain briefly.

Problem Solving Applications

READ • PLAN • SOLVE • CHECK

"The Wrangler," Fremont High School's newspaper, is published monthly by a group of students and their faculty sponsor. The paper's staff includes an advertising manager, a chief editor, a news editor, a sports editor, and a features editor. The other students work as reporters or photographers.

AS YOU READ

- cover—to report the news
- issue—an item or set of items given out at one time

Answer these questions about "The Wrangler."

1. The staff puts out 7 *issues* in a school year. The number of copies sold for each issue last year were 235, 248, 324, 275, 260, 281, and 357 copies. About how many copies more would need to be sold this year to increase sales 10%?

2. Gaylene is the advertising manager. Friday she deposited the following amounts in "The Wrangler" account: \$10.50, \$25.00, \$12.75, \$14.50, \$20.25, and \$18.00. How much was deposited altogether?

3. Thirty-five percent of the printing cost of the paper comes from the sale of ads. If it costs \$2,800 to print the paper, how much money must be earned from ads?

4. This year it will cost \$2,800 to print "The Wrangler." If there are 7 issues a year, what is the printing cost for each issue?

5. The two photographers for the paper were assigned to *cover* Homecoming Week. They spent 1 h at the pep assembly, $2\frac{1}{2}$ h at the football game, and 4 h at the Homecoming Dance. How many hours did they spend covering Homecoming Week?

***6.** Ads for the school paper are sold by the column inch. If an ad is 3 columns wide and 2 in. high, it is a 3×2 or 6-column-inch ad. Find how many column inches an ad would measure if it is 2 columns, $4\frac{1}{2}$ in. high.

Career: Communications Workers

READ • PLAN • SOLVE • CHECK

The three main areas in communications are radio, television, and printed publications such as newspapers. People work as photographers, typists, proofreaders, program planners, technical crew members, printers, and production managers. Some jobs require only a high school education, while others require a technical school or college education.

AS YOU READ

- contact sheet—a photographic print

Solve.

1. The programmer for a television station planned 9 commercials during a 1-hour program. The playing times for the commercials were as follows: 22.5 s, 30 s, 28.4 s, 20 s, 25.6 s, 29.1 s, 16 s, 18.5 s and 30 s. Find the total playing time for the commercials.

2. Bob Ridgeway is a newspaper photographer. After an assignment, it took him $1\frac{1}{2}$ h to process the film. It took another 15 min to make a *contact sheet*. It took 2 h more to prepare the photo for the news article. Find the total time it took to prepare the photo for use.

3. It costs \$950 to print the "Wiregrass Shopper," an 8-page weekly shopping paper. Find the average printing cost per page.

4. A photograph must be reduced to $\frac{2}{3}$ of its original size to fit a certain space in a newspaper. How long must the reduced photograph be if the original is 7 in. long?

5. A newspaper writer needs an article to fill a 15-column-inch space in the next edition. If it takes about 32 words to fill each column inch of space, about how many words long must the article be?

***6.** A printer has an order for 3,600 printed leaflets. If there is an 18% spoilage rate for this type of job, how many leaflets should actually be printed to allow for spoilage?

MAKING MATH COUNT

EXAMPLE

The check for Mr. and Mrs. Sim's dinner came to \$24.95. Find the amount of a 15% tip for service.

A \$3.75 **B** \$2.50
C \$2.00 **D** \$1.25

Think: $15\% = 10\% + 5\%$
$5\% = \frac{1}{2}$ of 10%
Round \$24.95 to \$25.00.
10% of \$25.00 = \$2.50
$\frac{1}{2}$ of \$2.50 = \$1.25
Add. **\$2.50 + \$1.25 = \$3.75**

So, **A** is the correct answer.

Choose the correct answer.

1. The dinner check is \$32.00. Find the amount of a 15% tip.
A \$1.60 **B** \$3.00
C \$3.20 **D** \$4.80

2. The lunch check is \$9.00. Find the amount of a 15% tip.
A \$0.45 **B** \$0.90
C \$1.35 **D** \$2.70

3. Edna had lunch for \$5.60 at a diner. About how much should she give the waitress for a 15% tip?
A \$0.56 **B** \$0.60
C \$0.75 **D** \$0.90

***4.** Mr. Barnes spent \$19.75 for dinner. What is the total cost including a 15% tip?
A \$21.75 **B** \$22.75
C \$24.00 **D** \$25.00

***5.** Pearl received a \$1.50 tip on a lunch check for \$10.00. What percent of the check was the tip?
A 10% **B** 15%
C 20% **D** 25%

***6.** Fred received a \$7.00 tip on a dinner check for \$35.00. What percent of the check was the tip?
A 5% **B** 10%
C 15% **D** 20%

CHAPTER TEST 19

Use the bank deposit slip below to answer the questions.

Bank No. 300100777	Acct. No. 671188730		
Date Apr. 5 19 84	Cash — Bills	138	00
Deposit to the account of:	Cash — Coins	4	50
Bill Harrell	Checks B. Tucker	75	89
102 Windham	I.R.S. Refund	168	91
Seaburg, Tx. 75660	Total		
Seaburg Commerce Bank	Less Cash		
Seaburg, TX 75660	Net Deposit		

1. What is Bill Harrell's bank account number?

2. How much cash did Bill Harrell deposit on April 5?

3. Find the total amount of checks deposited.

4. Find the net deposit.

You are planning to write a check to Thrifty Butcher Shop. In which blank on the check register do you write each of the following?

BE SURE TO DEDUCT FEES OR SERVICE CHARGES			DEBITS				CREDITS		BALANCE FORWARD	
ITEM NO. OR TRANSACTION CODE	DATE	DESCRIPTION OF TRANSACTION	AMOUNT OF PAYMENT/WITHDRAWAL (−)		√ TAX	FEE OR OTHER DEDUCT	AMOUNT OF DEPOSIT OR INTEREST (+)		G	
A	B	TO C	D				E		+ − F	
		FOR							BAL H	

5. today's date

6. $35.49

7. Thrifty Butcher Shop

8. the check number

9. If the old balance was $109.32, what is the new balance?

In which blank on the check does Robert write the following?

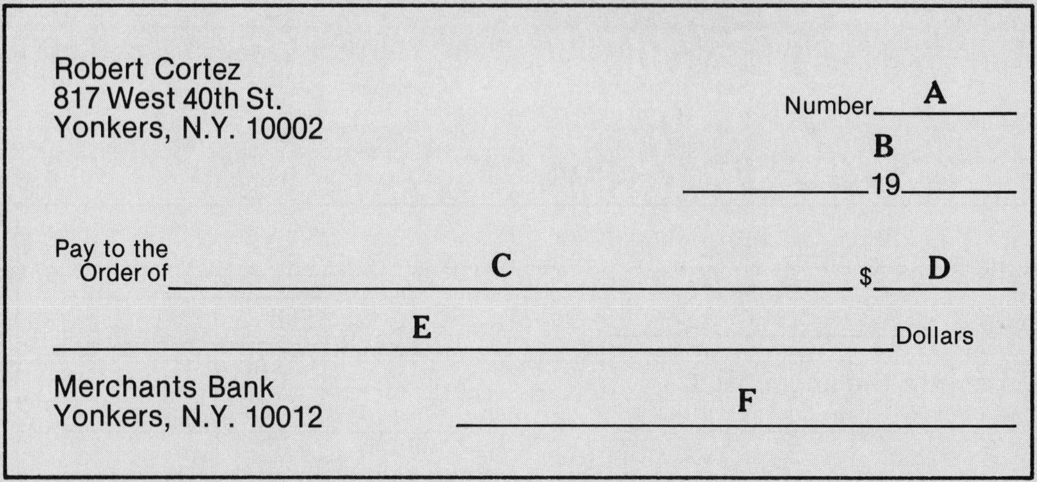

Robert Cortez
817 West 40th St.
Yonkers, N.Y. 10002

Number___A___

___B___19___

Pay to the Order of ___C___ $___D___

___E___ Dollars

Merchants Bank
Yonkers, N.Y. 10012

___F___

10. the date

11. check #22

12. Sun, Inc.

13. $35

Solve.

14. Marilyn loaned Marcus $1,200 at 12% simple interest for 3 y. How much interest did she receive?

15. Len borrowed $500 from his bank and paid it off 18 mo later at 10% simple interest. What was the total amount paid?

The Stephens family has an annual income of $34,500. This income is budgeted into several areas. Find the amount that is allowed annually for each budget area listed below.

16. Food, 25%

17. Utilities, 18%

18. Flora's bank statement shows a balance of $375.60. She has outstanding checks for $120.00, $69.35, and $132.95. What is her actual balance?

19. Ms. Chu's bank statement balance is $1,870.28. She has deposits of $500.00 and $273.20 and checks for $840.44 and $39.00 that are not shown. Find her actual balance.

20

Spending Money

CREDIT CARDS

To compute interest charges on credit card accounts

A credit card is issued to qualified consumers of goods and services who sign for the purchases and agree to pay within a certain period of time.

Each month, the cardholder receives a statement that lists all purchases and the total amount owed. The cardholder must make a minimum payment after receiving each monthly statement. Interest is charged on any unpaid balance and appears on the next month's bill.

EXAMPLE 1

Mr. Kahn's credit card bill was $420 last month. He made a payment of $40. Find the amount of interest added to next month's bill if the charge is $1\frac{1}{2}\%$ of the unpaid balance.

Subtract to find the unpaid balance.

$$\begin{array}{r l} \mathbf{\$420} & \longleftarrow \text{amount of bill} \\ \mathbf{-\ 40} & \longleftarrow \text{payment} \\ \hline \mathbf{\$380} & \longleftarrow \text{unpaid balance} \end{array}$$

Multiply to find the interest.
$1\frac{1}{2}\% = 1.5\% = 0.015$

$$\begin{array}{r l} \mathbf{\$380} & \\ \mathbf{\times 0.015} & \longleftarrow \text{rate of interest} \\ \hline \mathbf{\$5.700} & \longleftarrow \text{amount of interest} \end{array}$$

So, the interest charge is **$5.70.**

If no new charges are made the following month, the **payment due** will be $380 + $5.70, or $385.70.

EXAMPLE 2

Mr. Bates's credit card company granted him a $600 credit line. If he has charges of $370, how much credit is still available to him?

A **credit line** is the limit that a cardholder is allowed to charge.

Subtract the amount of charges from the credit line to find available credit.

$$\begin{array}{r l} \mathbf{\$600} & \longleftarrow \text{credit line} \\ \mathbf{-370} & \longleftarrow \text{amount of charges} \\ \hline \mathbf{\$230} & \longleftarrow \text{credit available} \end{array}$$

So, Mr. Bates has **$230** credit available.

GETTING READY

Solve.

1. Mr. Alger paid \$60 on his \$412 credit card bill. What amount of interest at $1\frac{1}{2}\%$ will be added to next month's bill?

2. Mrs. Bolt has a credit line of \$1,000 on her card. If she charges \$872, how much credit is still available?

3. Ms. Canten has an unpaid balance of \$385. What will be the $1\frac{1}{2}\%$ interest charge on next month's bill?

4. Mr. Delow has a credit line of \$750 and an unpaid balance of \$320. What is his available credit?

EXERCISES

Solve.

1. Mr. Gadd has a credit line of \$700. If he has charges of \$274, what is his available credit?

2. Mr. Hein has a credit line of \$800. If he has charges of \$453, what is his available credit?

3. Ms. Irving had to pay $1\frac{1}{2}\%$ interest on an unpaid balance of \$357. What interest did she pay?

4. Mrs. Johnson had to pay $1\frac{1}{2}\%$ interest on an unpaid balance of \$476.34. What interest did she pay?

5. Find the payment due on an unpaid balance of \$395 with a $1\frac{1}{2}\%$ interest charge.

6. Find the payment due on an unpaid balance of \$186 with a $1\frac{1}{2}\%$ interest charge.

Find the interest charges on the unpaid balance for the next billing period. (Use an interest rate of $1\frac{1}{2}\%$.)

7. Payment of \$15 made on a \$135 credit account

8. Payment of \$35 made on a \$265 credit account

9. Payment of \$65 made on a \$450 credit account

10. Payment of \$24 made on a \$175 credit account

11. Payment of \$45 made on a \$325 credit account.

12. Payment of \$120 made on a \$480 credit account

*13. Why do retail stores encourage consumer credit-card buying?

*14. List five disadvantages in using credit cards for making purchases.

COMPARING PRICES

To find the better buy and the amount of savings

- Find the cost of each purchase.
- Compare to find the better buy.
- Subtract to find the savings.

EXAMPLE 1

Which is the better buy: a \$170 stereo at 20% off or the same stereo priced at \$155 with 10% off?

Find the discount price for each item.

$$\begin{array}{r} \$170 \\ \times 0.80 \\ \hline \$136.00 \end{array} \qquad \begin{array}{r} \$155 \\ \times 0.90 \\ \hline \$139.50 \end{array}$$

Compare \$136 and \$139.50.
So, the **\$170 stereo at 20% off** is the better buy.

> 20% off means paying 80% of the original price. 10% off means paying 90% of the original price.

EXAMPLE 2

Store L and Store M both offer discounts on luggage. What is the savings if you buy the three pieces at Store M?

Luggage	Store L	Store M
26″ bag	\$ 95.94	\$ 74.87
29″ bag	110.96	112.94
Tote bag	39.24	31.60
Sale discount	15%	10%

> The **savings** is the difference between the two discount prices.

Find the total cost of the items.

$$\begin{array}{r} \$\ 95.94 \\ 110.96 \\ +\ \ 39.24 \\ \hline \$246.14 \end{array} \qquad \begin{array}{r} \$\ 74.87 \\ 112.94 \\ +\ \ 31.60 \\ \hline \$219.41 \end{array}$$

Find the discount prices.

$$\begin{array}{r} \$246.14 \\ \times 0.85 \\ \hline \$209.2190 \end{array} \qquad \begin{array}{r} \$219.41 \\ \times 0.90 \\ \hline \$197.4690 \end{array}$$

Round each discount price.

\$209.2190 ⟶ \$209.22
\$197.4690 ⟶ \$197.47

Subtract to find the savings. **\$209.22 − \$197.47 = \$11.75**
So, the savings at Store M is **\$11.75.**

GETTING READY

Which is the better buy?

1. An $89.95 picnic set at 10% off or the same set priced at $79.90

2. The items from Store A or Store B? What is the savings?

Item	Store A	Store B
Camera	$114.99	$82.40
Radio	22.99	17.60
12″ TV	125.92	86.34
Sale discount	5%	10%

3. A $450 boat at 10% off or the same boat priced at $410 and 5% off

EXERCISES

Which is the better buy?

	Store A		Store B	
1. Chair	$279.95	20% off	$265.99	5% off
2. Man's suit	$150.00	$\frac{1}{3}$ off	$110.00	10% off
3. Dishwasher	$339.00	$\frac{1}{3}$ off	$330.00	25% off
4. Watch	$85.00	10% off	$93.99	$\frac{1}{3}$ off
5. Table saw	$634.95	40% off	$595.95	$\frac{1}{3}$ off

Find the total amount of savings between the stores if you bought one of each item.

6.

Sleepwear	Al's	Bea's
Sleeper	$ 6.99	$ 5.77
Blanket	17.99	16.88
Gown	16.99	15.88
Pajamas	9.99	9.29
Robe	19.99	17.50
Slippers	11.95	10.75
Sale discount	25%	20%

7.

Camping Gear	Ray's	Flo's
Boots	$36.99	$37.50
Lantern	19.99	22.50
Camp stove	33.38	35.00
Thermal socks	6.75	7.50
Field jacket	69.99	72.50
Sleeping bag	37.75	42.50
Sale discount	10%	10%

***8.** At a sale, every item was sold at 25% off the regular price. If the sale price of a bicycle was $72, how much was the discount? (Hint: Find the regular price, then subtract.)

INSTALLMENT BUYING

To compute the carrying charges of installment buying

- Find the installment price.
- Subtract the cash price from the installment price.

When an item is paid for with cash, no interest or extra charges are added to the cost. With installment-plan buying, you are required to make a down payment; the rest of the cost is paid in regular weekly or monthly payments. The merchant charges you a fee, called the **carrying charge,** for all installment-plan purchases.

EXAMPLE

Find the carrying charge on a $275 TV purchased for $25 down and $26.50 a month for a year.

Find the installment price.

```
  $26.50  ←— monthly payment
    × 12  ←— number of payments
   53 00
  265 0
 $318.00  ←— total monthly payments
+  25.00  ←— down payment
 $343.00  ←— installment price
```

Subtract the cash price from the installment price to get the carrying charge.

```
 $343.00  ←— installment price
 −275.00  ←— cash price
  $68.00  ←— carrying charge
```

So, the carrying charge is **$68.**

The **installment price** is the total of regular payments plus the down payment.

The **carrying charge** is the difference between the installment price and the cash price.

GETTING READY

Find the carrying charge.

1. a $5,400 grand piano bought for $1,500 down and monthly payments of $130 for 3 y.

2. a $795 stereo system bought for $325 down and 12 monthly payments of $46.75

3. a $3,500 used car bought for $1,000 down and payments of $225 for 15 mo

4. a $600 sofa bought for $150 down and payments of $55.75 for 10 mo

EXERCISES

Solve.

1. Manuel bought a $650 refrigerator for $150 down and 12 monthly payments of $45. Find the carrying charge.

2. The Arnold family bought a $1,200 home music system for $200 down and 12 monthly payments of $90. Find the carrying charge.

3. Sheila bought a $725 stereo for $225 down and 12 monthly payments of $46.50. Find the carrying charge.

4. The Phillips family bought a $3,200 used car for $300 down and 18 monthly payments of $190. Find the carrying charge.

5. The Jackson family bought a set of furniture selling for $1,750 for $250 down and $76 per month for 2 y. Find the carrying charge.

6. A $3,600 living room set can be bought for $600 down and $176 per month for 2 y. Find the carrying charge.

7. A television sells for $745. It can be bought for $145 down and $22 per month for 3 y. Find the carrying charge.

8. A $945 refrigerator-freezer can be bought for $145 down and $31 a month for 3 y. Find the carrying charge.

*9. A washer-dryer can be bought for $155 down and $17.50 a month for 3 y. If the carrying charge is $130, find the cash price.

*10. List three *advantages* and three *disadvantages* of making purchases on the installment plan rather than paying cash.

ON YOUR OWN

1. Make a list of items that you feel are absolutely essential in a home.

2. A washing machine uses about 27 gal of hot water per load. How many gallons of water are used annually if 4 loads are done weekly?

3. A shower uses about 5 gal less than a bath. If a family of five persons takes a shower a day rather than a bath, how many gallons of water will be saved in a year?

4. The Meyer family moved from a single-family dwelling to a multiple-family dwelling. The annual energy cost of $1,260 was reduced by $\frac{1}{3}$ after the move. What were the energy costs for the multiple-family dwelling? At this rate, how much would be saved in a 10-year period?

RENTING AN APARTMENT

To compute annual expenses for renting an apartment

As much as 40% of a person's annual income may go for providing shelter. Renting an apartment is often a major concern for young adults. The costs of utilities (water, gas, electricity) and phone must be considered too.

EXAMPLE 1

Carlos earns \$10,000 per year. He has budgeted 30% of his income for housing. How much can he spend on housing per month?

Find 30% of \$10,000. **\$10,000 × 0.30 = \$3,000 per year**

Divide by 12. $\mathbf{\frac{\$3,000}{12} = \$250}$ **per month**

So, he can spend about **\$250** per month.

EXAMPLE 2

Maria rented a three-room furnished apartment for \$190/mo. What is her annual rent?

> **Annual rent** means rent for 1 y.

Multiply by 12.

```
  $190  <-- monthly rent
  × 12
   380
  1 90
$2,280  <-- annual rent
```

So, her annual rent is **\$2,280.**

EXAMPLE 3

Leroy rented a two-room furnished apartment for \$170/mo. He pays \$25/mo for utilities and \$12.50/mo for phone service. Find his annual expense.

> Include rent, utilities, and the cost of phone service when you figure annual rental expenses.

Add.

```
  $170.00  <-- rent
    25.00  <-- utilities
+   12.50  <-- phone
  $207.50  <-- monthly expense
```

Multiply by 12. **\$207.50 × 12 = \$2,490**

So, he pays **\$2,490** annually.

GETTING READY

Solve.

1. Jeff earns $14,500 per year. He allows $\frac{1}{4}$ of his income for shelter. About how much should he spend on living quarters per month?

2. Mandy rents a $2\frac{1}{2}$-room furnished apartment for $225/mo including utilities. What is her annual rent?

3. Quincy rents a 3-room furnished apartment for $210/mo. He pays $35/mo for utilities and $13/mo for phone service. Find his annual expense.

4. Ms. Martens pays $275 a month for her furnished apartment, including utilities. Her phone costs $14/mo. What is her annual expense for living quarters?

EXERCISES

Solve.

1. Mr. and Mrs. Kane rented a $275/mo furnished apartment. Utilities averaged $60/mo and the phone cost $15/mo. Find their annual expense.

2. Wendy Jordan rented a $1\frac{1}{2}$-room studio apartment for $195/mo and paid $12/mo for utilities and $12.75/mo for a phone. Find her annual expense.

3. Tony DeMeo rented a 3-room apartment for $235/mo including utilities. He paid $17.50 each month for phone service. Find his annual expense.

4. Ray Brown rents a $3\frac{1}{2}$-room apartment for $275/mo. His phone costs $12/mo and utilities are $55/mo. Find his annual expense.

5. The Daley family signed a 3-year lease on a $275/mo apartment. How much rent will the family pay over the 3 y-period?

6. The McCarthys signed a 2-year lease on a $375/mo furnished apartment. How much rent will they pay over the 2 y-period?

7. Jennifer Conney earns $15,000 per year. If she budgets 30% of her income for housing, about how much should she spend per month on living quarters?

8. Larry Giulini earns $18,000 per year and budgets 35% of his income for housing. About how much should he spend per month on living quarters?

ON YOUR OWN

Select six advertisements for apartments in a local paper and compute the annual rental expense.

BUYING A HOUSE

To compute the cost of buying a house

Most people cannot pay cash for a house. They make a down payment and usually sign a mortgage agreement for the remainder of the purchase price. Regular payments are made over a long period of time to pay off the mortgage. The cost of a mortgage depends on the interest rate and the length of time for repayment.

A **mortgage** is a long-term loan used for paying for a house.

MONTHLY REPAYMENT SCHEDULE FOR 15.75% MORTGAGE RATE

Loan	15 y	20 y	25 y	30 y	35 y	40 y
$40,000	580.53	549.02	535.72	529.85	527.21	526.01
$50,000	725.66	686.27	669.65	662.31	659.01	657.51
$60,000	870.79	823.53	803.58	794.78	790.81	789.01

EXAMPLE

The Allen family took out a $40,000 mortgage at 15.75% for 20 y. Use the chart above to find out how much they will pay in the 20 y. How much of this amount is interest?

Find how many months in 20 y.

$$12 \times 20 = 240 \text{ mo}$$

Multiply the monthly payment by the number of months.

```
    $549.02  ←— monthly payment
       ×240  ←— number of months
   21 960 80
   109 804
$131,764.80  ←— total payment
```

So, they will pay **$131,764.80** in the 20 y.

Now find the amount of interest.

Subtract the mortgage from the total payment.

```
 $131,764.80  ←— total payment
 − 40,000.00  ←— mortgage
  $91,764.80  ←— total interest paid
```

So, they will pay **$91,764.80** in interest.

GETTING READY

Solve, using the chart on page 386.

1. The Foster family took out a $40,000 mortgage at 15.75% for 30 y. How much will they repay by the end of the 30 y. How much of this is interest?

2. The Grant family took out a $60,000 mortgage at 15.75% for 25 y. How much will they repay at the end of 25 y? How much of this is interest?

EXERCISES

Solve, using the chart on page 386.

1. The Howells have a $50,000 mortgage at 15.75% for 30 y. How much will they pay by the end of the 30 y?

2. How much interest will the Howells pay in 30 y?

3. The Jenks' 15.75% mortgage of $60,000 is for 15 y. How much will they pay by the end of the 15 y?

4. How much interest will the Jenks pay in 15 y?

5. The Kenyons took out a 15.75% mortgage of $50,000 for 40 y. How much will they have paid at the end of 20 y?

6. The Fawcett family has a $60,000 mortgage of 15.75% for 35 y. How much will they have paid at the end of 15 y?

7. The Gunn family has a 15.75% mortgage for $40,000 for 25 y. How much interest will they pay in 25 y?

8. The Morrow family has a 15.75% mortgage for $40,000 for 35 y. How much interest will they pay in 35 y?

9. The Meyer family has a $50,000 mortgage at 15.75% for 25 y. How much will they pay by the end of 25 y?

*10. How much is saved by paying off a $50,000 mortgage at 15.75% in 20 y rather than 35 y?

*11. How much is saved by paying off a $60,000 mortgage at 15.75% in 25 y rather than 10 y?

BUYING A CAR

To compute the costs of buying a car

Many people borrow money to buy a car and repay the loan in monthly installments. A loan can be arranged through an automobile dealer or through a bank, credit union, or loan company.

EXAMPLE 1

Basil traded in his old car for a newer model. He signed an agreement to pay $185/mo for 12 mo in addition to $450 cash as a down payment. What did the purchase cost him?

Multiply.

$185 ⟵ monthly payment
× 12 ⟵ number of payments
370
1 85
$2,220 ⟵ total of payments

Add the down payment.

+ 450 ⟵ down payment
$2,670 ⟵ actual price

So, he paid **$2,670** in all for the newer car.

EXAMPLE 2

Mr. Steciak offers financing for a new car priced at $6,000 by charging $171.50/mo for 48 mo with 10% down. Compute the finance charge.

> The **finance charge** is the actual price minus the cash price.

Multiply.

$171.50 ⟵ monthly payment
× 48 ⟵ number of payments
137200
68600
$8,232.00 ⟵ total of payments

Find the down payment. **$6,000 × 0.10 = $600.00**

Add the down payment.

$8,232.00
+ 600.00 ⟵ down payment
$8,832.00 ⟵ actual price

Subtract.

$8,832.00 ⟵ actual price
− 6,000.00 ⟵ cash price
$2,832.00 ⟵ finance charge

So, the finance charge is **$2,832.00.**

GETTING READY

Solve.

1. Wendell Allen paid $196/mo for a year in buying a used car. He also paid $150 down. What did the purchase cost him?

2. Al Saunders offers financing for a new car priced at $8,000 by charging $230.65/mo for 48 mo with 10% down. Compute the finance charge.

3. Ray Salzman borrowed $7,000 to buy a new car. If the payments were $217 per month for 48 mo, compute the finance charge.

EXERCISES

Solve.

1. Ramón Alioto bought a used car for $400 down and $165/mo for 18 mo. Find the actual price.

2. Kathy Scott bought a car for $650 down and $217/mo for 18 mo. Find the actual price.

3. Jay Vacca borrowed $5,000 for a new car. If the payments were $198.50 per month for 36 mo, what was the finance charge?

4. Julia Gerrard bought an $8,700 car for 25% down and 48 monthly payments of $208. Find the finance charge.

5. Mr. Wheeler offers $8,000 financing by charging $243.45 per month for 48 mo. Compute the finance charge.

6. Mr. Huseck offers $8,000 financing by charging $213.82/mo for 48 mo. Compute the finance charge.

7. Mr. Rodman bought a $4,999 car for $1,200 down and $26.70/wk for 192 wk. Compute the finance charge.

8. Mrs. Weiss bought a $6,700 car for 25% down and $145/mo for 48 mo. Compute the finance charge.

9. Ida Estey paid $775 down and $227 a month for 36 mo for her car. Compute the total purchase price.

10. Candy Kroger paid $475 down and $235 a month for 48 mo for her car. Compute the total purchase price.

*11. After making a suitable down payment on a car, Newman Kopald needs to finance $3,400. Bank A offers an auto loan for 48 mo at 15%. Bank B offers a loan for 36 mo at 18%. Should Newman choose the bank with the lower rate of interest? Prove your answer is correct by showing the calculations.

FUEL COSTS

To compute miles per gallon (mpg) ratings and fuel costs

EXAMPLE 1

A car consumed 864 gal of gas in covering 21,600 mi. What is the car's mpg rating?

$$\text{mpg} = \frac{\text{distance (miles)}}{\text{gallons of fuel}}$$

Divide miles by gallons used. $\quad \mathbf{\frac{21{,}600}{864} = 25} \longleftarrow$ mpg rating

So, the mpg rating is **25 miles per gallon.**

EXAMPLE 2

If a car's mpg rating is 28 mpg, find the distance the car can cover on a 14.5-gal tankful of gas.

$$\text{mpg} \times \text{gal} = \text{distance}$$

Multiply mpg by the number of gallons.

$$\begin{array}{r} \mathbf{14.5} \\ \mathbf{\times 28} \\ \hline \mathbf{1160} \\ \mathbf{290} \\ \hline \mathbf{406.0} \end{array} \quad \begin{array}{l} \longleftarrow \text{gallons} \\ \longleftarrow \text{mpg rating} \\ \\ \\ \\ \longleftarrow \text{miles covered} \end{array}$$

So, the car can cover about **406 miles** on a tank of gas.

EXAMPLE 3

Gas costs $1.55 a gallon. If a car's mpg rating is 28 mpg, find the cost of gas to travel 4,200 mi.

$$\text{fuel used} = \frac{\text{distance}}{\text{mpg}}$$

Divide miles by mpg. $\quad \mathbf{\frac{4{,}200}{28} = 150\ gal}$

Multiply to find cost. $\quad \mathbf{\$1.55 \times 150 = \$232.50} \longleftarrow$ fuel cost

So, the cost of gas will be **$232.50.**

GETTING READY

Solve.

1. A car consumed 305 gal of gas to cover 8,325 mi. Find the mpg rating.
2. If a car has a rating of 16 mpg, how far can it travel on 13.5 gal of gas?
3. Gas costs $1.48 a gallon. Find the cost of gas to cover 960 mi in a car rated at 24 mpg.
4. Alice Anderson used 465 gal of gas while traveling on a 3,906-mile trip. What is her car's mpg rating?

EXERCISES

Solve.

1. The Q car gets 38.5 mpg. How far can it travel on 11.5 gal of gas?

2. The Z car gets 42.5 mpg. How far can it travel on 16 gal of gas?

3. Gas costs $1.46 a gallon. Find the cost of covering 5,408 mi in a car with a 26 mpg rating.

4. Gas costs $1.53 a gallon. Find the cost of covering 6,006 mi in a car rated at 19.5 mpg.

5. Find the mpg rating of a car that covered 8,600 mi on 430 gal of gas.

6. Find the mpg rating of a car that covered 3,570 mi on 204 gal of gas.

*7. Fran's new car gets 24 mpg. Her old car got 16 mpg. How many gallons of gas will the new car save in traveling 12,000 mi?

*8. Al's new car gets 30 mpg. His old car got 18 mpg. How much less gas will his new car use in traveling 16,200 mi?

*9. Gas sells for $1.47 a gallon. Carol's new car gets 32 mpg; her old car got 16 mpg. How much will she save on gasoline traveling 24,000 mi in her new car?

*10. Gas sells for $1.29 a gallon. Sarah's old car gets 12 mpg. If she gets a new car with a 36 mpg rating, how much will she save on gasoline driving 18,000 mi?

CALCULATOR

One liter is 0.227 gal. Use your calculator and this conversion factor to find the price per liter of these gasolines. Round to the nearest cent.

1. Premium costing $1.51 a gallon.
2. Regular costing $1.32 a gallon.
3. Unleaded costing $1.37 a gallon.

*4. What is the price per gallon of gasoline that costs $0.37 per liter?

Problem Solving Skill

To fill out a catalog order form

- Enter order numbers, prices, and weights.
- Use total weight to find shipping and handling charges.
- Use total price to find sales tax.
- Add for total cost.

EXAMPLE

Study the completed order form below.

NAME OF ITEM (1 or 2 words)	HOW MANY (Pkgs., etc.)	CATALOG NUMBER Letter(s)	Numbers	Letter(s)	PRICE FOR ONE (Pkg., Yd., Ea., etc.)	TOTAL PRICE	DELIVERY WTS.
Blank Tapes	3	A	858-2728	A	3.29	9.87	3.00
Cassett & Player	1	A	851-1776	A	49.95	49.95	6.50

DELIVERY WEIGHT IN LBS.	SHIPPING AND HANDLING CHARGE
0- .5 lb	1.12
.6- 1 lb	1.45
1.1- 2 lb	1.85
2.1- 3 lb	2.04
3.1- 4 lb	2.25
4.1- 5 lb	2.44
5.1- 6 lb	2.63
6.1- 7 lb	2.84
7.1- 8 lb	3.03
8.1- 9 lb	3.22
9.1-10 lb	3.32
10.1-11 lb	3.38

STATE SALES TAX RATES PER $1.00	
New Mexico	3½¢
New York	4¢
North Carolina	3¢
North Dakota	3¢
Ohio	4½¢
Oklahoma	2¢
Oregon	0

TOTAL WEIGHT	9.50
MERCHANDISE TOTAL	59.82
SHIPPING AND HANDLING	3.32
SALES TAXES	2.10
TOTAL CASH ENCLOSED	65.24

Here are the catalog entries for these items:

Cassette blank tapes	A 858-2728 A	wt. 1 lb	3.29
Cassette player with radio	A 851-1776 A	wt. 6.50 lb	49.95

Multiply the price for one tape by 3. **$3.29 × 3 = $9.87**

Find the merchandise total. **$9.87 + $49.95 = $59.82**

Multiply the weight of one tape by 3.

1 lb × 3 = 3 lb

Add the weights of tapes and cassette.

3.00 lb + 6.50 lb = 9.50 lb

Locate total weight in the shipping and handling chart.

9.50 lb is between 9.1 lb and 10 lb.

Find the shipping charge.

So, the charge is **$3.32.**

The customer is in New Mexico, so sales tax is $3\frac{1}{2}$¢ per $1.00.

Use 0.035 as the rate.

Multiply merchandise total by tax rate.

$59.82 × 0.035 = $2.0937

Round *up* to the next cent.

$2.0937 ⟶ $2.10

Add total price, shipping, and tax.

$59.82 + 3.32 + 2.10 = $65.24

So, the total cash enclosed is **$65.24.**

EXERCISES

Solve, using the form on page 392.

1. What is the tax in New York on a total order of $39.89?

2. What is the tax in Ohio on a total order of $112.50?

3. The two items in an order weigh 3.50 lb and 2.25 lb. What is the shipping and handling charge for the order?

4. An order contains two 1.5-pound items and a 1-pound item. What is the shipping and handling charge for the order?

Find the total cost of each order.

5. Ordered from North Carolina: one phonograph (weight 7 lb, price $32.95) and two blank tapes (weight 1 lb each, price $4.29 each)

6. Ordered from New Mexico: one AM/FM recorder-player (weight 3.50 lb, price $54.50) and two microphones (weight 2 lb each, price $19.95 each)

***7.** The following items are being ordered from Oklahoma:

Catalog listing

Quality recording microphone	A 858-1761 B wt. 2 lb	5.95
Cassette recorder-player	A 843-1502 B wt. 4.50 lb	39.50
Stereo headphones	C 921-1466 A wt. 2 lb	23.95

Copy the form on page 392 and fill it out completely with an order for one microphone, one cassette recorder-player, and two sets of headphones.

Problem Solving Applications

READ • PLAN • SOLVE • CHECK

The seniors at Jackson High School are preparing for a 4-day class trip to the World's Fair. The class officers are planning the details of the trip with the help of their class advisers.

AS YOU READ

- chaperone—an older person who goes with younger persons to supervise and to see that proper behavior is maintained
- charter—to hire

Answer these questions about the class trip.

1. Mrs. Parks estimates that 20% of the 287 seniors will not go on the trip. About how many students will go on the trip? Round to the nearest whole number.

2. The class officers have decided to *charter* buses. If each bus holds 50 people, how many buses are needed for 240 students and *chaperones*?

3. The Acme Bus Company charges $650/day for the deluxe coach and $595/day for the super coach. How much more would the deluxe coach cost for the 4 days?

4. The total cost of transportation, lodging, and admission is $157 per person. How much will it cost for 250 persons?

5. Easy Rest Motor Lodge charges $27.50/night per person. Highway 40 Motor Inn charges $30/night including breakfast. If breakfast costs about $4.75 per person, how much can be saved by staying at Highway 40 for 3 nights?

6. Mr. Jordan and Mrs. Parks estimated that 3 lunches and 3 dinners would cost about $50 per person and that spending money should be between $50 and $100. At most, how much should each person bring?

Career: Travel Agents

READ • PLAN • SOLVE • CHECK

Travel agents work for independent travel agencies and for corporate travel departments. Many of today's travel agents use computers for current information and reservations with major carriers. Training is received on the job and in vocational schools.

AS YOU READ

- book—to make reservations for travel and lodging
- cruise—a tour by ship
- villa—a country house

Solve.

1. A Virgin Islands special for summer is $541 per person for 4 days. A 40% discount is given to children under 10. How much would it cost for a couple and their 9-year-old twins to take this trip?

2. Ben Green works on commission. He *booked* a 7-day vacation to Puerto Rico for a family of 4 persons. The cost of the trip is $418 per person. How much does he earn for this booking if his commission is 5%?

3. Lucy Merrill owns Freedom Travel Agency. She booked a Nile *cruise* for a party of 14. The cost was $1,790 per person. What was the total cost of the cruise for the group?

4. Tom Jordan booked a yachting trip for a young, sporting goods salesman. The 10-day trip cost $755 plus $315 air fare. How much is the total cost per day?

5. Mara Greenfield booked a corporate business trip from New York to Los Angeles. The round-trip flight, hotel, and limousine totaled $1,782. Her commission was 2.5%. How much did she earn?

6. St. Maartens offers a visitor a *villa* vacation for $1,699. This includes air fare and car rental. The travel agency receives 8% commission on this vacation. How much does the agency earn booking 4 villa vacations?

CHAPTER TEST 20

Solve.

1. Miss Quinn earns about $13,000 a year. If 30% of her salary goes for housing, how much does she spend each month for housing?

2. Which is the better buy, a $340 stereo at Store A for 20% off or the same stereo for $310 at Store B for 10% off?

3. Mr. Ray ordered 32 yd^2 of outdoor carpet for his patio at $11.95 per square yard. If he paid 8% tax, what was the total cost?

4. Find the finance charge on a $3,600 used car bought for $600 down and $108 per month for 3 y.

5. Mr. Marten's credit card bill was $380 last month. He made a payment of $40. Find the amount of interest added to next month's bill if the charge is $1\frac{1}{2}$% of the unpaid balance.

6. Dawn rents a studio apartment for $350/mo. Her utilities are $20/mo and her phone bill averages $45/mo. Find her annual expense.

7. Mr. Rath bought a $3,850 living-room set for $800 down and $72.50 a month for 48 mo. Find the carrying charge.

8. Gas costs $1.48 a gallon. Find the cost of gas needed to travel 2,100 mi in a car rated at 28 mpg.

9. Find the mpg rating of a car that went 487 mi on 18 gal of gasoline.

10. Andy Cross pays $662.31 a month on his 30-year mortgage. How much will he spend to pay off the mortgage?

11. Find the better buy on mopeds.

    J.J.'s $340 40% off
    M.T.'s $300 25% off

12. Albert Longo borrowed $5,600 to buy a used car. He repaid the loan in 36 mo by paying $164 a month. Find the finance charge.

13. The monthly payment on a $70,000 mortgage at 14% is $829.41. What is the total payment over 30 y? How much of this is interest?

*14. A $5,500 loan has a monthly payment of $237/mo if the loan is for 2 y and $164/mo if the loan is for 3 y. How much more is the finance charge on the 3-year loan?

CUMULATIVE REVIEW 10-20

Add. Simplify when possible.

1. $\frac{3}{8} + \frac{2}{8}$ **2.** $\frac{2}{3} + \frac{1}{6}$ **3.** $\frac{2}{5} + \frac{4}{5}$ **4.** $1\frac{3}{5} + 2\frac{1}{2}$ **5.** $4\frac{7}{10} + 1\frac{2}{5}$

Subtract. Simplify when possible.

6. $\frac{7}{9} - \frac{1}{9}$ **7.** $\frac{11}{12} - \frac{3}{4}$ **8.** $2\frac{1}{4} - \frac{3}{4}$ **9.** $4\frac{3}{8} - 2\frac{4}{5}$ **10.** $7\frac{3}{100} - 3\frac{3}{10}$

Multiply. Simplify when possible.

11. $\frac{1}{6} \times \frac{2}{3}$ **12.** $7 \times \frac{4}{5}$ **13.** $3\frac{1}{5} \times 6$ **14.** $2\frac{3}{4} \times 4\frac{4}{5}$

Divide. Simplify when possible.

15. $10 \div \frac{5}{6}$ **16.** $\frac{2}{3} \div \frac{4}{9}$ **17.** $\frac{5}{6} \div 2\frac{1}{2}$ **18.** $2\frac{1}{5} \div 1\frac{3}{10}$

19. Team A won 26 games and lost 11 games. Write a ratio in fraction form for games won to games lost.

20. What is the probability of spinning a 5 on a spinner labeled 1 through 8?

21. Solve the proportion.
$\frac{3}{5} = \frac{6}{n}$

22. Solve by writing a proportion. Sara worked 5 h to earn $12. At that rate, how much will she earn for 15 h work?

23. Find the mean.
227, 846, 325, 432, 650

24. Find the median.
215, 235, 416, 565, 433, 347

25. Find the mode.
54, 56, 53, 42, 17, 45, 42, 32, 36

26. If $E = I \times R$ and $I = 8.4$ and $R = 178$, find E.

27. Solve $2n + 3 = 39$.

28. Solve $\frac{n}{5} + 12 = 17$.

29. Write 9% as a fraction.

30. Write 125% as a mixed number

31. Write 0.04% as a decimal.

32. Find 45% of 50.

33. 27 is what percent of 36?

34. 32 is 80% of what number?

Find the areas. (Use $\pi = 3.14$.)

35.

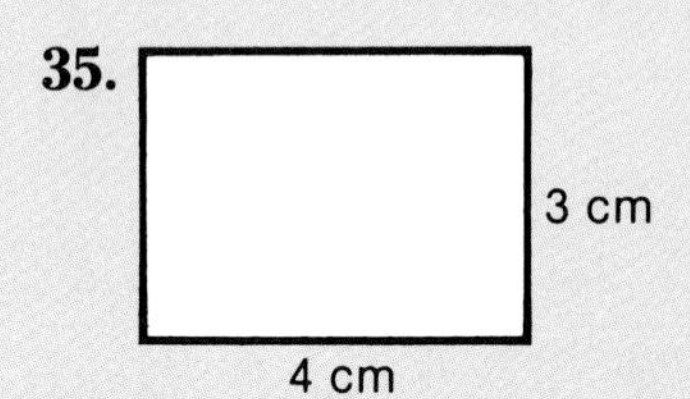

36.

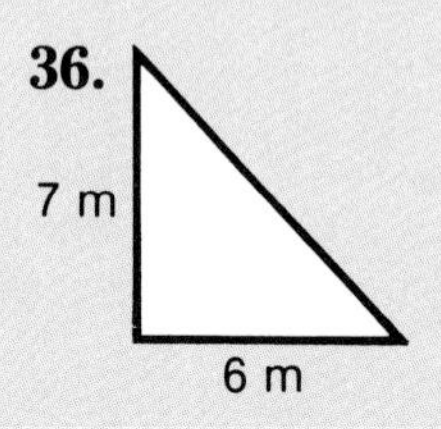

37.

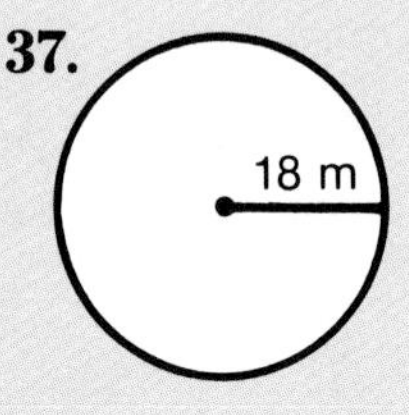

Classify the triangles by the measures of their angles.

38.

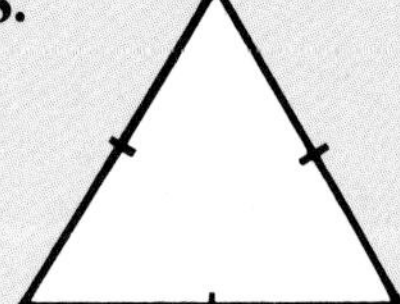

39.

40.

41. Classify the triangles in **38–40** by the measures of their sides.

42. Find the circumference of a plate with a diameter of 14 cm. (Use $\pi = 3.14$.)

43. Find the volume of a box that is 16 in. long, 14 in. wide, and 9 in. high.

Solve.

44. Bill was given a recipe that calls for $\frac{3}{4}$ c milk. He wants to make $\frac{1}{2}$ of the recipe. How much milk does he need?

45. The length of a room on a floor plan is $2\frac{3}{4}$ in. The scale is 1 in. = 6 ft. Find the actual length of the room.

46. Joe worked 40 hours for $7.75 per hour. His deductions were retirement, $18.60; Federal and State Tax, $38.15; FICA, $45.60; and credit union, $25. Find his net pay.

47. Jason's bank statement balance is $197.03. Deposits of $53.08 and $12.95 and a check for $29.87 are not shown. What is his actual balance?

48. Mr. Aaron sold a house for $70,000. He receives a commission of 7% on all sales. How much commission did he earn?

49. What is the interest on $4,500 for 2 y at 12%?

50. Find the carrying charge on a $4,500 piano bought for $700 down and monthly payments of $190 for 2 y.

51. Lisa rented an apartment for $375/mo. Her telephone cost $16/mo, and utilities were $90/mo. Find her annual expense.

Formulas and Tables

Perimeter	Equilateral Triangle	$P = 3 \times$ length of a side
	Polygon	$P =$ sum of lengths of sides
	Rectangle	$P = 2 \times$ length $+ 2 \times$ width
	Square	$P = 4 \times$ length of a side
Circumference	Circle	$C = \pi \times$ diameter or $C = \pi \times d$
Area	Circle	$A = \pi \times$ radius $\times$ radius or $A = \pi \times r \times r$
	Rectangle	$A =$ length $\times$ width or $A = l \times w$
	Square	$A =$ length of side $\times$ length of side or $A = s \times s$
	Triangle	$A = \frac{1}{2} \times$ base $\times$ height
Volume	Rectangular Solid	$V =$ length $\times$ width $\times$ height or $V = l \times w \times h$
Others	Carrying Charge	Carrying charge $=$ installment price $-$ cash price
	Diameter	Diameter $= 2 \times$ radius
	Distance	Distance $=$ rate $\times$ time
	Finance Charge	Finance charge $=$ actual price $-$ cash price
	Gross Pay	Gross pay $=$ hourly rate $\times$ number of hours
	Interest	Interest $=$ principal $\times$ rate (per year) $\times$ time (in years)
	Miles Per Gallon (mpg)	Miles per gallon $= \frac{\text{distance (in miles)}}{\text{gallons of fuel}}$
	Net pay	Net pay $=$ gross pay $-$ total deductions
	Radius	Radius $=$ diameter $\div 2$

METRIC

Length	Capacity	Mass
1 cm = 10 mm	1 L = 1,000 mL	1 g = 1,000 mg
1 m = 100 cm = 1,000 mm		1 kg = 1,000 g
1 km = 1,000 m		

CUSTOMARY

Length	Capacity	Weight
1 ft = 12 in.	1 c = 8 fl oz	1 lb = 16 oz
1 yd = 3 ft = 36 in.	1 pt = 2 c	1 T = 2,000 lb
1 mi = 5,280 ft = 1,760 yd	1 qt = 2 pt	
	1 gal = 4 qt	

Glossary

The explanations given in this glossary are intended to be brief descriptions of the terms. They are not necessarily definitions.

acute angle *(p. 280)* An angle that measures less than 90°.

area *(p. 164)* The number of square units needed to cover a region.

average *(p. 102)* Add the numbers and then divide this sum by the number of items in the set.

better buy *(p. 152)* An item with the lowest unit price.

carrying charge *(p. 382)* The difference between the installment price and the cash price.

centi *(p. 160)* A prefix meaning one hundredth.

circumference *(p. 296)* The distance around a circle.

credit line *(p. 378)* The limit a credit card holder may charge to an account.

degrees Celsius *(p. 174)* The unit of temperature in the metric system.

denominator *(p. 190)* The bottom number in a fraction. In the fraction $\frac{1}{2}$, 2 is the denominator.

diameter *(p. 296)* A chord that passes through the center of a circle.

difference *(p. 30)* The result of subtracting one number from another.

digit *(p. 8)* The ten symbols used to form numerals. 0,1,2,3,4,5,6,7,8,9 are digits.

discount rate *(p. 336)* The percent decrease in price.

dividend *(p. 96)* The number to be divided. In $18 \div 3 = 6$, the dividend is 18.

divisor *(p. 96)* The number that is used to divide another number. In $18 \div 3 = 6$, the divisor is 3.

equilateral triangle *(p. 284)* A triangle with three sides equal in length.

evaluating expressions *(p. 242)* To replace each variable with a given number and compute the value.

F.I.C.A. *(p. 350)* The amount contributed to Social Security. It stands for Federal Insurance Contributions Act.

formula *(p. 244)* An equation showing how quantities are related.

fraction *(p. 186)* This shows a part of a whole or a part of a set.

gram *(p. 171)* The basic unit of mass in the metric system.

gross pay *(p. 344)* The total amount earned before deductions are made.

hexagon *(p. 282)* A polygon with six sides.

improper fraction *(p. 208)* A fraction in which the numerator is greater than the denominator. The fraction $\frac{13}{5}$ is an improper fraction.

interest *(p. 366)* The amount of money the lender of money charges the borrower.

isosceles triangle *(p. 284)* A triangle with two sides equal in length.

kilo *(p. 166)* A prefix meaning one thousand.

least common denominator *(p. 210)* The smallest number that is a common multiple of a set of denominators.

like fractions *(p. 204)* Fractions that have the same denominator.

liter *(p. 168)* The basic unit of capacity in the metric system.

mean *(p. 102)* The average of the numbers in a set of data.

median *(p. 310)* The middle number in a set of numbers.

meter *(p. 159)* The basic unit of length in the metric system.

milli *(p. 166)* A prefix meaning one thousandth.

mode *(p. 309)* The number that occurs most often in a set of data.

net pay *(p. 352)* The amount of earnings after the total deductions are subtracted from gross pay.

numerator *(p. 190)* The top number in a fraction. In the fraction $\frac{1}{2}$, 1 is the numerator.

obtuse angle *(p. 280)* An angle that measures more than 90°.

octagon *(p. 282)* A polygon with eight sides.

parallelogram *(p. 286)* A quadrilateral whose opposite sides are equal and parallel.

pentagon *(p. 282)* A polygon with five sides.

percent *(p. 320)* Means *per hundred* or *out of a hundred.*

perimeter *(p. 288)* The distance around a polygon.

polygon *(p. 282)* A closed figure made up of line segments.

product *(p. 70)* The result of a multiplication problem.

proportion *(p. 272)* A sentence that two ratios are equal. $\frac{3}{4} = \frac{21}{28}$ is a proportion.

quadrilateral *(p. 282)* A polygon with four sides and four angles.

quotient *(p. 92)* The result of a division problem. In $18 \div 3 = 6$, 6 is the quotient.

radius *(p. 296)* The distance from the center of a circle to a point on the circle.

ratio *(p. 268)* A comparison of two numbers. The ratio $\frac{4}{5}$ is read 4 to 5.

rectangle *(p. 286)* A parallelogram with four right angles.

rhombus *(p. 286)* A parallelogram with four sides equal in length.

right angle *(p. 280)* An angle that measures exactly 90°.

scalene triangle *(p. 284)* A triangle with no sides equal in length.

square *(p. 286)* A rectangle with four sides equal in length.

sum *(p. 8)* The result of an addition problem.

triangle *(p. 282)* A polygon with three sides.

unlike fractions *(p. 212)* Fractions which do not have the same denominator.

variables *(p. 242)* Letters or symbols used to represent numbers.

volume *(p. 300)* The measure of space inside a solid figure.

Selected Answers

This section contains selected answers to the problems under **Getting Ready** *(all) and* **Exercises** *(odd-numbered).*

page 4

Getting Ready **1.** 13 **2.** 13 **3.** 11 **4.** 7 **5.** 12 **6.** 11 **7.** 6 **8.** 9 **9.** 4 **10.** 16 **11.** 11 **12.** 8
Exercises **1.** 7 **3.** 12 **5.** 17 **7.** 8 **9.** 5 **11.** 15 **13.** 10

page 5

Getting Ready **1.** 17 **2.** 11 **3.** 15 **4.** 13 **5.** 12 **6.** 11 **7.** 11
Exercises **1.** 12 **3.** 18 **5.** 13 **7.** 11 **9.** 12 **11.** 13 **13.** 16

page 6

Getting Ready **1.** 70 **2.** 800 **3.** 7,000 **4.** 170 **5.** 900
Exercises **1.** 70 **3.** 70 **5.** 110 **7.** 1,300 **9.** 12,000

page 9

Getting Ready **1.** 59 **2.** 82 **3.** 59 **4.** 42 **5.** 104 **Exercises** **1.** 28 **3.** 38 **5.** 17 **7.** 66 **9.** 47 **11.** 59 **13.** 49 **15.** 98 **17.** 64 **19.** 52 **21.** 34 **23.** 27 **25.** 83 **27.** 90 **29.** 40 **31.** 105 **33.** 102 **35.** 101 **37.** $101

page 11

Getting Ready **1.** 20 **2.** 20 **3.** 22 **4.** 23 **5.** 30 **6.** 31 **7.** 37 **8.** 36
Exercises **1.** 20 **3.** 20 **5.** 25 **7.** 26 **9.** 36 **11.** 30 **13.** 36 **15.** 36 **17.** 44 **19.** $18 **21.** 20 mi

page 12

Getting Ready **1.** 157 **2.** 114 **3.** 839 **4.** 1,175 **5.** 1,181
Exercises **1.** 76 **3.** 146 **5.** 191 **7.** 162 **9.** 108 **11.** 110 **13.** 84 **15.** 87 **17.** 122 **19.** 90 **21.** 122 **23.** 156 **25.** 162 **27.** 607 **29.** 1,489 **31.** 1,532 **33.** 638 **35.** 1,636 **37.** 15, 63; 70, 8; 78 **39.** 36, 36; 50, 22; 72

page 15

Getting Ready **1.** 161 **2.** 178 **3.** 237 **4.** 1,236
Exercises **1.** 273 **3.** 4,373 **5.** 6,161 **7.** 316 **9.** 4,007 **11.** 10,553 **13.** 2,640 **15.** $152

page 17

Getting Ready **1.** 80 **2.** 570 **3.** 390 **4.** 8,940 **5.** 300 **6.** 700 **7.** 900 **8.** 8,200 **Exercises** **1.** 80 **3.** 80 **5.** 470 **7.** 640 **9.** 90 **11.** 870 **13.** 100 **15.** 1,400 **17.** 900 **19.** 600 **21.** 300 **23.** 100 **25.** 5,800 **27.** 900 **29.** 1,000

page 19

Getting Ready **1.** 70 **2.** 900 **3.** 160 **4.** $5 **Exercises** **1.** 100 **3.** 110 **5.** 900 **7.** 300 **9.** $6 **11.** $10 **13.** $9 **15.** 1,200 **17.** 4,910 **19.** $150 **21.** 120 mi

page 21

Getting Ready **1.** 194 **2.** 88 **3.** $4.00 **4.** $1.30 **5.** $5.80 **6.** $5.22 **Exercises** **1.** 199 **3.** 122 **5.** 197 **7.** 146 **9.** 54 **11.** $198 **13.** $1.25 **15.** $5.20 **17.** $2.90 **19.** No

page 29

Getting Ready **1.** 9 **2.** 5 **3.** 9 **4.** 5 **Exercises** **1.** 9 **3.** 8 **5.** 5 **7.** 9 **9.** 7

page 31

Getting Ready **1.** 44 **2.** 72 **3.** 24 **4.** 32 **5.** 20 **6.** 28 **7.** 42 **8.** 60 **9.** 33 **10.** 55 **11.** 42 **12.** 54 **Exercises** **1.** 32 **3.** 92 **5.** 71 **7.** 94 **9.** 73 **11.** 80 **13.** 80 **15.** 80 **17.** 64 **19.** 73 **21.** 82 **23.** 80 **25.** 20 **27.** 73 **29.** 72 **31.** $44

page 32

Getting Ready **1.** 40 **2.** 300 **3.** 40 **4.** 800 **5.** 700 **Exercises** **1.** 60 **3.** 50 **5.** 90 **7.** 10 **9.** 60 **11.** 300 **13.** 900 **15.** 600 **17.** 6,000 **19.** 5,000

page 35

Getting Ready **1.** 74 **2.** 37 **3.** 77 **4.** 4 **5.** 63 **Exercises** **1.** 41 **3.** 37 **5.** 87 **7.** 45 **9.** 37 **11.** 28 **13.** 48 **15.** 49 **17.** 23 **19.** 48 **21.** 42 **23.** 37 **25.** 33 **27.** 57 **29.** 6 **31.** 19 **33.** 27 **35.** 9 **37.** 14 **39.** 19 **41.** 15 **43.** 29 **45.** 24 **47.** 49 **49.** 35 **51.** $25

page 37

Getting Ready **1.** 309 **2.** 668 **3.** 3,346 **4.** 2,859 **Exercises** **1.** 325 **3.** 606 **5.** 408 **7.** 774 **9.** 783 **11.** 327 **13.** 669 **15.** 491 **17.** 140 **19.** 749 **21.** 479 **23.** 3,737 **25.** 2,790 **27.** 26,881 **29.** 7,537 **31.** 33,519 **33.** 21,259 **35.** 42,000 **37.** 69,917 **39.** 41,995 **41.** 69,012 **43.** 57,432 **45.** 7,528 **47.** 23,912,123

page 38

Getting Ready **1.** $3.25 **2.** $5.02 **3.** $11.57 **4.** $4 **5.** $1.55 **6.** $7.20 **Exercises** **1.** $4.11 **3.** $3.66 **5.** $1.24 **7.** $0.02 **9.** $7.11 **11.** $5.32 **13.** $5 **15.** $7.50 **17.** $18.50

page 40

Getting Ready **1.** 22 **2.** 46 **3.** 66 **4.** $1.37 **5.** $2.23 **6.** $9.10 **Exercises** **1.** 50 **3.** 40 **5.** 160 **7.** 500 **9.** 230 **11.** $44 **13.** $4.50 **15.** $1.60

page 43

Getting Ready **1.** 500 **2.** 400 **3.** $500 **4.** 4,400 **5.** 1,400 **6.** 1,000 **7.** 13,000 **8.** $3 **9.** $10 **10.** $200 **Exercises** **1.** 400 **3.** 410 **5.** 8,000 **7.** 80 **9.** $10 **11.** $10 **13.** $10 **15.** $50 **17.** 1,000

page 51

Getting Ready **1.** Given: Time put in → 2 pm; Time to cook → 3 h 20 min Find: Time to remove

Exercises **1.** Given: mini-pennants → 261; standard pennants → 197 Find: number sold altogether **3.** Given: odometer at home → 18,931 mi; odometer in Chicago → 19,328 mi Find: how far Maria drove **5.** Given: weighs now → 128 lb; lost → 29 lb Find: how much weighed before diet **7.** probs. **1.** 458 **2.** 2,371 **3.** 397 mi **4.** $675 **5.** 157 lb **6.** $5.96

page 52

Getting Ready **a.** 2,146 **b.** homecoming football game **c.** 1,462 **d.** how many were not students
Exercises **1a.** 5 doz **1b.** George **1c.** 4 doz **1d.** how many dozen were left **3a.** before and after school **3b.** 45 **3c.** 28 **3d.** how many newspapers he delivers each day **5.** probs. **1.** 1 doz **2.** 13 **3.** 73 **4.** 1,221 kW·h

page 54

Getting Ready **1.** add **2.** multiply **3.** multiply **4.** divide **5.** divide
Exercises **1.** add **3.** subtract **5.** subtract **7.** multiply **9.** add **11.** divide **13.** divide **15.** Probs. **1.** 405 **2.** $437 **3.** $5.98 **4.** $2.25 **5.** $15\frac{1}{2}$ y old **6.** $11.87 **7.** $1\frac{1}{2}$ c **8.** $2\frac{1}{2}$ h **9.** 31 **10.** 42 **11.** 800 km/h **12.** $1.29 **13.** $5.12 **14.** $0.33

page 56

Getting Ready **1.** amount Randi gave **2.** 756 paperback books
Exercises **1.** price of 6 cans **3.** a one-week delay **5.** 9 y old cat; 1 y old puppy **7.** $60 commuting expense **9.** Probs. **3.** 6 da **4.** 38 **5.** 16 lb **6.** 1,250 **7.** $470 **8.** 6 h

page 58

Getting Ready **1.** 1,275 **2.** 86 **3.** 390,334 **4.** 54¢ **5.** $9.56
Exercises **1.** 450 m **3.** 1,374 mi **5.** $238 **7.** 147 **9.** $133.22

page 61

Getting Ready **1.** 354 **2.** 10,657 cubic units **3.** Est.: 3,000; Act.: 3,247
Exercises **1.** 8 doz **3.** 407 **5.** Est.: 1,200; Act.: 1,235 **7.** Est.: 300 km; Act.: 282 km

page 63

Getting Ready **1.** 45 **2.** 1 h 25 min
Exercises **1.** 9 **3.** $4.25 **5.** 170 **7.** $13

page 71

Getting Ready **1.** 45 **2.** 35 **3.** 24 **4.** 25 **5.** 8 **6.** 36 **7.** 40 **8.** 35 **9.** 48 **10.** 81 **11.** 32 **12.** 49 **13.** 56 **14.** 30 **Exercises** **1.** 27 **3.** 36 **5.** 21 **7.** 16 **9.** 24 **11.** 54 **13.** 63 **15.** 15 **17.** 64 **19.** 9 **21.** 0 **23.** 36

page 72

Getting Ready **1.** 160 **2.** 240 **3.** 800 **4.** 20,000 **5.** 4,200
Exercises **1.** 360 **3.** 640 **5.** 200 **7.** 280 **9.** 400 **11.** 5,600 **13.** 2,400 **15.** 63,000

page 73

Getting Ready **1.** 129 **2.** 368 **3.** 1,646 **4.** 45,055
Exercises **1.** 168 **3.** 909 **5.** 8,109 **7.** 24,884 **9.** 16,008 **11.** $93

page 75

Getting Ready **1.** 172 **2.** 834 **3.** 16,961 **4.** 3,822
Exercises **1.** 144 **3.** 483 **5.** 168 **7.** 729 **9.** 1,710 **11.** 3,157 **13.** 5,712 **15.** 2,488 **17.** 21,450 **19.** 39,258 **21.** $23,790

page 77

Getting Ready **1.** 468 **2.** 1,620 **3.** 3,304 **4.** 270,671
Exercises **1.** 1,127 **3.** 1,869 **5.** 1,792 **7.** 3,360 **9.** 2,880 **11.** 1,200 **13.** 2,304 **15.** 6,794 **17.** 21,672 **19.** 14,586 **21.** 22,575 **23.** 61,952 **25.** 54,929 **27.** 41,958 **29.** 42,050 **31.** 68,376 **33.** 220,039 **35.** 614,200 **37.** 30,512 **39.** 315,315 **41.** 17,150 lb

page 79

Getting Ready **1.** 84,832 **2.** 184,548 **3.** 249,798 **4.** 953,500
Exercises **1.** 80,256 **3.** 247,266 **5.** 142,043 **7.** 128,016 **9.** 185,991 **11.** 134,352 **13.** 393,300 **15.** 245,328 **17.** 394,638 **19.** 309,636 **21.** 750,438 **23.** 1,350,900 **25.** 4,395,132 **27.** 4,836,266 **29.** 6,074,622 **31.** 364,656 **33.** 748,996 **35.** 885,000 **37.** $72,048

page 81

Getting Ready **1.** 200 **2.** 7,000 **3.** 450,000 **4.** 5,600,000 **5.** 250; 2,500; 25,000 **6.** 3,260; 32,600; 326,000 **7.** 4,080; 40,800; 408,000 **8.** 12,970; 129,700; 1,297,000
Exercises **1.** 100 **3.** 10,000 **5.** 100,000 **7.** 90,000 **9.** 6,000 **11.** 2,400 **13.** 1,000 **15.** 490,000 **17.** 4,000 **19.** 2,000,000 **21.** 4,000,000 **23.** 6,300,000 **25.** 72,000,000 **27.** 3,600; 36,000; 360,000 **29.** 10,020; 100,200; 1,002,000 **31.** $10 **33.** 9

page 83

Getting Ready **1.** 12,000 **2.** 80,000 **3.** $100 **4.** $200
Exercises **1.** 900 **3.** 900 **5.** 20,000 **7.** 350,000 **9.** 54,000 **11.** $400 **13.** 5,000 **15.** $3.20

page 91

Getting Ready **1.** 4 **2.** 4 **3.** 7 **4.** 8 **5.** 6 **6.** 5 **7.** 8 **8.** 4
Exercises **1.** 8 **3.** 6 **5.** 9 **7.** 7 **9.** 8 **11.** 8

page 93

Getting Ready **1.** 80 **2.** 500 **3.** 40 **4.** 600 **Exercises** **1.** 10 **3.** 1,000 **5.** 60 **7.** 40 **9.** 600 **11.** 600 **13.** 80 **15.** 5 **17.** 30 **19.** 40 **21.** 90 **23.** 90 **25.** 10 **27.** 5 **29.** 4 **31.** 8 **33.** 7 **35.** 8 **37.** 30

page 95

Getting Ready **1.** 44 **2.** 19 **3.** 13 **4.** 114 **Exercises** **1.** 13 **3.** 15 **5.** 23 **7.** 118 **9.** 48

page 97

Getting Ready **1.** 5r6 **2.** 9r2 **3.** 132r2 **4.** 504r1
Exercises **1.** 4r2 **3.** 8r3 **5.** 6r2 **7.** 3r1 **9.** 36r3 **11.** 51r3 **13.** 31r3 **15.** 62r4 **17.** 1,472r5 **19.** 766r1 **21.** 15,638r3 **23.** 3,104r3 **25.** 61; 3

page 99

Getting Ready **1.** 3r10 **2.** 6r8 **3.** 27r7 **4.** 8r5 **5.** 220r11 **6.** 85r3 **7.** 1,569r23 **8.** 614r51
Exercises **1.** 2r12 **3.** 25 **5.** 22r26 **7.** 31r12 **9.** 302 **11.** 30r4 **13.** 703 **15.** 263 **17.** 206 **19.** 361r2 **21.** 5,065 **23.** 608 **25.** 189 **27.** 123r3 **29.** $2,428

page 100

Getting Ready **1.** 14 **2.** 242 **3.** 256r12 **Exercises** **1.** 35 **3.** 365 **5.** 563 **7.** 233r142 **9.** 340r843 **11.** $395

page 101

Getting Ready **1.** 211 **2.** 41 **3.** 53 **4.** 6,000 **Exercises** **1.** 201 **3.** 1,073 **5.** 137 **7.** 32 **9.** 159 **11.** 411

page 103

Getting Ready **1.** 50 **2.** 100 **3.** 87 **4.** 16 **Exercises** **1.** yes **3.** 91 **5.** 1,173

page 105

Getting Ready **1.** 10 **2.** 5 **3.** 20 **4.** $5 **5.** $1 **Exercises** **1.** 1,000 **3.** 900 **5.** 1,000 **7.** 1,000 **9.** 6 **11.** 60 **13.** 200 **15.** 1 **17.** 20 **19.** 2 **21.** 14 **23.** 4 **25.** $3 **27.** $10

page 116

Getting Ready **1.** five and six-tenths **2.** one hundred thirty-four dollars and six cents **3.** four hundred fifty-two and seven hundred eighty-three thousandths **4.** nineteen thousand, two hundred five and ninety-eight hundredths **5.** 201.08 **6.** 1,000.032 **Exercises** **1.** three and nine tenths **3.** six hundred and five hundredths **5.** 89.6 **7.** 305,600.19

page 117

Getting Ready **1.** 22 **2.** 88 **3.** 34 **4.** 157 **5.** 8 **6.** 19 **7.** 0 **8.** 400
Exercises **1.** 8 **3.** 101 **5.** 76 **7.** 0 **9.** 0 **11.** 608

page 119

Getting Ready **1.** 34.5 **2.** 99.1 **3.** 1136. **4.** 10.657 **5.** 420.54 **6.** 2097.2 **7.** 30.477 **8.** 647.9385 **Exercises** **1.** 69.2 **3.** 506.561 **5.** 39.99 **7.** 169.93 **9.** 468.2337 **11.** $205.91

page 121

Getting Ready **1.** 157.9 **2.** 98.93 **3.** 99.001 **4.** 499.84 **5.** 598.2 **6.** $869.04 **7.** 6.711 **8.** 4.979 **9.** 95.09 **10.** 195.97 **Exercises** **1.** 132.24 **3.** $716.25 **5.** $22.50 **7.** 33.44 **9.** 0.163

page 131

Getting Ready **1.** 63.6106 **2.** 7.1262 **3.** 18.050 **4.** 132.8540 **5.** 551.0421 **6.** 27.35271 **Exercises** **1.** 13.7067 **3.** 100.282 **5.** 1,265.28 **7.** 565.80 **9.** 108.3600 **11.** 5,352.50 **13.** 4,750.705 **15.** 39,294.2 **17.** 1,747.293 **19.** 64.635 **21.** 15,201.9 **23.** $5.10

page 132

Getting Ready **1.** 49.538 **2.** 1337.625 **3.** $18.672 or $18.67 **4.** 41.748 **5.** $6.36 **6.** $4.13
Exercises **1.** 3.105 **3.** 35.28 **5.** 517.92 **7.** $2.0493 or $2.05 **9.** $20.706 or $20.71 **11.** 74.52 **13.** 1.653 **15.** $8.874 or $8.87 **17.** 8.932 **19.** 66.88 **21.** $3,762.50 **23.** 0.4836 **25.** 43.47 **27.** $29.68 **29.** 0.0532 **31.** 131.1197 **33.** $89.10 **35.** $17.94 **37.** $279.96

page 134

Getting Ready **1.** 72; 720; 7,200 **2.** 6.5; 65; 650 **3.** $53.70; $537; $5,370 **4.** 60.08; 600.8; 6,008 **5.** 45.7 **6.** 34.6 **7.** 6,531 **8.** 854.3 **9.** 420 **10.** 5,400 **11.** 67.5 **12.** 5,135
Exercises **1.** 1.23; 12.3; 123 **3.** 0.3; 3; 30 **5.** 432.46; 4,324.6; 43,246 **7.** $18.30; $183; $1,830 **9.** 54.7 **11.** 5,613 **13.** 270 **15.** 70.5 **17.** 6,128 **19.** 9,251.8 **21.** $229

page 136

Getting Ready **1.** 191.8; 191.84 **2.** 0.5; 0.50 **3.** 55.9; 55.90
Exercises **1.** 5.8 **3.** 19.4 **5.** 108.7 **7.** 119.43 **9.** 32.80

page 143

Getting Ready **1.** 19.3 **2.** 5.80 **3.** 7.21 **4.** 89.2
Exercises **1.** 15.8 **3.** 145.6 **5.** 25.896 **7.** 1452 **9.** 330 **11.** 26.52 **13.** 978.5

page 144

Getting Ready **1.** 2.1 **2.** 1.09 **3.** 2.3 **4.** 8.46 **5.** 0.65 **6.** 0.45
Exercises **1.** 0.7 **3.** 2.07 **5.** 0.2125 **7.** 3.06 **9.** $3.04 **11.** 2.004 **13.** 0.192 **15.** $0.23 **17.** 1.25 **19.** $6.75 **21.** $17.88

page 146

Getting Ready **1.** 576 **2.** 3.8 **3.** 32 **4.** 360 **5.** 3.7 **6.** 430
Exercises **1.** 81 **3.** 10.7 **5.** 30 **7.** 510 **9.** 161.1 **11.** 290 **13.** 3 **15.** 25.3 **17.** 6.34 **19.** 5.32 **21.** 5,000 **23.** 2 **25.** 694 **27.** 1.5 **29.** 15.5 **31.** 211

page 148

Getting Ready **1.** 0.678; 0.0678; 0.00678 **2.** 0.7654; 0.07654; 0.007654 **3.** 0.0465; 0.00465; 0.000465 **4.** 0.087 **5.** 6.85 **6.** 0.9543 **7.** 0.6486 **8.** 0.456 **9.** 53.822 **10.** 0.546 **11.** 9.56
Exercises **1.** 9.27 **3.** 0.0735 **5.** 0.09 **7.** 0.0078 **9.** 0.495 **11.** 9.756 **13.** 0.000776 **15.** 0.2376 **17.** 47.29 **19.** 5 **21.** 80 **23.** 59.86 **25.** 8.8 **27.** 6.752

page 150

Getting Ready **1.** 244.7; 244.66 **2.** 66.8; 66.79 **3.** 2.4; 2.41
Exercises **1.** 33.7 **3.** 9.5 **5.** 14.0 **7.** 3.27 **9.** 3.09

page 159

Getting Ready **1.** 3 cm **2.** 4 mm
Exercises **1.** 8 km **3.** 1 mm **5.** 4 m

page 160

Getting Ready **1.** 93 mm, 9.3 cm **2.** 78 mm, 7.8 cm **3.** 70 mm, 7.0 cm **4.** 83 mm, 8.3 cm **5.** 99 mm, 9.9 cm **Exercises** **1.** 84 mm, 8.4 cm **3.** 133 mm, 13.3 cm **5.** 97 mm, 9.7 cm **7.** 67 mm, 6.7 cm **9.** 75 mm, 7.5 cm **15.** 35 mm, 3.5 cm **17.** 49 mm, 4.9 cm **19.** 60 mm, 6.0 cm **21.** 82 mm, 8.2 cm **23.** 72 mm, 7.2 cm

page 162

Getting Ready **1.** 2.7 cm, 2.8 cm. 3.7 cm, 2.7 cm, P = 11.9 cm **2.** 2.7 cm, 3.6 cm, 4.6 cm, P = 10.9 cm **3.** 3.2 cm, 1.0 cm, 4.2 cm, 2.0 cm, 2.5 cm, P = 12.9 cm **4.** 1.5 cm, 3.9 cm, 1.8 cm, 2.6 cm, P = 9.8 cm **Exercises** **1.** 2.7 cm, 3.0 cm, 4.1 cm, 2.6 cm, P = 12.4 cm **3.** 2.1 cm, 1.3 cm, 3.2 cm, 2.1 cm, 2.0 cm, P = 10.7 cm **5.** 5.8 cm, 2.9 cm, 5.8 cm, 2.9 cm, P = 17.4 cm **7.** 2.4 cm, 1.8 cm, 1.6 cm, 3.6 cm, 2.6 cm, 3.3 cm, P = 15.3 cm **9.** Perimeter is doubled

page 165

Getting Ready **1.** 6 cm^2 **2.** 10 cm^2 **3.** 3.5 cm^2 **Exercises** **1.** 8 cm^2 **3.** 9.5 cm^2 **5.** 8.5 cm^2 **7.** 12 cm^2

page 166

Getting Ready **1.** 800 **2.** 24 **3.** 0.8 **4.** 5,420 **5.** 82,400 **6.** 83.28 **Exercises** **1.** 2,000 **3.** 7,200 **5.** 3,000,000 **7.** 2,890,000 **9.** 7.5 **11.** 1.1 **13.** 35 **15.** 45,000 **17.** 1.7 **19.** 0.365 **21.** 975,000 **23.** 56,800,000,000

page 168

Getting Ready **1.** 2 L **2.** 250 mL **Exercises** **1.** 19 L **3.** 946 mL **5.** 250 L **7.** 4,000 mL

page 169

Getting Ready **1.** 16,000 **2.** 9.867 **3.** 0.00872 **Exercises** **1.** 5,000 **3.** 12,000 **5.** 5.24 **7.** 9,500 **9.** 490 **11.** 0.099 **13.** 0.00209 **15.** 0.0055

page 171

Getting Ready **1.** 500 mg **2.** 1 kg **Exercises** **1.** 1 kg **3.** 85 mg **5.** 25 kg **7.** 2 g

page 172

Getting Ready **1.** 72,000 **2.** 52,000 **3.** 876 **4.** 4 **5.** 650 **6.** 0.00158 **7.** 1.832 **8.** 0.228 **Exercises** **1.** 24,000 **3.** 1 **5.** 3,000 **7.** 0.1 **9.** 743 **11.** 0.001 **13.** 1,480 **15.** 1.742 **17.** 310 **19.** 1.346 **21.** 0.473 **23.** 0.00374 **25.** 0.000004872

page 174

Getting Ready **1.** Too cold **2.** Too hot **3.** Comfortable **4.** Too cold **5.** 15°C **6.** 10°C **7.** 55°C **Exercises** **1.** Comfortable **3.** Too cold **5.** 0°C **7.** −1°C **9.** Dropped 20°C **11.** Rose 6°C **13.** Rose 14°C **15.** 16°C

page 177

Getting Ready **1.** 3,520 **2.** 9 **3.** 48 **4.** 8 **5.** 192 **6.** 4,000 **Exercises** **1.** 5,280 **3.** 4 **5.** 4 **7.** 5 **9.** 32 **11.** 16 **13.** 2 **15.** 240 **17.** 80 **19.** $2\frac{1}{2}$ **21.** 37,889.28

page 187

Getting Ready **1.** $\frac{4}{6}$ or $\frac{2}{3}$ **2.** $\frac{5}{9}$ **3.** $\frac{5}{8}$ **4.** $\frac{3}{6}$ or $\frac{1}{2}$ **5.** $\frac{2}{3}$ **Exercises** **1.** $\frac{2}{7}$ **3.** $\frac{4}{6}$ or $\frac{2}{3}$ **5.** $\frac{3}{5}$

page 189

Getting Ready **5.** $\frac{4}{6}, \frac{2}{3}$ **6.** $\frac{8}{10}, \frac{4}{5}$ **7.** True **8.** True **9.** False **10.** True **Exercises** **5.** $\frac{2}{12}, \frac{1}{6}$ **7.** True **9.** False **11.** True **13.** False **15.** True **17.** True **19.** False **21.** True

page 191

Getting Ready **1.** $\frac{2}{6}, \frac{3}{9}, \frac{4}{12}, \frac{5}{15}$ **2.** $\frac{4}{10}, \frac{6}{15}, \frac{8}{20}, \frac{10}{25}$ **3.** $\frac{6}{8}, \frac{9}{12}, \frac{12}{16}, \frac{15}{20}$ **4.** $\frac{2}{20}, \frac{3}{30}, \frac{4}{40}, \frac{5}{50}$ **5.** 3 **6.** 15 **7.** 35 **8.** 30 **9.** 70 **10.** $\frac{8}{40}, \frac{4}{20}, \frac{2}{10}, \frac{1}{5}$ **11.** $\frac{10}{50}, \frac{5}{25}, \frac{4}{20}, \frac{2}{10}$ **12.** $\frac{20}{60}, \frac{10}{30}, \frac{4}{12}, \frac{1}{3}$ **13.** $\frac{24}{36}, \frac{16}{24}, \frac{12}{18}, \frac{6}{9}$ **14.** 9 **15.** 1 **16.** 5 **17.** 7 **18.** 4

Exercises **1.** $\frac{2}{12}, \frac{3}{18}, \frac{4}{24}, \frac{5}{30}$ **3.** $\frac{14}{20}, \frac{21}{30}, \frac{28}{40}, \frac{35}{50}$ **5.** 9 **7.** 18 **9.** 9 **11.** 24 **13.** 15 **15.** $\frac{24}{32}, \frac{12}{16}, \frac{6}{8}, \frac{3}{4}$ **17.** $\frac{30}{50}, \frac{15}{25}, \frac{6}{10}, \frac{3}{5}$ **19.** 1 **21.** 2 **23.** 5 **25.** 5 **27.** 4 **29.** 8

page 192

Getting Ready **1.** $\frac{2}{3}$ **2.** $\frac{1}{3}$ **3.** $\frac{1}{3}$ **4.** $\frac{1}{2}$ **5.** $\frac{3}{4}$ **6.** $\frac{5}{8}$ **7.** $\frac{4}{5}$ **8.** $\frac{3}{7}$
Exercises **1.** $\frac{1}{3}$ **3.** $\frac{1}{4}$ **5.** $\frac{8}{15}$ **7.** $\frac{5}{7}$ **9.** $\frac{2}{5}$ **11.** $\frac{2}{3}$ **13.** $\frac{3}{5}$ **15.** $\frac{2}{7}$ **17.** $\frac{1}{4}$ **19.** $\frac{5}{7}$ **21.** $\frac{3}{4}$ **23.** $\frac{1}{2}$ **25.** $\frac{3}{8}$ **27.** $\frac{10}{11}$ **29.** $\frac{6}{7}$ **31.** $\frac{3}{4}$

page 194

Getting Ready **1.** $>$ **2.** $<$ **3.** $>$ **4.** $<$ **Exercises** **1.** $>$ **3.** $<$ **5.** $>$ **7.** $>$ **9.** $<$ **11.** $>$ **13.** $>$ **15.** $<$

page 204

Getting Ready **2.** $\frac{2}{3}$ **3.** $\frac{1}{2}$ **4.** 1 **5.** $\frac{5}{8}$ **Exercises** **5.** $\frac{7}{9}$ **7.** $\frac{15}{17}$ **9.** $\frac{4}{5}$ **11.** $\frac{7}{8}$ **13.** $\frac{1}{2}$ **15.** 1 **17.** $\frac{4}{5}$ **19.** 1 **21.** $\frac{4}{7}$ **23.** $\frac{11}{12}$ **25.** $\frac{7}{8}$ **27.** $\frac{4}{5}$ **29.** $\frac{3}{8}$ in.

page 207

Getting Ready **2.** $\frac{3}{11}$ **3.** 0 **4.** $\frac{2}{3}$ **5.** $\frac{3}{5}$ **Exercises** **5.** $\frac{6}{11}$ **7.** $\frac{1}{6}$ **9.** 0 **11.** $\frac{2}{5}$ **13.** $\frac{2}{11}$ **15.** $\frac{1}{3}$ **17.** $\frac{3}{7}$ **19.** $\frac{2}{3}$ **21.** $\frac{1}{2}$ **23.** $\frac{2}{5}$ **25.** $\frac{1}{10}$ **27.** $\frac{1}{5}$ mi

page 208

Getting Ready **1.** $4\frac{1}{2}$ **2.** $1\frac{6}{7}$ **3.** $4\frac{5}{6}$ **4.** $\frac{8}{5}$ **5.** $\frac{26}{7}$ **6.** $\frac{159}{8}$ **Exercises** **1.** $1\frac{3}{5}$ **3.** $9\frac{1}{2}$ **5.** $2\frac{1}{9}$ **7.** $\frac{5}{3}$ **9.** $\frac{17}{7}$ **11.** $\frac{83}{10}$

page 209

Getting Ready **1.** $1\frac{2}{5}$ **2.** $1\frac{1}{2}$ **3.** $2\frac{3}{4}$ **4.** $7\frac{7}{9}$ **5.** $9\frac{4}{15}$ **Exercises** **1.** $1\frac{1}{5}$ **3.** $11\frac{1}{2}$ **5.** $3\frac{1}{5}$ **7.** $12\frac{1}{4}$ **9.** $14\frac{1}{2}$

page 211

Getting Ready **1.** $\frac{3}{9}, \frac{4}{9}$ **2.** $\frac{15}{20}, \frac{2}{20}$ **3.** $\frac{15}{24}, \frac{14}{24}$ **4.** $\frac{2}{6}, \frac{5}{6}$ **5.** $\frac{6}{8}, \frac{5}{8}$ **6.** $\frac{5}{35}, \frac{21}{35}$ **7.** $\frac{4}{20}, \frac{15}{20}, \frac{2}{20}$ **8.** $\frac{20}{60}, \frac{45}{60}, \frac{16}{60}$ **Exercises** **1.** $\frac{2}{4}, \frac{1}{4}$ **3.** $\frac{5}{10}, \frac{4}{10}$ **5.** $\frac{3}{12}, \frac{2}{12}$ **7.** $\frac{5}{15}, \frac{9}{15}$ **9.** $\frac{5}{20}, \frac{14}{20}$ **11.** $\frac{4}{8}, \frac{5}{8}$ **13.** $\frac{6}{10}, \frac{7}{10}$ **15.** $\frac{5}{15}, \frac{8}{15}$ **17.** $\frac{8}{10}, \frac{7}{10}$ **19.** $\frac{12}{16}, \frac{5}{16}$ **21.** $\frac{4}{12}, \frac{3}{12}, \frac{1}{12}$ **23.** $\frac{21}{24}, \frac{16}{24}, \frac{18}{24}$ **25.** $\frac{6}{42}, \frac{9}{42}, \frac{10}{42}$ **27.** Tuesday

page 213

Getting Ready **1.** $\frac{11}{12}$ **2.** $1\frac{1}{9}$ **3.** $\frac{19}{90}$ **4.** $1\frac{1}{8}$ **5.** $3\frac{13}{15}$ **6.** $10\frac{11}{12}$
Exercises **1.** $\frac{13}{15}$ **3.** $1\frac{17}{30}$ **5.** $\frac{17}{30}$ **7.** $9\frac{37}{40}$ **9.** $4\frac{7}{8}$ **11.** $10\frac{11}{12}$ **13.** 20 **15.** $9\frac{79}{100}$ **17.** $20\frac{11}{40}$ **19.** $3\frac{1}{4}$ yd

page 215

Getting Ready **1.** $\frac{7}{12}$ **2.** $1\frac{1}{10}$ **3.** $1\frac{3}{4}$ **4.** $3\frac{1}{2}$ **5.** $5\frac{1}{2}$ **6.** $3\frac{7}{12}$ **Exercises** **1.** $\frac{3}{8}$ **3.** $\frac{1}{15}$ **5.** $3\frac{4}{9}$ **7.** $3\frac{2}{3}$ **9.** $2\frac{5}{6}$ **11.** $4\frac{2}{5}$ **13.** $5\frac{1}{4}$ **15.** $4\frac{1}{6}$ **17.** $5\frac{1}{10}$ **19.** $5\frac{1}{2}$ **21.** $1\frac{1}{2}$ **23.** $4\frac{1}{3}$ **25.** $2\frac{3}{4}$ **27.** $5\frac{7}{24}$ **29.** $12\frac{9}{10}$ **31.** $76\frac{1}{4}$ ft

page 223

Getting Ready **1.** $1\frac{1}{2}$ **2.** $1\frac{2}{3}$ **3.** $3\frac{1}{3}$ **4.** 2 **5.** $1\frac{1}{5}$ **Exercises** **1.** $3\frac{3}{4}$ **3.** $1\frac{1}{5}$ **5.** $1\frac{3}{5}$ **7.** $4\frac{1}{2}$ **9.** $4\frac{1}{2}$ **11.** $2\frac{4}{5}$ **13.** $2\frac{1}{4}$ **15.** 2 **17.** 6 **19.** 2 **21.** $\frac{1}{2}$ in.

page 224

Getting Ready **1.** $\frac{5}{12}$ **2.** $\frac{3}{20}$ **3.** $\frac{1}{5}$ **4.** $\frac{2}{5}$ **5.** $\frac{1}{2}$ **6.** $\frac{5}{16}$ **7.** $\frac{1}{6}$ **8.** $\frac{5}{8}$ **9.** $\frac{1}{2}$ **10.** $\frac{18}{35}$ **Exercises** **1.** $\frac{1}{10}$ **3.** $\frac{3}{10}$ **5.** $\frac{7}{20}$ **7.** $\frac{1}{10}$ **9.** $\frac{5}{16}$ **11.** $\frac{5}{8}$ **13.** $\frac{5}{18}$ **15.** $\frac{1}{27}$ **17.** $\frac{1}{12}$ **19.** $\frac{2}{5}$ **21.** $\frac{25}{36}$ **23.** $\frac{5}{21}$ **25.** $\frac{2}{27}$ **27.** $\frac{1}{3}$ lb

page 226

Getting Ready **1.** $2\frac{1}{2}$ **2.** $1\frac{2}{7}$ **3.** $14\frac{1}{4}$ **44.** 26 **Exercises** **1.** $1\frac{3}{5}$ **3.** $1\frac{31}{32}$ **5.** $\frac{27}{35}$ **7.** $1\frac{19}{21}$ **9.** $16\frac{1}{2}$ **11.** $9\frac{1}{3}$ **13.** $2\frac{2}{5}$ **15.** $1\frac{1}{5}$ **17.** $2\frac{4}{5}$ **19.** $25\frac{2}{3}$ **21.** $10\frac{1}{8}$ **23.** $2\frac{2}{3}$ **25.** $4\frac{1}{2}$ **27.** $1\frac{3}{4}$ **29.** $7\frac{1}{2}$ **31.** $9\frac{4}{5}$ **33.** $4\frac{3}{8}$c **35.** 147 mi

page 228

Getting Ready **1.** $\frac{1}{7}$ **2.** $\frac{9}{2}$ **3.** $\frac{5}{6}$ **4.** $\frac{3}{17}$ **5.** $\frac{3}{10}$ **6.** $\frac{8}{15}$ **Exercises** **1.** $\frac{16}{3}$ **3.** $\frac{7}{3}$ **5.** $\frac{11}{2}$ **7.** $\frac{23}{6}$ **9.** $\frac{1}{12}$ **11.** $\frac{5}{18}$ **13.** $\frac{8}{13}$ **15.** $\frac{1}{33}$ **17.** $\frac{4}{9}$

page 229

Getting Ready **1.** $\frac{5}{48}$ **2.** $\frac{7}{50}$ **3.** $\frac{3}{40}$ **4.** $\frac{23}{112}$ **5.** $\frac{19}{40}$ **Exercises** **1.** $\frac{7}{80}$ **3.** $\frac{1}{6}$ **5.** $\frac{1}{9}$ **7.** $\frac{11}{100}$ **9.** $\frac{11}{21}$ **11.** $\frac{9}{100}$ **13.** $1\frac{13}{15}$ **15.** $1\frac{8}{15}$ **17.** $\frac{19}{36}$ **19.** $\frac{7}{30}$ **21.** $\frac{11}{12}$ yd or 33 in.

page 231

Getting Ready **1.** $\frac{9}{16}$ **2.** $\frac{7}{12}$ **3.** $9\frac{3}{5}$ **4.** $5\frac{5}{9}$ **5.** $\frac{7}{10}$ **6.** $\frac{1}{14}$ **7.** $4\frac{2}{3}$ **8.** $1\frac{7}{93}$ **Exercises** **1.** $\frac{5}{9}$ **3.** $\frac{35}{48}$ **5.** $\frac{1}{16}$ **7.** $\frac{1}{2}$ **9.** $\frac{4}{7}$ **11.** $1\frac{1}{15}$ **13.** $\frac{1}{9}$ **15.** 6 **17.** $\frac{1}{4}$ **19.** $\frac{5}{48}$ **21.** $\frac{1}{5}$ **23.** 2 **25.** $1\frac{7}{29}$ **27.** 5 **29.** 40

page 232

Getting Ready **1.** 19 **2.** 54 **3.** $\frac{39}{56}$ **4.** $1\frac{1}{2}$ **Exercises** **1.** 18 **3.** 16 **5.** 110 **7.** $52\frac{1}{2}$ **9.** $4\frac{7}{12}$ **11.** $2\frac{2}{5}$ **13.** $\frac{26}{45}$ **15.** $\frac{63}{80}$ **17.** 18 **19.** $4\frac{5}{32}$ **21.** \$40 **23.** \$6 **25.** 8 **27.** $\frac{17}{40}$ **29.** 420 **31.** 783 **33.** 8 **35.** $3\frac{3}{8}$ mi

page 235

Getting Ready **1.** $\frac{2}{5}$ **2.** $\frac{1}{4}$ **3.** $1\frac{3}{10}$ **4.** $1\frac{1}{25}$ **5.** $2\frac{13}{20}$ **6.** 0.5 **7.** 0.6 **8.** 0.833 **9.** 3.9 **10.** 2.667

Exercises **1.** $\frac{9}{10}$ **3.** $\frac{16}{25}$ **5.** $\frac{7}{10}$ **7.** $\frac{5}{8}$ **9.** $\frac{3}{4}$ **11.** $\frac{17}{100}$ **13.** $1\frac{1}{2}$ **15.** $4\frac{3}{50}$ **17.** $3\frac{3}{25}$ **19.** $2\frac{1}{8}$ **21.** 0.25 **23.** 0.375 **25.** 0.625 **27.** 0.1 **29.** 0.125 **31.** 0.06 **33.** 0.18 **35.** 0.405 **37.** 3.75 **39.** 2.286 **41.** 35

Page 243

Getting Ready **1.** 11 **2.** 9 **3.** 9 **4.** $\frac{3}{10}$ **5.** 63 **6.** \$35.20 **7.** 7 **8.** 9 **9.** $\frac{1}{7}$ **Exercises** **1.** 13 **3.** $7\frac{3}{4}$ **5.** 0.9 **7.** \$235 **9.** 24 **11.** 8 **13.** 121.5 **15.** 32 **17.** 8 **19.** 32 **21.** 14

Page 245

Getting Ready **1.** 46 **2.** 485 **3.** $\frac{1}{3}$ **4.** $192\frac{1}{2}$ **5.** \$1.20 **6.** 1.1
Exercises **1.** 5,160 **3.** \$187 **5.** 3.72 **7.** 7 **9.** 121 **11.** 17.4

page 247

Getting Ready **1.** $z = 21 + 19$ **2.** $w = 3.9 - 1.7$ **3.** $m = 50 \div 4$ **4.** $r = 6.5 \times 1.1$ **5.** $d = \$40.20 - \$19.95$ **6.** $f = 30 \div \frac{2}{3}$ **7.** $n = 65 \times 5$ **8.** $s = \frac{7}{2} \div 2$ **9.** $a = 3.62 + 1.89$
Exercises **1.** $s = 21 - 9$ **3.** $u = 8.3 - 5.4$ **5.** $r = 72 \div 8$ **7.** $m = 1.1 \times 3.3$ **9.** $p = \$18.80 + \$20.12$ **11.** $z = 3.8 \times 10$ **13.** $q = 6\frac{1}{3} \div 5$ **15.** $p = \frac{7}{8} - \frac{3}{4}$ **17.** $y = 3 \div 18$

page 249

Getting Ready **1.** 38 **2.** 28 **3.** 57 **4.** 1.3 **5.** $2\frac{3}{4}$ **6.** \$62.39
Exercises **1.** 4 **3.** 35 **5.** 3.4 **7.** \$5.25 **9.** $15\frac{5}{6}$ **11.** $4\frac{1}{3}$ **13.** \$102.75 **15.** $3\frac{3}{4}$ **17.** -271

page 251

Getting Ready **1.** 6 **2.** 175 **3.** \$83.36 **4.** 154
Exercises **1.** 22 **3.** 56 **5.** 3,380 **7.** 3 **9.** \$143 **11.** 480 **13.** 7,602 **15.** $\frac{5}{3}$ **17.** $\frac{1}{2}$

page 253

Getting Ready **1.** $3w$ **2.** $s + \$500$ **3.** $\frac{1}{2}p$ or $\frac{p}{2}$ **4.** $p - \$70$
Exercises **1.** $2n$ **3.** $\frac{a}{3}$ or $\frac{1}{3}a$ **5.** $s + \$50$ **7.** $h - 2$ **9.** $1\frac{1}{2}s$ or $s + \frac{1}{2}s$

page 254

Getting Ready **1.** $c = 15 + 1.50y$ **2.** $A = \frac{x + y}{2}$ or $A = \frac{1}{2}(x + y)$
Exercises **1.** $T = P + I$ **3.** $p = r + 0.1s$ **5.** $E = \frac{9r}{p}$ **7.** \$112

page 257

Getting Ready **1.** \$6/h **2.** 41°F **3.** $f = h + s$; 205
Exercises **1.** \$8/h **3.** 32°F **5.** 5h **7.** 20*s*

page 262

Getting Ready **1.** $\frac{12}{8}$ **2.** $\frac{2}{16}$ **3.** $\frac{500}{9}$ **4.** $\frac{2}{5}$ **Exercises** **1.** $\frac{300}{8}$ **3.** $\frac{1{,}000}{2{,}000}$ **5.** $\frac{8}{19}$ **7.** $\frac{6}{100}$ or $\frac{0.06}{1.00}$ **9.** $\frac{30}{41}$ **11.** $\frac{19}{2}$ **13.** $\frac{2}{1}$ **15.** $\frac{8}{3}$ **17.** $\frac{21}{59}$ **19.** $\frac{50}{9}$

page 264

Getting Ready **1.** 2 to 5 **2.** $\frac{20}{9}$ **3.** 3 to 4 **4.** $\frac{17}{20}$ **Exercises** **1.** 1 to 4 **3.** 2 to 3 **5.** $\frac{5}{18}$ **7.** $\frac{3}{7}$ **9.** $\frac{2}{5}$

page 266

Getting Ready **1.** 0.625 **2.** 0.727 **3.** 0.444 **4.** 0.188 **5.** 0.325
Exercises **1.** Edith **3.** Sue **5.** Tom **7.** OJ **9.** Lou

page 269

Getting Ready **1.** Walters **2.** B **3.** 4 prs. for \$8.75 **4.** 32 oz for \$1.28 **Exercises** **1.** Eagles **3.** 204 km in 4 h **5.** 6 pencils for \$0.60 **7.** Joan **9.** Teri; Phil

page 271

Getting Ready **1.** $\frac{1}{4}$ **2.** $\frac{1}{6}$ **3.** $\frac{1}{6}$ **4.** $\frac{1}{3}$ **Exercises** **1.** $\frac{1}{2}$ **3.** $\frac{1}{2}$ **5.** $\frac{5}{9}$; $\frac{1}{900}$ **7.** $\frac{1}{4}$

page 273

Getting Ready **1.** Yes **2.** No **3.** Yes **4.** 8 **5.** 3 **6.** 9 **Exercises** **1.** Yes **3.** Yes **5.** 8 **7.** $7\frac{1}{2}$ or 7.5 **9.** 20 **11.** 150 kg

page 281

Getting Ready **1.** Right, 90° **2.** Obtuse, 129° **3.** Acute, 31° **4.** Acute, 57° **Exercises** **1.** Obtuse, 123° **3.** Acute, 19°

page 282

Getting Ready **1.** Pentagon **2.** Triangle **3.** Hexagon **Exercises** **1.** Triangle **3.** Hexagon **5.** Pentagon

page 285

Getting Ready **1.** Right, isosceles **2.** Obtuse, scalene **3.** Acute, equilateral **Exercises** **1.** Acute **3.** Right **5.** Scalene **7.** True

page 286

Getting Ready **1.** Rectangle **2.** Parallelogram **3.** Square **Exercises** **1.** Parallelogram **3.** Square

page 289

Getting Ready **1.** 24 cm **2.** 14 cm **3.** 19 cm **Exercises** **1.** 88 mm **3.** 27.8 **5.** 12.9 cm **7.** 446 cm **9.** 8.6 m

page 291

Getting Ready **1.** 10.92 ft^2 **2.** 9 cm^2 **3.** 57.4 m^2 **4.** 525 mm^2 **Exercises** **1.** 2,460 mm^2 **3.** 56.25 cm^2 **5.** 336 ft^2 **7.** 13.5 cm^2 **9.** 54 cm^2 **11.** 1,060 cm^2 **13.** 468 cm^2 **15.** 4,500 cm or 45 m

page 292

Getting Ready **1.** 10 cm^2 **2.** 1.44 ft^2 **3.** 11.1 cm^2 **4.** 525 mm^2 **Exercises** **1.** 132 sq. units **3.** 1,248 sq. units **5.** 3.64 sq. units **7.** 132 mm^2 **9.** 140 in^2. **11.** 63.75 m^2 **13.** $7\frac{1}{2}$ m^2 **15.** 0.0025 m^2

page 295

Getting Ready **1.** 12 ft^2 **2.** 9.5 cm^2 **3.** 9 cm^2 **Exercises** **1.** 12.25 cm^2 **3.** 13 cm^2 **5.** 9.5 cm^2 **7.** 1,350 mm^2 **9.** 4.2 cm^2

page 297

Getting Ready **1.** 37.68 cm **2.** 35.168 in. **3.** 87.92 mm **4.** 43.96 m **5.** 21.98 cm **6.** 373.66 km **Exercises** **1.** 37.68 cm **3.** 75.36 ft **5.** 351.68 in. **7.** 439.6 mm **9.** 314 m **11.** 31.4 m **13.** 6.28 m

page 298

Getting Ready **1.** 452.16 mm^2 **2.** 176.625 ft^2 **3.** 254.34 cm^2 **4.** 1,384.74 m^2 **5.** 803.84 in^2. **6.** 124.6266 cm^2
Exercises **1.** 452.16 $in.^2$ **3.** 1,808.64 yd^2 **5.** 254.34 m^2 **7.** 18.0864 cm^2 **9.** 0.1256 m^2 **11.** 132.665 mm^2 **13.** 301.5656 m^2 **15.** 13.8474 m^2 **17.** 254.34 mm^2 **19.** 452.16 mm^2 **21.** 28.5 cm^2

page 301

Getting Ready **1.** 64 ft^3 **2.** 384 m^3 **3.** 5.184 m^3 **4.** 84 cm^3
Exercises **1.** 336 ft^3 **3.** 48 $in.^3$ **5.** 1,188 $in.^3$ **7.** 288 m^3 **9.** 24.192 m^3 **11.** 8.448 m^3 **13.** B **15.** 6 cm

page 309

Getting Ready **1.** R: 50, M: 45 **2.** R: 47, M: 53
Exercises **1.** R: 21, M: 39 **3.** R: 126, M: 222

page 310

Getting Ready **1.** 75 **2.** 1.645 **Exercises** **1.** 250 **3.** 39.9 **5.** $29.00 **7.** R: 0.27, Med: 1.69, M: 1.64 **9.** R: 30, Med: 95, M: 79

page 312

Getting Ready **1.** 112.6 **2.** $3.31 **3.** 3.75 **Exercises** **1.** 85.5 **3.** $32.10 **5.** 26.1

page 320

Getting Ready **1.** $\frac{28}{100}$ **2.** $\frac{85.3}{100}$ **3.** $\frac{205}{100}$ **4.** $\frac{0.5}{100}$ **Exercises** **1.** $\frac{45}{100}$ **3.** $\frac{100}{100}$ **5.** $\frac{24.6}{100}$ **7.** $\frac{3}{100}$ **9.** $\frac{1}{100}$ **11.** $\frac{1.8}{100}$ **13.** $\frac{3}{100}$

page 323

Getting Ready **1.** $\frac{17}{100}$ **2.** $\frac{1}{2}$ **3.** $\frac{1}{6}$ **4.** $\frac{47}{100}$ **5.** $2\frac{1}{5}$ **6.** 4
Exercises **1.** $\frac{1}{4}$ **3.** $\frac{5}{8}$ **5.** $\frac{3}{5}$ **7.** $\frac{7}{20}$ **9.** $\frac{12}{25}$ **11.** $\frac{5}{6}$ **13.** $1\frac{1}{4}$ **15.** $\frac{1}{100}$ **17.** $\frac{1}{5}$ **19.** 1 **21.** $\frac{1}{20}$ **23.** $\frac{61}{200}$

page 325

Getting Ready **1.** 0.25 **2.** 0.08 **3.** 1.32 **4.** 0.001 **5.** 0.125 **6.** 0.863
Exercises **1.** 0.10 **3.** 0.04 **5.** 0.1225 **7.** 5 **9.** 0.02 **11.** 3.75 **13.** 0.009 **15.** $0.33\frac{1}{3}$ **17.** 0.08 **19.** 0.39 **21.** 3.90 **23.** 0.048 **25.** 0.035 **27.** 0.006 **29.** 3.25

page 326

Getting Ready **1.** 78% **2.** 30% **3.** 453% **4.** 180% **5.** 9%
Exercises **1.** 7% **3.** 38% **5.** 70% **7.** 10% **9.** 17.6% **11.** 150%

page 329

Getting Ready **1.** 75% **2.** 20% **3.** 87.5% **4.** 11.1% **5.** 380%
Exercises **1.** 40% **3.** 30% **5.** 90% **7.** 49% **9.** 125% **11.** 420% **13.** 330% **15.** 250% **17.** 140% **19.** 8% **21.** 175% **23.** 2.5% **25.** 200%

page 331

Getting Ready **1.** 25; 5; 20 **2.** 21 **3.** $34.23 **4.** 9.6 **5.** $831.17
Exercises **1.** 24 **3.** 35.2 **5.** $24 **7.** $6.90 **9.** 744 **11.** $34.25 **13.** $473

page 333

Getting Ready **1.** 73.3% **2.** 30% **3.** 42.9% **4.** 200%
Exercises **1.** 50% **3.** 25% **5.** 80% **7.** 25% **9.** 33.3% **11.** 4% **13.** 29%

page 334

Getting Ready **1.** 52 **2.** $58.00 **3.** 110 **4.** 55 **5.** $520
Exercises **1.** 20 **3.** 120 **5.** 200 **7.** 20 **9.** 64 **11.** 24 **13.** $5,500 **15.** $130 **17.** $1,825 **19.** 76 played; 19 lost

page 337

Getting Ready **1.** 25% **2.** 12.7% **3.** 100% **4.** 50% **5.** 125%
Exercises **1.** 55% **3.** 40% **5.** 20.8% **7.** 20% **9.** 33.3%

page 339

Getting Ready **1.** $1.50 **2.** $3.10 **3.** 18 **4.** 30 **5.** $4.05 **6.** $7.00
Exercises **1.** $0.57 **3.** 97 **5.** $17.20 **7.** $0.09 **9.** $1.25 **11.** $2.00 **13.** $4 **15.** $1.80

page 345

Getting Ready **1.** 58 h **2.** 52 h **3.** $56\frac{1}{2}$ h **4.** 35 h **5.** $231.80 **6.** $735 **Exercises** **1.** 38 h **3.** 49 h **5.** 73 h **7.** $53\frac{1}{2}$ h **9.** $62\frac{1}{2}$ h **11.** $123.50 **13.** $241.50 **15.** $172.96 **17.** $382.70 **19.** $656.25 **21.** $200.25

page 346

Getting Ready **1.** $585 **2.** $368.20 **3.** $460 **4.** $240
Exercises **1.** $78 **3.** $137.50 **5.** $351.80 **7.** $529.90 **9.** $6,500

page 348

Getting Ready **1.** $1,695 **2.** $2,550 **3.** $2,580 **Exercises** **1.** $348 **3.** $1,450 **5.** $892.20 **7.** $187.50 **9.** $1,120 **11.** $734

page 353

Getting Ready **1.** $72.27 **2.** $223.99 **Exercises** **1.** $156.06 **3.** $173.61 **5.** $120.58

page 355

Getting Ready **1.** TD: $69.19 NP: $100.81 **2.** NP: $74.05
Exercises **1.** TD: $75.08 NP: $130.92 **3.** GP: $224 NP: $141.54

page 360

Getting Ready **1.** $667.54 **2.** $184.31
Exercises **1.** 304090504 **3.** 3 **5.** $225.00 **7.** $385.70 **9.** $10.55 **11.** $763.41

page 363

Getting Ready **1.** May 4, 1984 **2.** $48.98 **3.** $115.00 **4.** $835.69 **5.** May 8 **6.** $822.89; $707.89
Exercises **7.** $160.83 **9.** $652.85

page 364

Getting Ready **1.** $485.00 **2.** $574.89 **Exercises** **1.** $479.18 **3.** $347.66 **5.** $400.05 **7.** $317.40 **9.** Yes, by $1.99 **11.** $36

page 367

Getting Ready **1.** $150 **2.** $63.75 **3.** $19.50 **4.** $2,370 **5.** $540 **6.** $663 **Exercises** **1.** $105 **3.** $39.90 **5.** $1,065 **7.** $1,428.75 **9.** $575 **11.** $2,500 **13.** $70 **15.** 9.5%

page 369

Getting Ready **1.** 12 **2.** $114.49 **3.** $66.04 **4.** $14.49; $604 **Exercises** **1.** $561.80 **3.** $12.24 **5.** $2.53 **7.** $101.26

page 379

Getting Ready **1.** $5.28 **2.** $128 **3.** $5.78 **4.** $430 **Exercises** **1.** $426 **3.** $5.36 **5.** $400.93 **7.** $1.80 **9.** $5.78 **11.** $4.20

page 381

Getting Ready **1.** $79.90 **2.** Store B, $82.99 **3.** $410 and 5% off **Exercises** **1.** Store A **3.** Store A **5.** Store A **7.** $11.38 at Flo's

page 382

Getting Ready **1.** $780 **2.** $91 **3.** $875 **4.** $107.50 **Exercises** **1.** $40 **3.** $58 **5.** $324 **7.** $192 **9.** $655

page 385

Getting Ready **1.** $302.08 **2.** $2,700 **3.** $3,096 **4.** $3,468 **Exercises** **1.** $4,200 **3.** $3,030 **5.** $9,900 **7.** $375

page 387

Getting Ready **1.** $190,746; $150,746 **2.** $241,074; $181,074 **Exercises** **1.** $238,431.60 **3.** $156,742.20 **5.** $157,802.40 **7.** $150,746 **9.** $200,895 **11.** $137,650.80

page 389

Getting Ready **1.** $2,502 **2.** $3,871.20 **3.** $3,416 **Exercises** **1.** $3,370 **3.** $2,146 **5.** $3,685.60 **7.** $1,327.40 **9.** $8,947 **11.** No Bank A: $0.15 \times \$3,400 \times 4 = \$2,040$; Bank B: $0.18 \times \$3,400 \times 3 = \$1,836$

page 390

Getting Ready **1.** 27.3 mpg, or 27 mpg **2.** 216 mi **3.** $59.20 **4.** 8.4 mpg, or 8 mpg **Exercises** **1.** 442.75 mi **3.** $303.68 **5.** 20 mpg **7.** 250 gal **9.** $1,102.50

Index

PHOTO CREDITS

Page 1—Lou Jones, Image Bank; **11**—Alon Reininger, DPI; **15**—Erika Stone, Peter Arnold; **24**—Richard Hutchings; **25**—HRW Photo by Russell Dian; **27**—Steve Dunwell, Image Bank; **39**—Raoul Hackel, Stock, Boston; **43**—HRW Photo by Ken Karp; **44**—HRW Photo by Louis Fernandez; **45**—Thomas Ives; **47**—Roberto Valladares, Image Bank; **54**—Tom O'Brien, International Stock Photo; **59**—Educational Dimensions; **61**—Doris S. Baum, DPI; **64**—Gale Brunswick, Peter Arnold; **65**—Pam Hasegawa, Taurus; **67**—Allen Green; **68**—HRW Photo by Ken Lax; **79**—HRW Photo by Vadnai; **81**—HRW Photo by Michal Heron; **86**—Wil Blanche, DPI; **87**—Porterfield-Chickering, Photo Researchers; **89**—John Zoiner, International Stock Photo; **105**—HRW Photo by Richard Haynes; **106**—Scott Hyde, International Stock Photo; **107**—HRW Photo by Richard Hutchings; **109**—Mickey Palmer, Focus on Sports; **111**—Bill Gallery, Stock, Boston; **113**—Gregory Edwards, International Stock Photo; **122**—Phiz Mezey, DPI; **123**—L. L. T. Rhodes, Taurus; **125**—HRW Photo by Richard Haynes; **127**—George Hall, Woodfin Camp; **131**—HRW Photo by Ken Karp; **138**—Sybil Shackman, Monkmeyer; **139**—George Dodge, DPI; **141**—HRW Photo by Yoav Levy; **153**—John Zoiner, International Stock Photo; **154**—Rhoda Sidney, Monkmeyer; **155**—Mike Malyszko, Stock, Boston; **157**—Bill Gallery, Stock, Boston; **178**—Yoav Levy, Phototake; **179**—Michal Heron; **183**—Phillip A. Harrington, Peter Arnold; **198**—Jim Cartier, Photo Researchers; **199**—Michal Heron; **201**—Robert Isear, DPI; **211**—Joel Gordon, DPI; **213**—Emil Javorsky, DPI; **218**—Erika Stone; **219**—Bruce Roberts, Photo Researchers; **221**—Stephen Myers, International Stock Photo; **225**—Lea, Omni-Photo Communications; **227**—Pellegrini's Studio, International Stock Photo; **231**—Van Bucher, Photo Researchers; **233**—Ann Hagen, DPI; **236**—Eric Carle, Shostal Associates; **237**—Bonnie Freer, Peter Arnold; **241**—Larry Dale Gordon, Image Bank; **246**—Mimi Forsyth, Monkmeyer; **258**—HRW Photo by Louis Fernandez; **259**—HRW Photo by Russell Dian; **261**—Horst Schäder, Peter Arnold; **271**—Freda Leinwand, Monkmeyer; **276**—James H. Karales, Peter Arnold; **277**—Michal Heron; **279**—John Blaustein, Woodfin Camp; **287**—Wil Blanche, DPI; **302**—Donald L. Miller, International Stock Photo; **305**—Ray Ellis, Kay Reese & Associates; **307**—Dan McCoy, Rainbow; **315**—U.S. Department of Energy; **316**—HRW Photo by Louis Fernandez; **317**—HRW Photo by George Tames; **319**—George Hall, Woodfin Camp; **321**—G. Zimbel, Monkmeyer; **327**—Alvis Upitis, Shostal Associates; **335**—Bill Stanton, International Stock Photo; **339**—Freda Leinwand, Monkmeyer; **343**—Dan McCoy, Rainbow; **349**—Randa Bishop, DPI; **351**—HRW Photo by Louis Fernandez; **353**—James Broderick, International Stock Photo; **356**—William J. Jahoda, Photo Researchers; **359**—Sepp Seitz, Woodfin Camp; **372**—Pam Hasegawa, Taurus; **373**—Irving Schild, DPI; **374**—HRW Photo by Ken Karp; **377**—Tom Bieber, Image Bank; **387**—Vic Cox, Peter Arnold; **391**—Donald L. Miller, International Stock Photo; **394**—Ida Wyman, International Stock Photo; **395**—Richard Hutchings, Photo Researchers.